ANODIC PROTECTION

Theory and Practice in
the Prevention of Corrosion

ANODIC PROTECTION

Theory and Practice in the Prevention of Corrosion

OLEN L. RIGGS, Jr.

Kerr – McGee Technical Center
Oklahoma City, Oklahoma

and

CARL E. LOCKE

University of Oklahoma
Norman, Oklahoma

Consulting editor
NORMAN E. HAMNER

PLENUM PRESS · NEW YORK AND LONDON

Library of Congress Cataloging in Publication Data

Riggs, Olen L
  Anodic protection.

  Bibliography: p.
  Includes index.
  1. Metals–Anodic oxidation. I. Locke, Carl E., joint author. II. Hamner, Norman E.
III. Title.
TA462.R46   620.1'623   80-20412
ISBN 0-306-40597-0

Information herein credited to the Russian journal *Protection of Metals* (Zashchita Metallov) is reprinted with permission of Plenum Publishing Corp., 227 W. 17th St., New York, N.Y. 10011 and is copyrighted by Plenum. Information herein credited to the *Journal of The Electrochemical Society* is reprinted by permission of the publisher, The Electrochemical Society, Inc., P.O. Box 2071, Princeton, N.J. 08540.

© 1981 Plenum Press, New York
A Division of Plenum Publishing Corporation
227 West 17th Street, New York, N.Y. 10011

In appreciation for the thoughtful support,
we dedicate this book to
our understanding wives: *Ann* and *Sammie*

Preface

The objectives of this book are to give technical information about anodic protection, explain how economic analyses are made to determine whether or not it should be used, and describe some of the applications and equipment. Limitations of the technique will be pointed out.

Technological changes that have resulted in higher temperatures, pressures, and velocities increase corrosion rates and markedly influence materials selection and design decisions. Continuous cycle systems impose increased demands on system reliability. New processes require more sophisticated equipment made of costlier metals which are often in short supply and subject to the vagaries of international commerce. The impact of continuing inflation influences decisions related to capital expenditures and maintenance costs.

Some problems caused by these considerations can be solved, or solutions simplified, by the use of anodic protection. Technical and scientific information is presented on applications to industrial equipment, economics, design and installation, operation and maintenance, electrochemical principles, laboratory tests and procedures. A historical summary, patent list, glossary of terms, and a subject index are included.

It is important to acknowledge that much of the information has been from the original work of others, including the publications of many friends. Many of the applications come from developments pioneered by scientists in Russia, whose innovative work involves corrosive solutions to which anodic protection has not been applied elsewhere. Results of original research and engineering applications by the authors are also included. Manufacturers of equipment and system engineers have cooperated in supplying information and their cooperation is gratefully acknowledged.

There is every reason to believe that an increasing number of anodic-protection installations will be made in environments where it has been successful in the past, as well as in others in which it is not commonly used now. Better understanding of reactions accumulated from actual working equipment and from laboratory and pilot plant experimentation will permit its use in systems where it has not been employed frequently so far.

Techniques and applications described in this book may encourage increased applications of anodic protection. When this happens, the authors will appreciate information about new installations and techniques that may be used in revisions of this book.

Particular thanks is due to the Kerr-McGee Corporation for its continued support and to the University of Oklahoma for its cooperation in making time available to prepare the book. The authors also appreciate the cooperation of publishers who have authorized use of many excerpts from their books. Without this cooperation, preparation of the book would have been much more difficult.

The ready cooperation of industry, authors, and publishers that characterizes the technical and scientific community in the United States has been instrumental in helping to make available information about ways to do things better, easier, and more economically.

Olen Riggs, Jr.,
Kerr-McGee Corporation

Carl E. Locke,
University of Oklahoma

Norman E. Hamner,
Editorial Consultant

Contents

APPENDIXES

Anodic Protection of Metals—A Technique Whose Time Has Come

The twin compulsions of materials scarcity and escalating costs have brought about a marked increase in interest among designers, materials specialists, and engineers in effective methods to control wear and corrosion. This interest has resulted in the most intense studies of materials properties that has ever been experienced in the industrial nations of the world. Additional impetus has been given to this trend by the impending necessity to find and develop ways in which fuels can be produced to take the place of diminishing supplies of petroleum and natural gas, not to mention the financial drain of buying petroleum from abroad at escalated prices.

Ecological restraints imposed on industry have increased the cost of production and control methods formerly used. Companies seeking capital by selling equity in the stock market cannot find the ready buyers available a decade ago, so managers resort to conservation of existing plants and equipment when formerly they would have scrapped old machinery and replaced it with new.

Many Protective Methods Are Used

Equipment is protected by the best available methods. Liquid-phase environments may call for the application of inhibitors and detergents which improve heat-exchange efficiences and also reduce corrosion. In appropriate environments and conditions, electrochemical methods—anodic or cathodic protection—are applied, often with dramatic benefits and economies. Metallurgists are exploring the benefits of alloying, which may significantly reduce attack rates by additions of very small percentages of elements. They also apply heat-treatment regimes which prevent or reduce the probability of stress-corrosion cracking or accelerated local attack. Design changes and protective coatings are also among the frequently employed methods to reduce corrosion.

Whether one or more of the available protective methods is used involves weighing the economical as well as the technological alternatives and other considerations, including the ability of maintenance workers to cope with control techniques. Prerequisite to decisions is also the necessity of avoiding ecological contamination such as that involved in discharging blowdown from a cooling water system.

Cathodic protection should not be applied to underground structures in such a way that it damages other buried metal in the vicinity. When an effluent is changed from the liquid to the gaseous phase, discharge of the gas into the atmosphere is rigidly controlled. Even the terminal disposition of dangerous wastes is hedged about with restrictions that increasingly limit options.

Under these circumstances it is refreshing to learn about a technique which, when properly applied in appropriate environments and on specific materials, imposes minimum burdens on the external environment, produces low levels of effluent and usually only innocuous quantities of reaction products. Energy consumption is low, and installation and maintenance costs are modest. This technique is anodic protection, which, when applied properly, has reduced corrosion damage up to 100% on some materials in some environments.

Anodic Protection Used Effectively

For more than 15 years anodic protection has been effectively used to control corrosion by sulfuric acid—more than 100 sulfuric acid storage tanks of significant size have been under anodic protection for many years. It has been used effectively to protect carbon steel against 10–45% solutions of sodium hydroxide at 25–60°C. When sodium chlorate, chloride, and sulfate were added to the solution, protection was not adversely affected. The technique has also been used with success in ammonia–ammonium nitrate and nitrogen fertilizer solutions, among others. Anodic potentials protect titanium from chlorides up to 170°C, but not from bromides or iodides.[1] Other experimental evidence indicates that they will protect Nichrome and Inconel 600 against molten sodium sulfate.

In Russia, tanks containing hydroxylamine solutions contaminated with sulfate and sulfur dioxide have been protected. Protection has also been provided against attack by chloride solutions containing nitrates.

Laboratory tests indicate that anodic polarization of carbon steels produces a protective film in 500°C, 103 MPa, high-purity water which resists activation by Co^{60}. Producing a more protective film during hot conditioning should keep down subsequent activation rates.[2] This could extend the safe operational life of nuclear reactors producing steam.

Protection of Alloy Steel Important

While carbon steel can be protected against acid and some alkaline solutions by anodic protection, what is probably more important is its function in protecting steels containing chromium, nickel, and molybdenum. Passivation of these alloys is easier than it is for carbon steels. This may be significant beyond the effect it has on the immediate reduction of corrosion rates.

The urgent necessity to develop energy sources independent of petroleum has underlined the importance of corrosion-resistant alloys. Evidence is accumulating that the best alternatives available to the United States and other industrial nations will require a high order of performance by materials because of the aggressive environments common to the utilization of alternative energy sources. For example, it has been established that 17% or more chromium is required for steels used to resist geothermal brines at 250°C that contain less than 20% NaCl.[3] A protective chromium oxide layer is essential for iron- and nickel-base alloys in coal gasification systems.[4]

Because anodic protection's effectiveness is enhanced by the chromium content of alloys, with efficiencies approaching 100%, as is described elsewhere herein, it acquires importance that goes beyond the immediate goals of extending the service life of equipment. As important as this is from the standpoint of engineering economics, there are other considerations that make it even more significant.

Strategic and Absolute Factors Bearing on Materials

The task of the designer, materials specialist, and engineer is complicated by the bare facts of the strategic availability of materials, especially with respect to chromium and cobalt. There is also reason to be concerned about the absolute availability of some metals. These effects have been intensely studied in the United States and elsewhere because there is some reason to believe that modern industrial processes that depend on alloyed metals face critical supply problems, some of which may be difficult, if not impossible, to resolve.

Chromium is the most important of the strategic alloys to the United States at the present time.[5-8] The consequences of chronic materials shortages have been considered and special attention has been given to chromium. This metal is probably the most important for corrosion- and heat-resistant alloys. In 1976, 69% of the chromium imported into the United States was used in engineering alloys. This consumed 248,000 tons. Of the known world resources of chromium, 62% is in South Africa, 33% in Rhodesia, 1.2% in the USSR,

0.3% in the Philippines, 0.1% in Turkey, and all others, 3.3%. The United States has none.[9]

Cobalt, an essential alloying element for high-temperature environments where it is used to resist oxidation at 600–1200°C, in gas turbines, for example, and specified for use against hydrochloric acid (and as the main element in prosthetic implants, in which chromium is also essential), is in limited supply in the United States.[10–12] Sources of cobalt (1971–1973) are Zaire, 67%; Zambia, 9.6%; Morocco, 6.7%; U.S. and Canada, 8.9%; Finland, 4.3%; and others, 3.1%.

Prices on chromium increased rapidly after a ban on imports from Rhodesia when Russia became a principal supply source. In 1978 cobalt increased in price from $7 to $20 a pound after the Zaire conflict. During the months preceding hostilities there, the Soviet Union, Poland, and East Germany bought between 500 and 2000 tons of cobalt. Cartel-like agreements among countries supplying bauxite, copper, iron, and tungsten ore following the 1973 Near-East petroleum embargo put supplies of these and other essential metals into the political arena.[13] These developments have increased the necessity for controlling the consumption of strategic metals.[14]

Substitute Alloys for Chromium

"At this time there is no satisfactory chromium substitute as an alloying addition to resist high-temperature oxidation and sulfide corrosion," according to the National Academy of Sciences. Although considerable attention is given to various conservation, substitution, and design measures to reduce chromium consumption, little attention has been given to the possible impact of higher prices.[14,15]

When tests were made to identify alloying additions that will permit reducing the chromium content of AISI Type 304 steels, it was discovered that satisfactory performance in some corrosives could be attained by addition of aluminum, molybdenum, and silicon, but that the chromium content could not be reduced below 12%.

Prices of Substitute Alloys Increase

Ferritic steels containing 2% molybdenum (of which 18,000 tons were produced in 1976) can be substituted for chromium alloys in many applications. However, the price of molybdenum increased from 8 to 11% in each of three recent years. Alloys (except for stainless and tool steels) consumed 50.7% and stainless steels, 13.4% of molybdenum during 1976.[15]

While aluminum is useful as an alloy and as a coating for CrNi and other heat- and corrosion-resistant steels and may be an alternative material in some cases, it is an energy-intensive metal. A ton of aluminum requires 16,000 kW hr of electricity and 90×10^6 Btu of thermal energy.[16] Most of the bauxite and alumina used in the US for aluminum production is imported. It takes 45,000 Btu to produce a pound of aluminum castings, while only 9,000 Btu/lb (0.4536 kg) is needed for iron.[16a]

Corrosion Protection Is Necessary

Corrosion is not the only reason for the failure of industrial equipment, but it is a common, if not the most prevalent, cause of leaks in pipes, tanks, and other equipment. These leaks are a critical matter when they involve hot, poisonous, combustible, explosive, or ecologically objectionable materials. In large-scale liquid process systems, a small leak can cause operational interruptions that can result in thousands of dollars a day in production losses.

A railroad tank car or a highway truck leaking ammonia, a combustible, corrosive acid, or alkali solution creates a health and property-damage hazard and can contaminate the environment in the vicinity of the leak. Huge awards for injuries, loss of life, or property damage may be the result. Criminal charges are possible also.

Contamination Control Important

As described in detail in Chapter 2, control of product contamination is a major problem with many commodities, such as is the case with sulfuric acid. When 98% protection of vessels handling sulfuric acid concentrations of 93 to 105% can be achieved, this is a major factor in meeting the quality standards required by a user. Even when high-alloy materials are used to reduce corrosion that adulterates a product requiring high purity, anodic protection not only reduces the likelihood of contamination, but extends the life of the equipment, a significant savings in the case of expensive alloys.

The most spectacular successes with anodic protection have been achieved on sulfuric acid heat exchangers. Figure 1-1 shows the effects of sulfuric acid attack on a stainless-steel cooler attributed to turbulence erosion, as well as corrosion. Anodic protection of heat exchangers also permits better heat transfer by allowing greater velocities and reduces first cost by making it possible to use smaller exchangers and related components. This reduces the cost of equipment, which is further reduced by decreasing the cost of metal in a "cor-

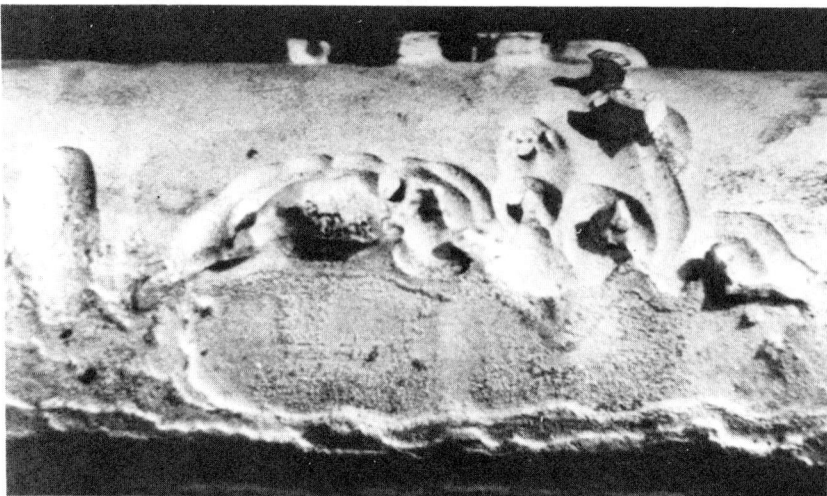

Figure 1-1. Heat-exchanger tubes in sulfuric acid. Top: protected tubes after exposure in absorbing acid service. Bottom: local activation attack during unprotected operation (J. M. Stammen, *Anodic Protection: State of the Art, Corrosion* 71/81, NACE, Houston, TX).

rosion allowance." These benefits persist throughout the useful life of equipment.[17]

   Anodic protection can also be used in hot and cold alkaline solutions,[18–20] such as caustics, for protection against hydroxylamine sulfate,[21] nitrate, and chloride solutions, and as a standby to protect titanium–palladium heat-exchanger tubes exposed to residues from zinc reduction.[22] This latter application is particularly important because process upsets could cause catastrophic failure of equipment without standby protection.

Experiments in molten sodium sulfate indicate that anodic protection may be useful in protecting Nichrome and Inconel 600.[23] Good results have been achieved in the protection of titanium.[24] Recent tests in desalination environments indicate that properly alloyed arsenical brasses that are amenable to anodic protection in hot seawater may be produced.[25]

Further details on some of these applications and numerous others will be found elsewhere in this book.

Ecological Considerations Are Important

Because anodic protection does not usually produce any effluent, it has ecological advantages over some other control methods. It may reduce the need for inhibitors or eliminate their use entirely. There is evidence of synergism between anodic protection and inhibitors in some environments. When inhibitors are used, they may cause problems when they must be discharged into the environment.

When galvanic anodes must be exposed in a solution to provide cathodic protection, the corrosion products they generate may create problems. Anodic protection does not usually introduce significant quantities of corrosion products or reaction species into process streams.

Energy-Conservation Values

When the effective service life of alloys containing scarce or expensive metals, especially boron, chromium, or molybdenum, can be extended indefinitely, there is a substantial energy savings. The amount of energy required to produce corrosion-resistant alloys is important. When the potential energy savings possible with other steels and with titanium, for example, is added, the technique gives a net savings many orders of magnitude more than the net energy component of anodic-protection equipment and the energy consumed by the system.

If corrosion attrition of expensive equipment can be reduced, as it can be by anodic protection, the scrap or recycled value of the equipment is enhanced. When corrosion is reduced to a small fraction of the unprotected rate, equipment may be used until it becomes obsolete. The effect this has on amortization and tax schedules can be important. Preservation of equipment is a certain way of conserving energy that has been expended to mine, refine, and fabricate it.

Reclamation of materials is one way to effect energy savings.[26] In the case of chromium–nickel, cobalt, or titanium alloys there are, however, some difficulties that may or may not make reclamation relatively ineffective. In recycled CrNi alloys, relatively small amounts of metals may adversely affect corrosion rates. A somewhat similar situation prevails with respect to other corrosion-resistant metals, especially those for high-temperature service.

Metals recycled for castings, for example, may have properties and characteristics that influence their use for this purpose. It is also likely that similar reservations can be made with respect to recycled wrought alloys.[27]

These limitations on the use of recycled materials increase the importance of protecting equipment by every practical and economic method. Proper selection and design, as well as electrochemical or other protective techniques, become imperative.

Applications and Limitations

Applications to control corrosion by a wide range of corrosive solutions are covered in detail elsewhere in this book. Nevertheless, it is useful to cover some aspects of application in general so that the potentials and limitations of anodic protection can be better understood. Among the advantages of anodic protection is its ability to protect surfaces at a distance from the cathode (Figure 1-2), including surfaces inside crevices which cannot be protected by presently available techniques.

Figure 1-3 shows potentiostatic curves indicating that an extremely narrow crevice (0.009 in., 0.023 cm) in a chromium–nickel casting steel (CF-8 or UNS 92600) cannot be protected in sulfuric acid at 25°C. This is in contrast to the situation depicted in Figure 1-4, which shows that a similar crevice can be protected in AISI 304 steel. The reasons for this difference in behavior reside mainly in the potentials required to passivate the crevices in the two alloys. Current required to passivate CF-8 is several orders of magnitude greater than that required for AISI 304 steel.

Elsewhere, another authority[28] says, "It is much more difficult to protect surface irregularities, such as recessions around sharp slots, grooves, or crevices." These differences of opinion and contradictions, or apparent contradiction by experimental evidence, result from the fact that successful application of anodic protection requires a precise match among environment, materials, and protective currents.

Equipment to be anodically protected should be designed with a minimum of irregularities because "incomplete passivation can have catastrophic consequences.[11] In some environments, specifically in producing rayon fibers, titanium may be alloyed with 0.1% palladium, which effectively protects crevices.

An example of anomalies sometimes found when applying electrochemical

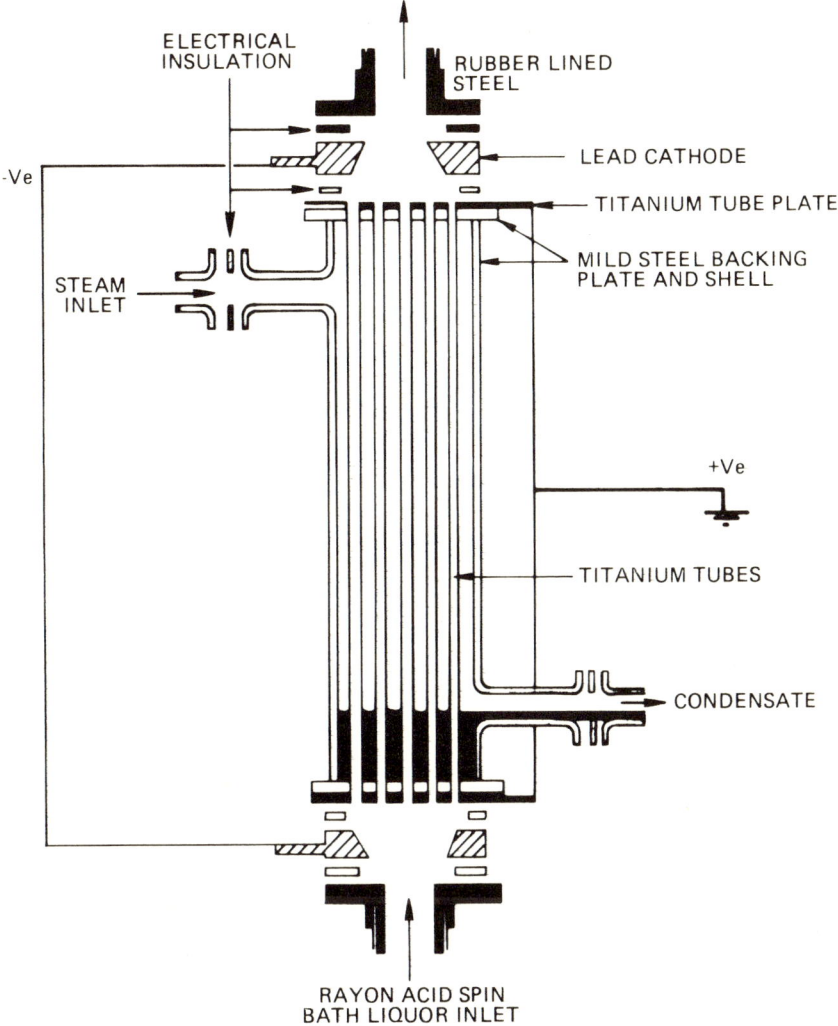

Figure 1-2. Schematic of anodic-protection installation on a titanium heat exchanger with 25 mm outside diameter by 2.75-m-long tubes. The whole bundle was in a mild-steel shell, 254 mm in diameter. Note the location of cathodes (Figure 7, Reference 24).

protection to materials is provided by the reaction of titanium in rayon spin bath solutions (10% sulfuric acid + sodium and zinc sulfates, hydrogen sulfide, and carbon bisulfide) as reported by Evans *et al.*[24] While the alloy TiPd is effective in preventing crevice attack, this solution will dissolve TiPd at a rate four times faster than it attacks commercial-grade titanium when both are

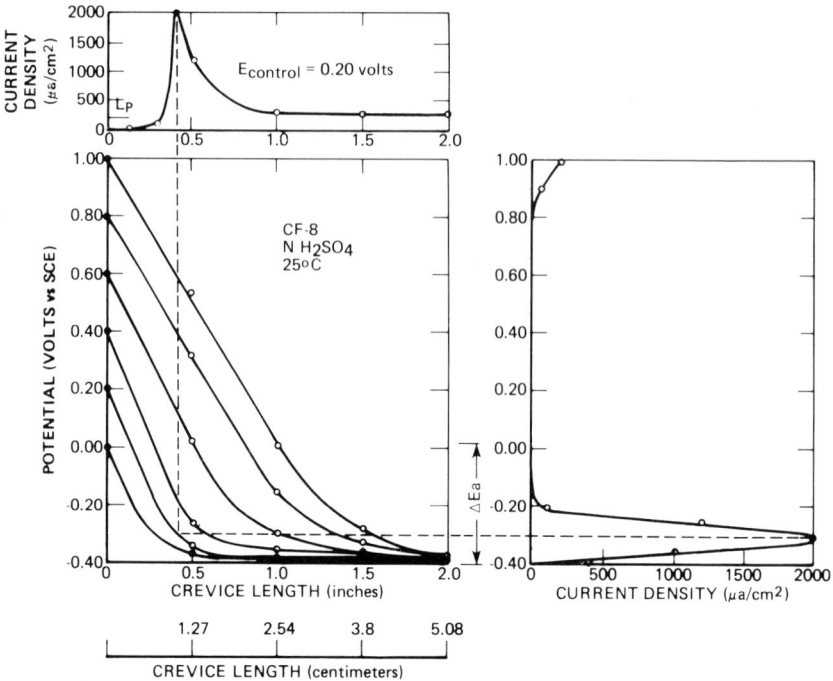

Figure 1-3. Anodic polarization of a narrow crevice. Crevice potential variations of CF-8 steel under anodic polarization, in hydrogen-saturated sulfuric acid at 25°C. Crevice width is 0.009 in. (0.23 mm). W. D. France, Jr. and N. D. Greene, Jr, *Passivation of Crevices during Anodic Protection, Corrosion* **24**(8), 1968 pp. 247–251.

anodically protected. Other anomalies are reported by these authors in their experiments with rayon spin bath solutions.

Limitations Can Be Anticipated

In contrast to cathodic protection, where historically a substantial amount of trial and error is involved, anodic protection's limitations can be determined in laboratory tests in advance of the actual installation. Results of these tests usually establish the exact conditions that must be maintained for effective protection.

Similarly, laboratory tests can be used to determine in advance what effects process variables will have on protection. Consequently, to the extent that these variables can be anticipated, the success of an installation can be assured in advance. Nevertheless, expert surveillance is desirable in many cases because minor changes in system parameters that are not anticipated in design can result in a shift of potentials into an active region with a resulting acceleration of attack.

Furthermore, it is obvious that there should be an uninterrupted supply of electricity to the system. Interruption of the protective current is not invariably productive of attack, however, because in some systems, passivation may persist for extended times. Loss of passivity may require application of massive currents to reestablish it, as is explained elsewhere. Application of these currents on demand by the system controller is well within the capability of existing technology, so hazards from this effect are diminishing as control systems become more sophisticated. As an example of what may happen, in the case of titanium heat exchangers in rayon spin baths, passivity persisted for several hours after current was turned off. Laboratory tests can determine, with respect to a given combination of corrosives and materials, whether or not such persistence can be expected and if so, for what interval.

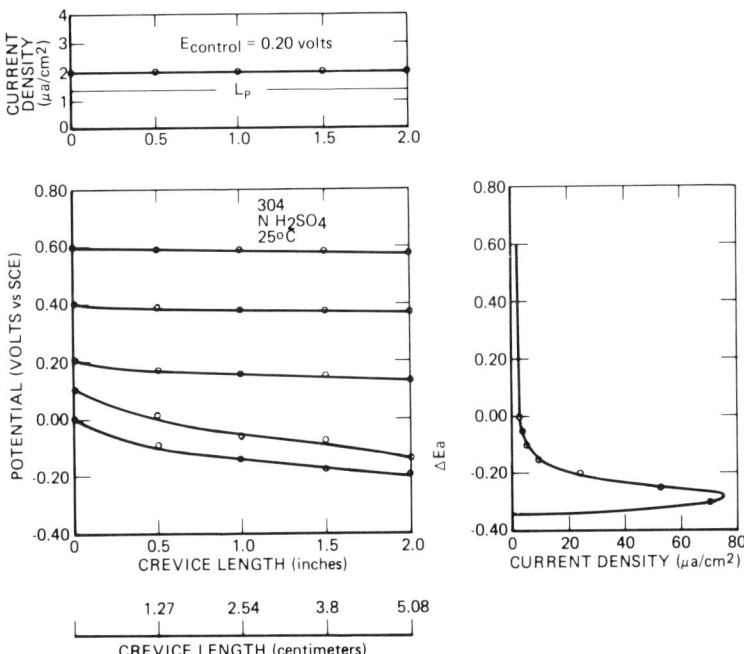

Figure 1-4. Anodic polarization of a narrow crevice in AISI 304 steel. Note the difference in current density compared to that in a similar crevice in Figure 3.

Installation Expense Factors

The sometimes considerable cost of initial installation can often be minimized by the technique of switching the controlling circuits among a number of protective installations. Because of the passivitation–persistence factor, which is characteristic of some installations, one set of control instruments can serve a number of tanks, heat exchangers, or other pieces of equipment. As each installation is switched into the circuit, its condition is analyzed and a sufficient increment of current is directed to the system to bring passivity back to design levels.

Recently developed large-capacity potentiostats have increased the range of the switching installations and enhanced the capability of the systems to automatically compensate for upsets.

Rapid Development Anticipated

Rapid development of anodic protection is anticipated. One acknowledged authority[29] on corrosion control says,

> In the future, anodic protection will probably revolutionize many current practices of corrosion engineering. Utilizing this technique, it is possible to reduce the alloy requirements for a particular corrosion service. Anodic protection can be classed as one of the most significant advances in the entire history of corrosion science.

Table 1-1 lists parameters of some recent installations of anodic protection. Many more are discussed elsewhere herein.

The scientific and engineering principles applicable are presented in sufficient detail to permit a clear understanding of the potential of the technique. Also, sufficient information is provided concerning the equipment, engineering applications, and other details so that the applicability of anodic protection to a wide variety of corrosive solutions can be determined with reasonable accuracy. It is not contended, however, that anodic protection can be applied offhand without careful preliminary work and expert assistance. The advantages to be gained, however, justify serious consideration.

This book provides the essential fundamental information concerning the electrochemistry and the laboratory procedures that can be used to establish the required application parameters in advance of installation. It also presents examples of engineering, details of equipment circuitry, and selection of materials that have been used successfully.

TABLE 1-1. Results of Recent Applications of Anodic Protection

Corrosive solution	°C	Material					Unprotected		Protected		Passive range v	Reference
		Cr	Ni	Mo	Ti	Other	mpy	mm/y	mpy	mm/yr		
Sulfuric acid	40–80	13–18	12–10	3			7.9	0.2	0.32	0.008	0.45–0.81	30
93%ᵃ	70	17	12	2.25		0.05 C, 68 Fe	29.0	0.74	1.8	0.05	+0.53–+0.71	
98%	100	AISI 316					500.7	12.7	0.26	0.007		31
93%ᵃ	100	22	42	3		1.8 Cu, 0.04 C	303.0	7.7	16.3	0.41		
98%ᵇ	100	Incoloy 825					23.6	0.6	5.4	0.14	+0.68–+0.71	
Nitric acid + hydroxyl amine, Na sulfate + hydrazine hydrate	17–18	10					435.4	11.0	4.4	0.12	+1.85–2.0	32
10–45% NaOH + other alkaline solutions + NaCl, sulfate + chlorate	25	Carbon Steel									0.1ᶜ	33
less than 3% NaOH only	60											

ᵃ +4 ppm Fe, 800 ppm sulfur dioxide.
ᵇ +4 ppm Fe, 75 ppm sulfur dioxide.
ᶜ Range in concentrated solutions only.

References

1. J. B. Cotton, *A Perspective View of Localized Corrosion of Titanium,* in *Localized Corrosion,* edited by Staehle, Brown, Kruger, and Agrawal, NACE, Katy, TX pp. 676–679, (1969).
2. D. H. Liste and R. S. Pathania, *Radioactive Contamination Under Applied Potentials in High Temperature Water, Corrosion* **34**(7) July (1978).
3. D. W. Shannon, *Role of Chemical Components in Geothermal Brines on Corrosion, Corrosion* 78/57, NACE, Katy, TX.
4. K. Natesan, *High Temperature Corrosion,* National Bureau of Standards Special Publication **468,** pp. 159–171, April (1977).
5. *Contingency Plans for Chromium Utilization, Report of Committee,* National Materials Advisory Board, National Academy of Science, Washington, DC 1978.
6. N. H. Henniker, *Engineering Implications of Chronic Materials Scarcity,* Office of Technology Assessment, Washington, DC, 1976.
7. *Rational Use of Potentially Scarce Metals,* NATO Scientific Affairs Division, Brussels, Belgium, 1976.
8. *Engineering Implications of Chronic Materials Scarcity,* Office of Technology Assessment, Washington, DC.
9. A. G. Gray, *Assessing Needs for Critical Materials, Metal Progress,* p. 23, August (1977).
10. D. Chatterji *et al., Protection of Superalloys for Turbine Applications, Advances in Corrosion Science Vol. 6,* edited by Staehle and Fontana, Plenum, New York, pp. 1–87 (1977).
11. E. Jackson and M. J. Wilkinson, *Effect of Thiourea and Some of Its Derivatives on the Corrosion Behavior of Cobalt in Concentrated Hydrochloric Acid, Br. Corros. J.* **11**(4), pp. 208–211 (1976).
12. T. M. Devine and J. Wulff, Comparative Crevice Corrosion Resistance of CoCr Base Surgical Implant Alloys. *JECS* **123,** pp. 1433–1437 (1976).
13. W. Casper, *Raw Materials Supply: Challenge and Chance for Economy and State, Metallgeschaft,* 1976.
14. J. R. Stephens and C. A. Barrett, *Oxidation and Corrosion Behavior of Modified Composition, Low Cr 304 SS Alloys,* Washington, DC, May 1977; *Environmental Degradation of Engineering Materials,* Blacksburg, VA, pp. 257–266 (1977).
15. *Materials Engineering,* p. 102, September 1977.
16. S. W. Boercker, *Energy Use in the Production of Primary Aluminum,* Oak Ridge, TN, August 1978.
16a. J. B. Wachtman, Jr. *et al., Progress Report: Energy Related Aspects of Materials,* Dearborn, MI, 1977.
17 D. Fyfe *et al., Anodic Protection of Sulfuric Acid Plant Cooling Equipment, Corrosion* 75/68, NACE, Katy, TX.
18. A. I. Tsinman and L. A. Danielyan, *Anodic Protection of Carbon Steel in Alkaline Solutions, Zasch. Met.* **12**(4), pp. 450–452, July–August (1976).
19. G. Herbsleb and W. Schwenk, *Potential Dependence of Intercrystalline Stress Corrosion of Unalloyed and Low Alloys Steels in Hot Alkaline Solution, Stahl Eisen* **90,** p. 903 (1970).
20. H. Grafen and D. Kuron, *On the Intercrystalline SCC of Mild Steels in Caustic Solutions,* Archiv fur Eisenhuttenvesen, p. 36 (1965).
21. V. S. Kuzub and A. L. Anokhin, *Investigation of Electrochemical and Corrosion Behavior of a Cathode of Type OKn23N28M3D3T Steel With Intermittent Polarization in an Industrial Solution of Hydroxylamine Sulfate, Zasch. Met.* **12**(3), pp. 282–284, May–June (1976).
21. V. S. Kuzub and V. S. Novitskii, *Effects of Temperature and Concentration on the Ratio*

of Nitrate and Chloride on the Pitting of 18Cr10NiTi Steel, Zasch. Met. **15**(5), pp. 560–562, September–October (1975).

22. R. W. Allan and C. J. Haight, *Prevention of Scaling and Corrosion of Ti Alloy Heat Exchanger Tubes, Corros. Austral.* **2**(3), pp. 12–15 (1977).

23. J. B. Berkowitz and W. D. Lee, *Investigation of the Possibilities for Electrochemical Control of Hot Corrosion,* Final Report, 1972–1973, ADL-75526-FR, NTIS. Cambridge, MA.

24. L. S. Evans *et al., Anodic Protection of Titanium in the Rayon Industry, Proceedings of the 4th International Congress on Metallic Corrosion,* NACE, Houston, TX, pp. 625–635 (1972).

25. F. H. Cocke *et al., Development of Low Cost, Corrosion Resistant Alloys and of Electrochemical Techniques for Their Protection in Desalination Systems,* OSW R & D Progress Report Number 801, NTIS, Waltham, MA.

26. *Resource Conservation, Resource Recovery and Solid Waste Disposal,* Committee on Public Works, Washington, DC, p. 31, November (1973).

27. R. W. Lobenhofer, *Something More to Worry About, Modern Casting,* p. 35, May (1978).

28. *Corrosion,* Vol. 2, Second edition, edited by L. L. Shrier (Newnes-Butterworth's, London, pp. 11–118, (1977).

29. M. G. Fontana and N. D. Greene, *Corrosion Engineering,* Second edition, McGraw-Hill, New York, p. 214 (1978).

30. Yu M. Mironov and V. A. Makarov, *Anodic Protection of Steels 18Cr10NiTi and 13Cr12Ni3MoTi in Sulfuric Acid, Zasch. Met.* **12**(1), pp. 50, 51, January–February (1976).

31. R. M. Kain and P. F. Morris, *Anodic Protection of FeCrNiMo Alloys in Concentrated Sulfuric Acid, Corrosion* 76/149, NACE, Katy, TX (1976).

32. I. V. Oknin and S. G. Gabriel, *(Experimental) Corrosion of Stainless Steels During Polarization in Nitric Acid Solutions with Added Sodium Sulfite, Hydroxylamine Sulfate, Hydrazine Hydrate, Zasch. Met.* **12**(3), pp. 264–268, May–June (1976).

33. A. I. Tsinman and L. A. Danielyan, *Anodic Protection of Carbon Steel in Alkali Solutions, Zasch. Met.* **12**(4), pp. 450–452, July–August (1976).

2

Anodic Protection of Industrial Equipment

Many methods of corrosion control were first applied in an empirical manner, often many years before the scientific principles underlying the methods were identified. Anodic protection, however, is an exception to this procedure in that an understanding of the basic mechanisms was developed in the laboratory first. These discoveries were followed by a period of pilot-scale development, which included several trial installations in chemical plants. Finally came the commercial development of the technique. This development procedure is very satisfying to the scientist and engineer.

Anodic protection is a corrosion-control method with very definite limitations in that the metal–environment system must have anodic-polarization behavior characteristic of active–passive materials. The polarization curve typifying this system and its significance from practical and theoretical viewpoints are fully discussed in Chapters 4 and 7. For systems involving active–passive materials, anodic protection can be a successful method of controlling corrosion. The technique has been applied in a wide variety of metal–environment systems, especially during the last 18 years. While most of these applications have been successful, some have not.

This chapter is a review of a wide variety of corrosion-control problems to which anodic protection has been applied. Discussion of the equipment involved is given in Chapter 3.

Sulfuric Acid Applications

Sulfuric acid is the chemical manufactured in greatest volume worldwide. Carbon and stainless steels are the most common construction materials for equipment used in the manufacture, storage, and transportation of this commodity which is used in concentrations of 77 to 100% in the United States. Fuming sulfuric acid, or oleum, is also manufactured. Oleum is 100% sulfuric acid in which additional SO_3 has been dissolved. Anodic protection can be used to control these acids over a wide temperature range.

Applications to control corrosion by sulfuric acid have been the most common uses of anodic protection and have been the mainstay of its commercial success. This section includes descriptions of applications of anodic protection to control corrosion in plants using sulfuric acid. Because this segment of the chemical industry is so important to anodic protection, a brief summary of the extent of corrosion problems associated with sulfuric acid is presented.

Sulfuric Acid Corrosion

Sulfuric acid is manufactured and sold in a wide range of concentrations in the United States and abroad. Because most storage and transportation equipment is made of carbon steel, most of this section will be devoted to discussions of carbon steel corrosion, particularly in acid concentrations ranging from 77 to 100%. In addition, problems associated with fuming acid are discussed. Also included is a brief summary of problems encountered when austenitic stainless steels are used.

While corrosion rates of carbon steels in sulfuric acid are mainly functions of temperature, velocity, and acid concentration and purity, rates are greatly affected by minor alloying elements in steel, particularly copper. This alloying factor is also discussed.

Fontana[1] has published data in which the effects of temperature and concentration on corrosion rates are summarized as shown in Figure 2-1. This agrees with other data which show that the maximum rate of attack on steel

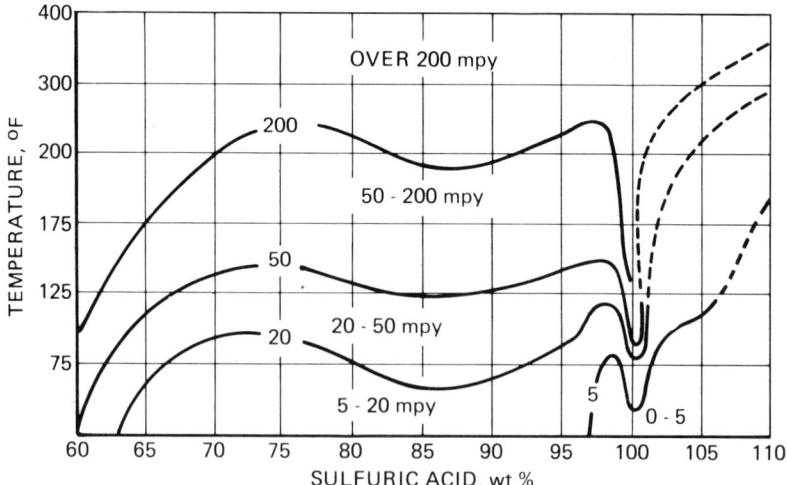

Figure 2-1. Corrosion of steel by sulfuric acid as a function of temperature.[1]

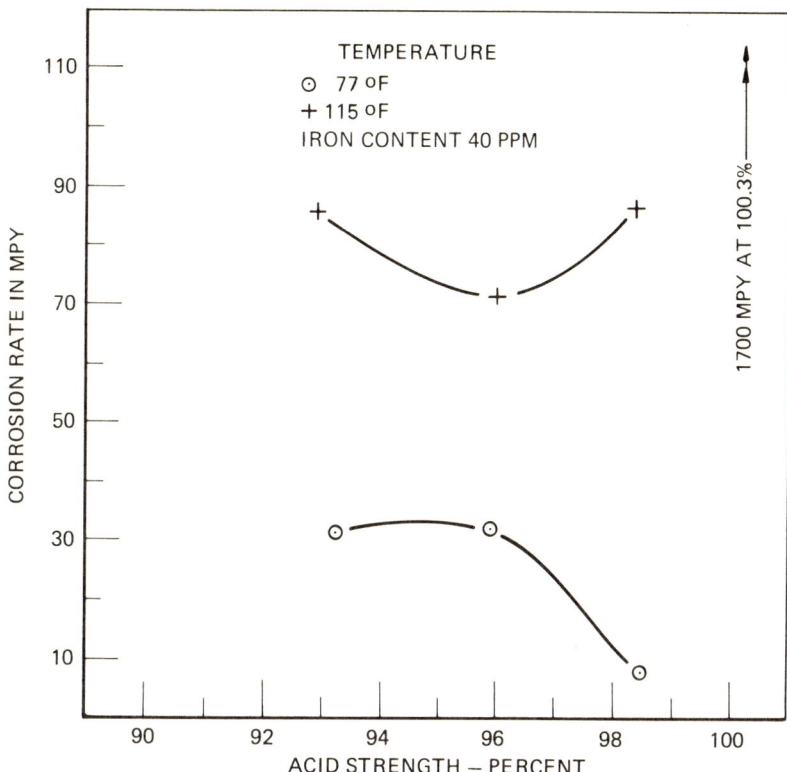

Figure 2-2. Corrosion rate of carbon steel in concentrated sulfuric acid as a function of acid strength.[5]

occurs in 83% acid.[2] Other published data do not agree with this. Foroulis[3] indicates that the 40% concentration is the most aggressive to carbon steel.

Corrosion rates of steel in 77–100% concentrations are in the range 20–40-mils (thousandths of an inch) per year (or mpy) or 508–1016 microns per year (μm/yr) at 24°C. The increase in corrosion rates at 100% concentrations is shown in Figure 2-1. Fisher and Brady[4] found rates of 27 mpy (686 μm/yr) in a steel tank at 100% concentration and 40°C. This is in close agreement with the data shown in Figure 2-1.

Fyfe, Vanderland, and Rodda[5] found that the corrosion rate of steel in concentrated sulfuric acid was a function of concentration, temperature, and iron content. Figure 2-2 illustrates laboratory data which indicate that a 21°C change in temperature greatly increases the corrosion rate. In addition, these data show that the corrosion rate in 98.5% acid is affected by a temperature

increase to a greater extent than it is by changes in concentration. Fyfe *et al.*
found that the iron content of the acid also affected the rate, as shown in Figure
2-3.

Corrosion rates of steel in sulfuric acid are characteristically low because
of the ferrous sulfate corrosion product in the acid which, when deposited on
surfaces, acts as a diffusion barrier and thus lowers the rate. This relatively
soft film can be easily swept away at low velocities, so equipment such as
pumps have much higher attack rates. This erosion–corrosion effect also can
be a problem, as pointed out by Fyfe *et al.,*[5] under inlet nozzles located near
a vessel sidewall. Splashing of acid during the filling of a tank can thin the wall
severely.

Austenitic stainless steels behave as active–passive metals in sulfuric acid,
and corrosion rates of passive steels are low. In many cases rates of less than
1 mpy (0.254 μm/yr) have been listed for AISI 300 series steels. However, if
the steels are in the active state, the rate can be well over 100 mpy (25.40 μm/

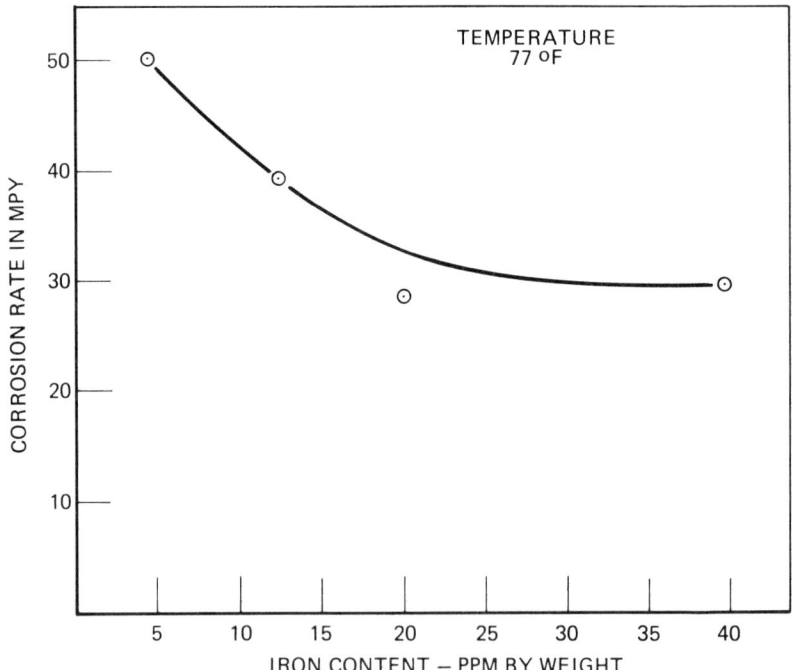

Figure 2-3. Influence of iron contamination in 93.5% sulfuric acid on its corrosivity to carbon
steel.[5]

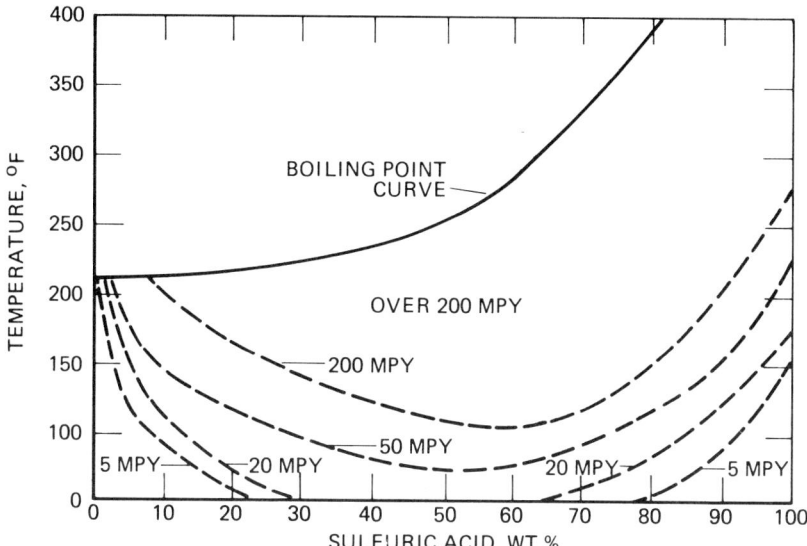

Figure 2-4. Corrosion of type 316 stainless steel by air-free sulfuric acid as a function of temperature.[1]

yr). Figures 2-4 and 2-5 illustrate corrosion rates of 300 series steels in air-free and aerated sulfuric acid.

Fyfe *et al.*[5] recently discussed a survey of sulfuric acid storage tanks used by Canadian Industries, Ltd. They cited a catastrophic rupture of a 3000-ton (2722-tonne) tank containing 93% acid. The failure resulted from the thinning of a wall near an inlet nozzle from which acid splashed during filling.

This same paper also pointed out that a small amount of copper (0.25%) in steel halved the corrosion rate. The steel in tanks about 25 years old usually contains about 0.4% copper, while most steel plate now being produced contains about 0.1% copper, a difference that portends possible trouble due to corrosion in sulfuric acid storage tanks in the next few years. Many plants and storage vessels have been made of steels with low copper content, owing to the rapid expansion of the fertilizer industry in the last 10 to 15 years.

Hydrogen Attack Is Also a Problem

Additional problems, such as the grooving of steel by hydrogen evolution, are documented. The grooves are commonly found in side manholes and side-mounted tank outlets, where they run circumferentially from 9- to 3-o'clock positions. Laboratory investigations confirm that the grooves result from hydro-

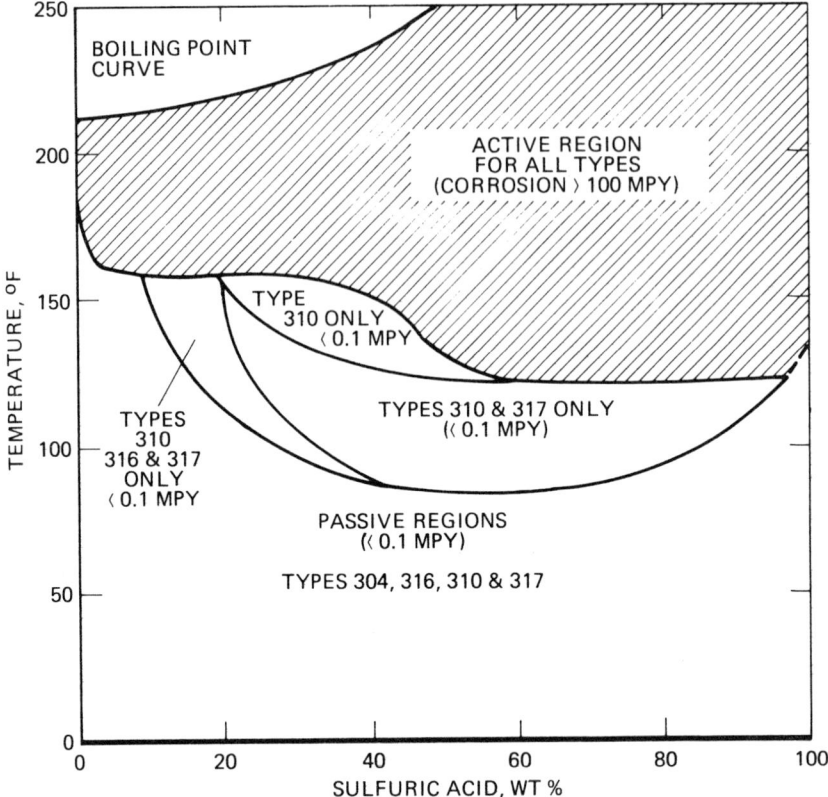

Figure 2-5. Passivity regions for stainless steels in aerated sulfuric acid.[1]

gen gas streaming along lines and disrupting the soft, protective iron sulfate film.

Blistering is also a common problem in sulfuric acid storage vesssels. As is well known, these blisters are caused by the accumulation of hydrogen in laminations, inclusions, or minute voids within a metal plate. The accumulation is often the result of diffusion of atomic hydrogen generated by corrosion. The hydrogen atoms penetrate the metal matrix until they reach a discontinuity, where they combine to form molecular hydrogen gas. Molecular hydrogen cannot readily diffuse through the metal.

The equilibrium constant for the reaction

$$2H^0 \rightarrow H_2$$

is 4×10^{36}.[6] This means the tendency for the formation of molecular hydrogen from the two atoms of hydrogen is very great. Therefore, for all practical purposes, molecular hydrogen will be the species in any void within the metal. Some think the pressure can build to extremely large values. One source says the pressure can be several hundreds of thousands of atmospheres. Schuyler has published interesting calculations concerning the pressures necessary to deform steel to produce blisters of various sizes.[7] Schuyler used the tensile strengths of the steels and the dimensions of the blisters observed in field storage tanks. The pressures necessary to form these blisters ranged up to 240 atmospheres (324.5 kg/cm^2). Therefore it may be assumed that other estimates of possible pressures are seriously in error.

Product Purity Is Important

Many attempts to control corrosion in sulfuric acid vessels are made to preserve product purity. Because there are many end uses and reactions for sulfuric acid in which a high iron content can be troublesome, a manufacturer can command a higher price for acid low in dissolved iron than he can for contaminated acid. As an example, electrolytic-grade sulfuric acid must not have over 50 ppm iron. It is almost impossible to store acid for any reasonable time in a bare or unprotected steel tank and maintain this level.

Several investigators have reported on the rates of iron pickup by various concentrations of acid. Riggs, Hutchinson, and Conger[8] described an oleum storage tank in which 105% acid had as high as 1273 ppm iron, which caused severe processing problems in a sulfonation plant owing to iron soap formation. Fisher and Brady[4] described a tank used for 100% acid in which the iron content increased from 31 to a total of 145 ppm. Fresh acid equivalent to about one-third the capacity of the tank was put into the vessel every four to six days. The acid produced a white iron hydroxide precipitate which gave it a cloudy to milky color due to the high iron content. This contamination caused problems in the production of various chemicals.

Sudbury, Locke, and Coldiron[9] told of similar iron pickup in a 98% acid storage tank from which the acid dissolved iron at the rate of 10 ppm/day to reach a maximum of 140 ppm/day. When this acid was used to produce linear alcohols, the iron content was a deterrent to acceptable product purity.

Kolotyrkin *et al.*[10] reported iron pickup in a storage vessel containing 94% sulfuric acid at 4–30°C. They found that the iron content increased from 110 to 280 ppm during 28 days storage.

It can be concluded from the above examples that the iron content may increase at rates of 5–20 ppm/day in unprotected storage tanks. It is obvious that the rate depends on various factors including acid concentration, vessel size, configuration or orientation (vertical or horizontal), acid residence time, and storage temperature. Consequently, to maintain a low iron content in acid, it is necessary to use some sort of corrosion control in storage vessels.

Protection of Sulfuric Acid Storage Equipment

The most common use of anodic protection has been to control corrosion of sulfuric acid storage tanks made of carbon steel. There are approximately 100 tanks now under protection in the United States and an unknown number elsewhere. Justification for corrosion controls on tanks has been given in the preceding section. Economic aspects of anodic protection are discussed in Chapter 5. Details of design, installation, equipment selection, and maintenance procedures follow in Chapters 3 and 4. This section concerns several anodic protection installations and emphasizes the results.

Table 2-1 lists literature references to the installation of anodic protection for sulfuric acid storage tanks, in most of which protection was applied to reduce iron pickup. Results of all installations have been favorable. As an example of this success, Figure 2-6 is a plot of the iron content of oleum stored in an anodically protected vessel for a year. The incoming oleum had an iron content of approximately 40 ppm. Table 2-2 lists the average iron content of 100% sulfuric acid stored in an anodically protected steel tank. These data were averaged over the months before and after the installation. Figure 2-7 is a plot of the iron content of 94% acid stored in a carbon steel tank with and without anodic protection. These data also indicate that, at a temperature of

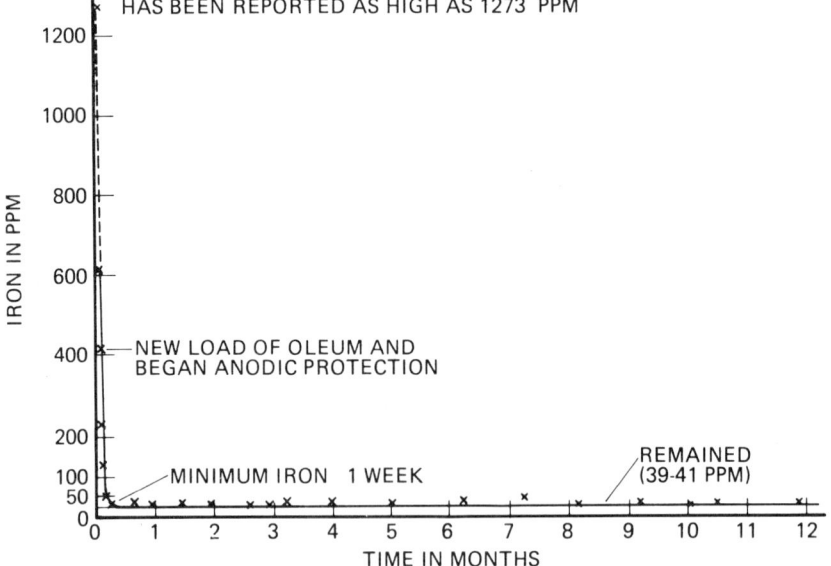

Figure 2-6. Effectiveness of anodic protection of oleum storage tank.[8]

TABLE 2-1. Anodic Protection of Steel Storage Tanks Containing Sulfuric Acid at Ambient Temperatures

Concentration (%)	References
Oleum	11,12
100[a]	4
98	10,12–15
93	3,5,16,17
90–94	16–25
77–oleum	19

[a]At 30°C.

TABLE 2-2. Average Iron Contents Before and After Anodic Protection of Storage Tanks, 100 % Sulfuric Acid[4]

	Iron (ppm[a])	
Location	Before	After
Discharge	145	35
Feed	31	26
Iron pickup	114	9

[a]Parts per million.

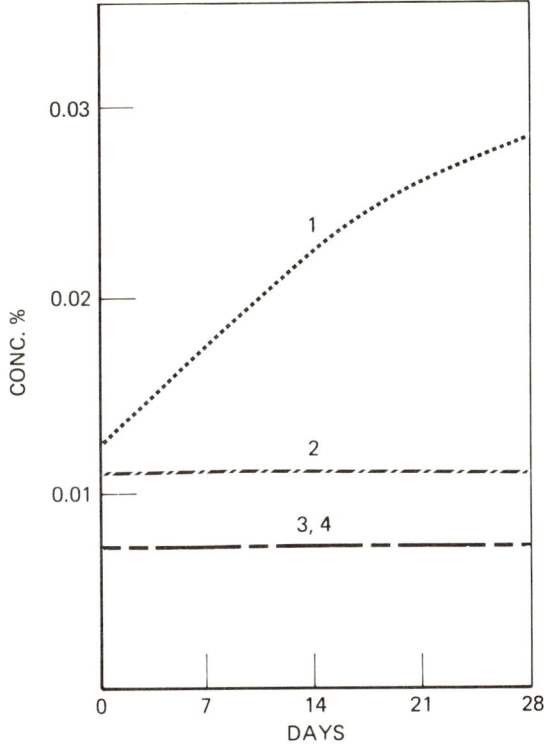

Figure 2-7. Rate of accumulation of iron in sulfuric acid: (1) At 4–30°C without protection; (2) At 4–30°C under protection; (3) Below 0°C under protection; (4) Below 0°C without protection.[10]

0°C or below, iron did not accumulate in the acid and at ambient temperatures (4–30°C), anodic protection drastically decreased the amount of iron pickup, as compared to that in the unprotected tanks.

In addition to the iron-content data cited, some investigators have calculated corrosion rates. Figure 2-8 illustrates data taken using corrosion coupons in a 10,000-ton (9072-tonne) 93% sulfuric acid tank in which corrosion rates in the protected case are approximately the same at all levels. This is important because in the upper zones specimens are alternately exposed and immersed, owing to fluctuating levels. These results may be interpreted as an indication that protection persists even though portions of the wall are not continuously submerged in fluid. Table 2-3 lists corrosion rates for 50 days from a storage tank containing 100% acid. The coupon at the three-foot level on the unpro-

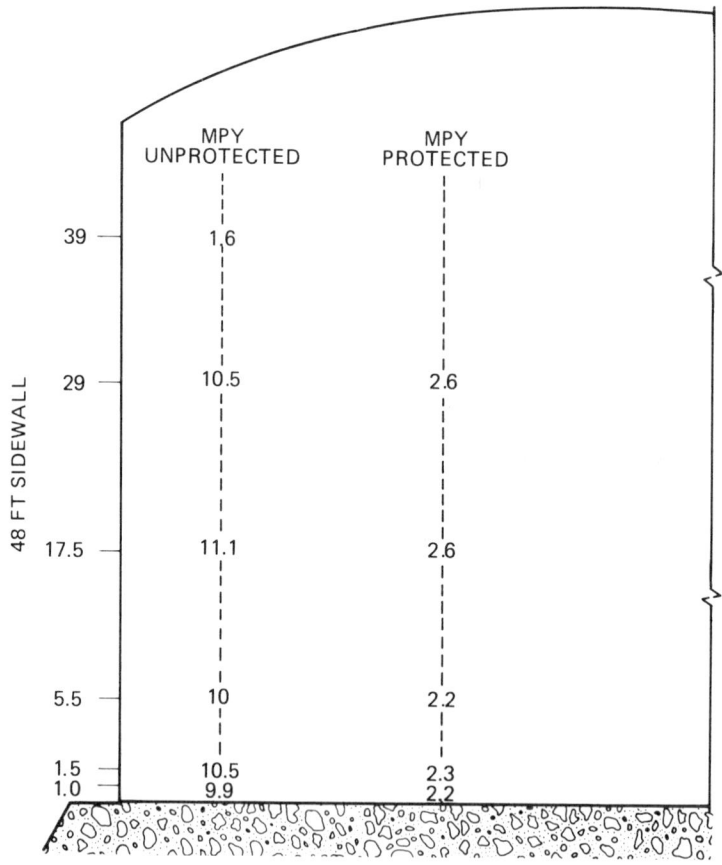

Figure 2-8. Effectiveness of anodic protection in a 93% sulfuric acid storage tank.[5]

TABLE 2-3. Results of Test on Anodically Protected and
Unprotected Coupons Exposed 50 Days in 100% Sulfuric Acid
Storage Tank[4]

Distance of coupon from tank bottom		Corrosion rates			
		Unprotected		Protected	
feet	cm	mpy	μm/yr	mpy	μm/yr
0	0.00	35.3	896.6	3.6	91.44
1	30.48	34.1	866.1	3.4	86.36
2	60.96	31.2	729.5	3.1	78.74
3	91.44	2.0[a]	50.8	3.2	81.26
4	121.92	29.9	759.5	3.5	88.90
5	152.40	22.1	561.3	5.8	147.3
6	182.88	4.3	109.2	5.4	137.2
7	213.36	6.3	160.0	5.3	134.6
8	243.84	3.8	96.5	4.1	104.11
9	274.32	2.2	55.9	1.4	35.66
10	304.80	0.8	20.3	0.9	22.88
11	335.28	0.8	20.3	0.8	20.33

[a]Coupon inadvertently connected electrically to the anodically protected tank wall.

tected tank was inadvertently connected electrically to the protected tank wall, so the coupon was thus also protected, resulting in a reduction of its corrosion rate by a factor of 10. Kolotyrkin *et al.*[10] reported that the corrosion rate of anodically protected steel in 94% acid with a mean temperature of 10°C was 10.5 times lower than that of unprotected steel.

These data indicate that anodic protection is an efficient means of controlling the corrosion of steel storage tanks in sulfuric acid service. This has been confirmed by the successful use of anodic protection for as long as 17 years on many vessels in the sulfuric acid industry.

An interesting citation concerning a system for protection of sulfuric acid "cisterns" was found in Polish corrosion literature.[18] This system, as described in an abstract, did not use a potential controller.

Application Details in Storage Tanks

Installation of anodic protection in a storage tank is shown schematically in Figure 2-9. There are two electrodes, control and dc power-supply circuitry, and a connection to the tank. A cathode (or multiple parallel cathodes for large tanks) is connected to the negative pole of the power supply, the positive pole

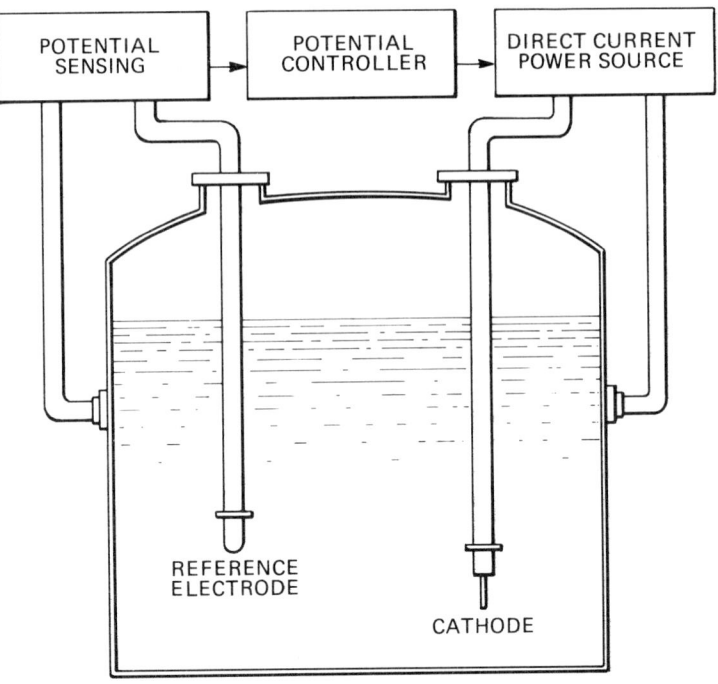

Figure 2-9. Anodic protection system schematic.

is connected to the tank wall, and control circuitry is connected to both the reference electrode and the tank wall. The controller monitors the potential between the reference electrode and the tank wall and adjusts the power output in such a way that the potential is maintained between preset–predetermined limits. Each of these elements will be discussed in detail later.

It is possible to protect several sulfuric acid storage tanks with one controller and one power supply. This is accomplished by switching the control and power circuitry between vessels in a time sequential method. This has been done in many installations,[19] beginning with a field application around 1962. Two similar systems were discussed in 1966 literature[12] and a patent was issued for this concept in 1969.[20] While holders of the patent were not the first to use the method, they were obviously the first to seek a patent. The switching mode improves the economic feasibility of anodic protection because, although electrodes are needed for each vessel, one set of control and power-supply circuits suffices for several installations. Up to four vessels have been protected by one controller.[19]

More dilute concentrations of acid are usually contained in stainless steel, concerning which two interesting examples have been mentioned in the Russian literature. A stainless steel (1Kh18N10T, containing 18% Cr and 10% Ni) measuring tank containing 50% sulfuric acid was anodically protected to reduce corrosion rates from 9.1 $g/m^2/hr$ to 0.003 $g/m^2/hr$ (400 to 0.13 in./yr).[21] Other investigators describe synergism resulting from the use of an inhibitor (Katapin A, reported to be C_nH_{2n+1}—C_6H_4—N—C_5H_5)[22] and anodic protection to control corrosion of the same type of stainless steel in 6.4% sulfuric acid at 10°C. The corrosion rate was lowered by a factor of 1900 by this system.[23]

Spent Sulfuric Acid

Many companies recycle sulfuric acid after it has been used. The recycled acid solutions have a wide range of concentrations of organic material and water and can be very corrosive to carbon steel.

Corrosion characteristics of some typical recycled sulfuric acids and the effectiveness of anodic protection against them are shown in Table 2-4. Other literature also contains information concerning experience with anodic protection of storage vessels and process tanks handling these corrosive solutions. Redden[24,25] described an anodic protection installation designed to control corrosion of a steel tank used for blending two types of spent acid from an alkylation process. Hayes[26] discussed multiple installations to protect steel vessels handling an 86% spent acid in which a switching mechanism was used to cycle one controller and one power supply to protect three vessels. Another system described involved one controller to protect two tanks handling 86% spent acid. The published results indicate that corrosion rates were lowered by 90% in the summer and 98% in the winter. Problems encountered included insufficient power-supply sizes, the effects of which were aggravated by the switching system.

Heat Exchangers for Sulfuric Acid

There have been interesting applications of anodic protection to processing equipment such as heat exchangers. The literature describes several instances in which heat exchangers in sulfuric acid service have been anodically protected, as summarized in Table 2-5. The most common use has been for acid coolers.

Sulfuric acid is manufactured by converting sulfur dioxide to sulfur trioxide in a catalytic-conversion step. The sulfur trioxide is then absorbed into concentrated sulfuric acid, in most cases, 98% solutions. The absorption is highly exothermic, so the acid must be cooled to control operating temperatures. Figure 2-10 is a flow diagram of this absorption–cooling system.[29]

TABLE 2-4. Corrosion Rates of Unprotected and Anodically Protected Mild Steel Coupons in Black Acid Solutions[2]

Type of acid	Solution composition, wt.%			°C	Liquid phase[b] (rate/yr)				Polarization potential (mV)	Current density[a]		Percent protection	Film life[c] (minutes)
	H₂SO₄	H₂O	Organic acid in water		Unprotected		Protected			mA/ft²	mA/m²		
					mpy	mm/yr	mpy	mm/yr					
Reprocessed alcohol acid[d]			97	60	21	0.53	3.4	0.9	+900	4.0	43.04	84	>15
Spent alcohol[d]			up to 91	26–60	45–77	1.14–1.96	8.3–1.6	0.21–0.04	+900–+1000	7.0–0.5	75.3–0.54	84	>15
Spent alkylation acid[e]	25–72.6	4.0–36.4	4.8–62.6	27–71	14–253	0.36–6.43	1.4–127	0.04–3.23	+800–+1300	3–187	32.3–2012.1	81–99	<1–5

[a] Near end of test
[b] Approximately 24-hr tests
[c] With coupons passivated at least 1 hr
[d] Quiescent
[e] Quiescent and stirred
[f] Some tests showed increased attack under anodic protection × 9.7.

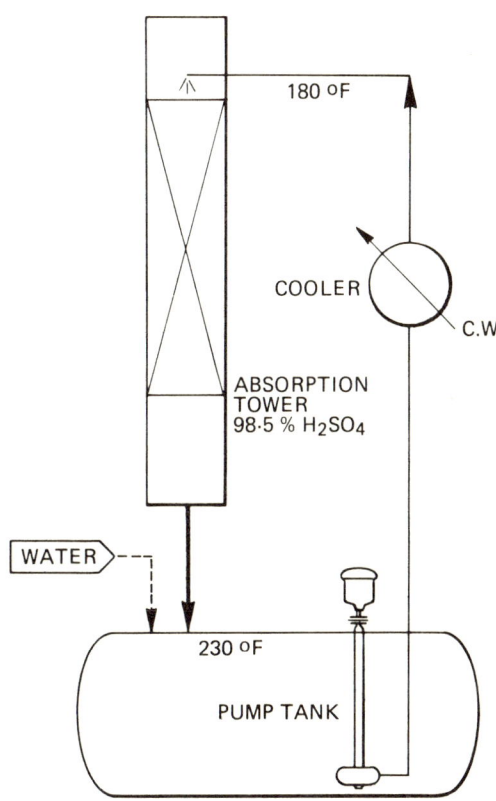

Figure 2-10. Typical arrangement of cooler, pump tank, and absorption tower.[29]

For many years the coolers were made of cast iron with acid inside and water trickling over the outside surfaces. These "trombone-like" coolers were a continuing source of maintenance problems, so most acid plants had a stockpile of spare coolers. In addition, iron pickup from the coolers was substantial.

Two companies have had heat exchangers designed with anodic protection

TABLE 2-5. Heat-Exchanger Applications in Sulfuric Acid

Concentration (%)	Metal	Temperature (°C)	References
77	1Kh18N9T	100–120	27
93, 98	AISI 316	115.6	28
93	AISI 316L	65.6	30
98	AISI 316L	110.0	29,31
3 + sodium sulfate, hydrogen sulfide	Titanium	<24	35,36

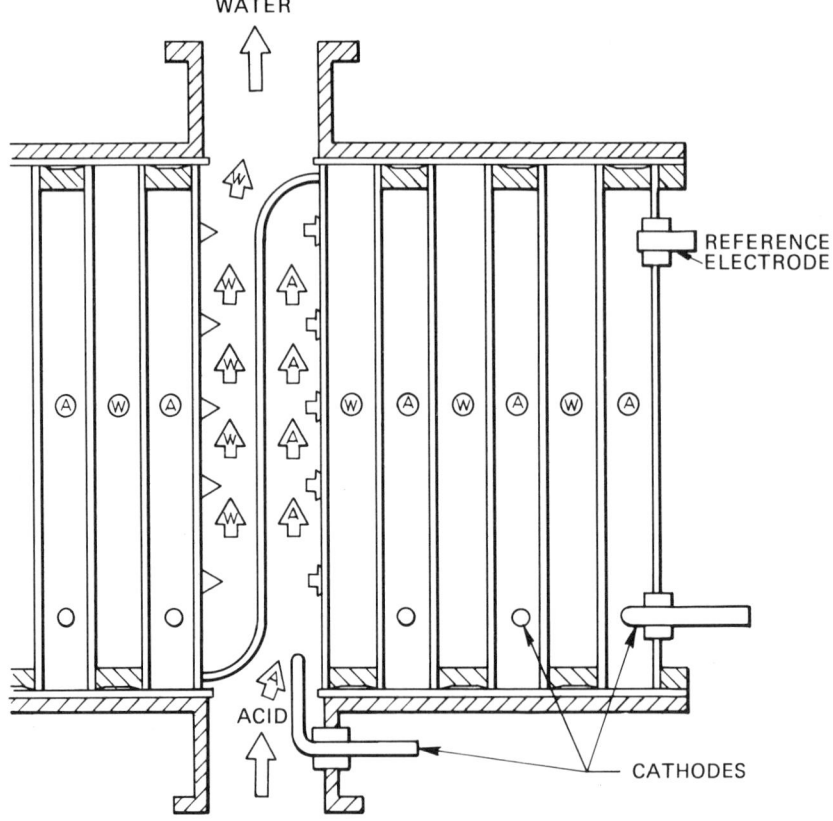

Figure 2-11. Cross section of spiral heat exchanger with anodic protection electrodes installed.[28]

systems as integral parts of the equipment to replace cast-iron coolers.[28,29] The first installation of this type was made in 1966 from AISI 304 steel to be used in cooling 93% acid at 66°C. The anodic protection system worked satisfactorily to protect the stainless steel from attack by the acid. Because the plant was located on the shores of the Great Salt Lake and water from the lake was used for cooling, the exchanger failed owing to stress–corrosion cracking caused by chlorides in the cooling water. This is a good example of a successful operation where the patient died.

The unit discussed above was a spiral-type exchanger, as shown in Figure 2-11. It has a helical spiral channel through which the acid flows, adjacent to which is another helical spiral channel through which the cooling water flows. The wire-type cathodes are discussed in Chapter 3. The operating parameters and results of using the anodic protection system are given in Table 2-6.

TABLE 2-6. American Heat
Reclaiming Spiral Heat Exchanger for
Sulfuric Acid

Metal	AISI 316
Type	Spiral
Concentration	93–98%
Temperature	<240°C
Corrosion rate	1 mpy (25.4 μm/yr)
Cathode electrode	Wire
Reference electrode	Proprietary

A brief mention of protection of a spiral cooler in Poland has been made.[30] These exchangers were constructed from a 23Cr–28Ni–3Mo stainless steel and were used for 93% sulfuric acid at 90°C. No description of the corrosion rates was given.

Another company has been very active in the commercialization of anodically protected shell and tube heat exchangers, which are shown in Figure 2-12. Acid is circulated on the shell side, and water on the tube side. Table 2-7 is a summary describing several similar applications. More than 50 plants around the world have exchangers like this now in use, cooling about 40×10^6 tons of acid a year.[31] Figure 2-13 is a photograph of three heat exchangers under anodic protection in a large sulfuric acid plant.

Tubes are made of various alloys resistant to the cooling water. If the water is not brackish, AISI 316L is used for tubes. In brackish water Incoloy 825, 904L, Alloy 20Cb3, or Hastelloy C is used for tubing, tube sheets, and waterbox liners. The shell side is made of 316L or 304L, depending on the acid strength.[29]

The exchangers have been designed carefully and constructed to minimize problems due to acid leaks, excessive acid turbulence, and stress corrosion on the cooling-water side.

TABLE 2-7. Chemetics Heat Exchanger for Sulfuric Acid[a]

	Temperature		Corrosion rates[b]				
			Unprotected		Protected		
Concentration (%)	°C	°F	mpy	mm/yr	mpy	μm/yr	Cathodes[c]
93	60	140 ⎱	200–400	5.08–10.16	1	25.4	5
99	110	230 ⎰					

[a]Metal: AISI 316 and others. Type: Shell and tube. Flow pattern: Shell side, acid; tube side, water.
[b]316 L.
[c]Proprietary.

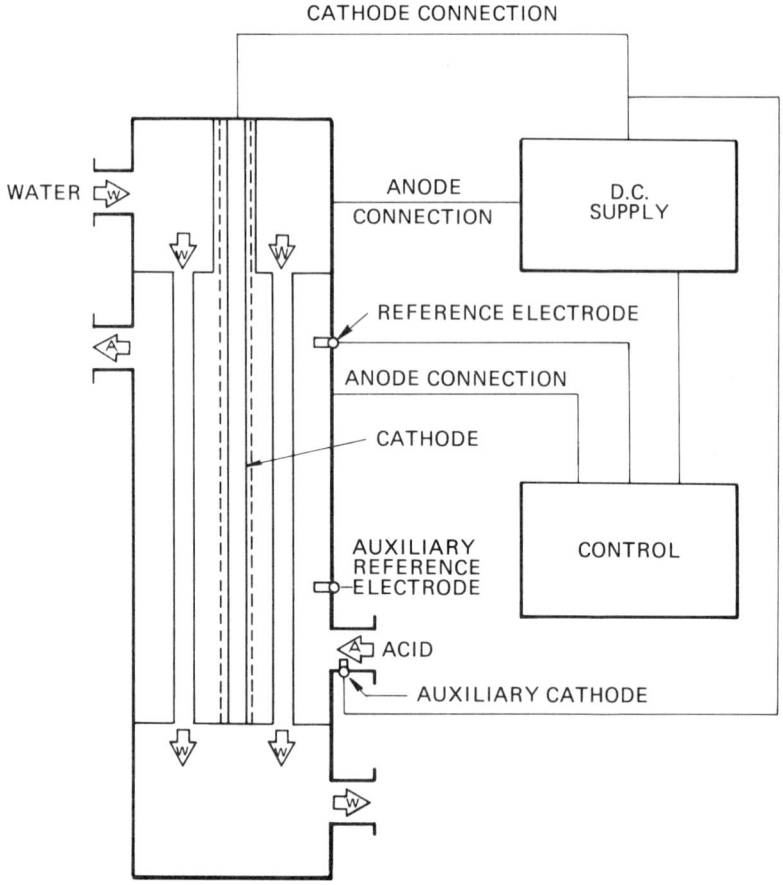

Figure 2-12. Anodic protection installation on shell and tube heat exchanger for sulfuric acid cooling.[29]

A description of another interesting heat exchanger in sulfuric acid service has been given in the Polish literature.[32] Double-pipe heat exchangers made of 1H18N9T and H17N13M2T were anodically protected from corrosion by concentrated sulfuric acid. The brief description available discloses that the lives of these coolers can be extended up to 12 years.

Two citations in *Chemical Abstracts* describe anodic protection of shell and tube exchangers in sulfuric acid service in France.[33,34] Protection of these heat exchangers, made of Z2CND17-13 and 2CN18-12 stainless steels, was reported.

Titanium tubes in an exchanger handling dilute sulfuric acid from a rayon-fiber spinning bath have been anodically protected.[35,36] The success of this application has led one company to construct an anodically protected evaporater for this service.[36] A recent publication from the USSR[37] disclosed work on a model heat exchanger for 93.6% acid. The investigators suggested that the heat exchanger be passivated in cold acid, then heated to operating conditions at 100°C. They observed no corrosion of the protected model in this test.

Recently an interesting application of anodic protection was made to large cast-iron containers holding boiling sulfuric acid.[38] These pots are used in the concentrations of 70–95% acid at 154°C. Corrosion rates up to 35 mm/yr (0.73 in./yr) have been reported in practice. Anodic protection was applied to

Figure 2-13. Anodically protected stainless-steel shell and tube sulfuric acid coolers in a large sulfuric acid plant. Chemetics International, Inc.[31]

these pots (which are 2.5 m in diameter by 2 m deep) after laboratory investigation proved that the corrosion rate could be lowered to 2.5 mm/yr. The stirrer was used as the cathode after it was properly insulated from the pot. A pyrometer pocket made of silicon cast iron is used as the reference electrode.

The pot was passivated initially when filled with 60% acid. Results of these installations have been encouraging, but the equipment has not been in service long enough to give a valid indication of extended life. Earlier laboratory work was reported on a similar system.[39]

Transport of Sulfuric Acid

Trial applications of anodic protection to control the corrosion of tanks used to move sulfuric acid from manufacturer to customers have been made. One of the authors of this book was involved with installations on steel-tank trucks for oleum and 93% acid transport. The trials were terminated when problems developed with the maintenance of the systems. The truck for 93% acid used a portable control and power system and required a larger, stationary, power supply. The latter was used to passivate the tank after it was filled, while the portable unit was used to maintain protection during the delivery trip. Good results from this system did not have sufficient impact to induce plant personnel to allocate time to the truck after it was filled or to the anodic protection system.

A successful tank-truck application is described in which a stainless steel tank was used for 93% and 99% acid. Anodic protection reduced iron pickup by 80–90%.[40]

Another brief mention of anodic protection of railroad tank cars in sulfuric acid service was reported elsewhere.[41] No details were given about the system or the results.

Protection Provided for Barge

A report describes anodic protection of a barge transporting sulfuric acid.[42] Additional information about this application revealed it contained several small carbon steel tanks with a capacity of 3000 tons (2721.6 tonnes) of 93% sulfuric acid. Before installation of anodic protection there was an increase in the iron content to 200 ppm during an eight-day haul. After the first two trips following the installation of protection, iron pickup dropped to 8 ppm for the trip and the acid was clear rather than milky, as it had been when iron pickup was 200 ppm. The barge was used for two years before being removed from service due to acid-marketing considerations.

Nitrate Fertilizer Container Protection

Increases in crop yields in the United States over the past few years have been due in large part to the augmented use of chemical fertilizers. Large-scale users of chemical fertilizers apply solutions containing various forms of nitrogen to the soil, including ammonium nitrate and other ammonia and urea combinations. These solutions can be corrosive to mild steel, so in many cases aluminum or certain chromium–nickel steels are used.

Anodic protection is very effective in controlling the corrosion of steel in these solutions.[43] Tables 2-8 and 2-9 give corrosion data on protected and unprotected coupons obtained during laboratory studies, indicating corrosion rates as high as 200–300 mpy (5000–7600 μm/yr). Table 2-9B contains corrosion rates of steel suspended in the vapor space above these fertilizer solutions. Many of these systems have zero corrosion rates in the liquid phase because steels in the solutions have two stable electrode potentials. One of them is the active state in which the electrode will corrode and the other potential is within the passive potential region. Once the potential is shifted to the passive range it will not move unless some process upset occurs to force the potential back to the active range. So no current is required and the steel does not corrode. It is important to note that 83% ammonium nitrate with a pH of 3 will severely corrode steel when it is anodically polarized, but that other combinations can be protected anodically.

An unpublished survey of anodic protection installations issued during 1969 listed seven vessels as large as 92 ft (28 m) in diameter in which nitrogen fertilizer solutions are handled under anodic protection. Table 2-10 lists installations of anodic protection on fertilizer-handling equipment and the sources of information about them.

Kuzub *et al.*[45] describe results when using anodic protection on a steel vessel containing a solution of NH_4NO_3, NH_3, $(NH_4)_2CO_3$, and urea for one year. An identical unprotected tank was used for comparison. Corrosion coupons were placed in both. Coupons from the protected vessel did not lose weight, while coupons from the unprotected vessel had a 1.8-mm/yr corrosion rate. The authors said that the unprotected vessel developed leaks and "the solution became dark brown in color from the corrosion products." Windows were cut in both vessels and the interiors were inspected. The protected vessel's wall was clean and free of indications of corrosion. Welds appeared to be in their original state. Internal surfaces of the unprotected vessel were corrosion-damaged, welds were split, and leaks had occurred in the joints. This installation was preceded by an earlier study in which smaller vessels were successfully anodically protected.[46]

Sudbury, Banks, and Locke[43] reported on the protection of an 11,000-gal

TABLE 2-8. Results of Corrosion and Polarization Studies of Carbon Steel in Liquid Fertilizer Solutions[a]

Solution	Concentration (%)	pH	°C	Liquid-phase rate (24 hr) Unprotected mpy	mm/yr	Protected mpy	mm/yr	Polarized potential (mV)	Current density mA/ft²	mA/m²	Vapor-phase rate mpy	µ/yr
Ammonia	28–30	13–14	27	0–350	0–8.89	0.1–0.5	0.00254–0.0727	−200–+200	0.3	3.228		
Ammonium nitrate ± ammonium hydroxide	67	4.5–8	27	0.3–350	0.00762–8.89	0–0.5	0–0.0727	+500	0.6–0.7	6.456–7.532	0.85–66	22–1676
	83	3–7	74–93	1.3	51.18	0–35	0–0.889	+600–+800	0.5–95	5.38–1022.2	0.3–20	7.6–508
Commercial solutions[b]		11	27	59–101	1.4986–3976.4	0	0	+100–−100	0.4–1.0	4.304–10.76	9.6–11	244–279

[a]Summarized data from Table 1 in Reference 43.
[b]%: 19–25 ammonia; 65–68 ammonium nitrate; + or − 6 urea; 6 to 9 water.

TABLE 2-9. AISI 1020 Steel Coupons in Nitrogen Fertilizer Solutions (27°C)[a,b]

Solution components	Wt. %	pH	Liquid phase[f]				Potential[g] (mV)	
			Unprotected		Protected		Unprotected	Protected
			mpy	μm/yr	mpy	μm/yr		
Ammonia	Trace–30.6							
Ammonium nitrate	0–69	7–10	0–230	0–5842	0	0	−180 − −1200	−100 − −250
Urea	0–43							
Water	6–30							

[a]Summarized from Table 1 in Reference 44.
[b]Inhibitor concentration: 0–0.1%. Current density[c] (mA/ft²): 0–0.5 (5.38 mA/m²) cathodic. Percent inhibition: −7– +100. Flade current density[d] (mA/ft²): 1–100 (0.01082–1.08 A/m²). Passive film life, days[e]: 9–30.
[c]At end of 24-hr test.
[d]From polarization curves.
[e]From 30-day tests with coupons initially passivated for one hour.
[f]Approximately 24-hr tests.
[g]Active to saturated calomel electrode at 27°C.

TABLE 2-9B. Vapor Phase Corrosion Rates above Solutions Listed in Table 2-9A[a]

	Suspended in vapor		Barely touching solution	
Unit	24 hrs	7 days	24 hrs	7 days
mpy	0–36	1–8	8–135	19–94
μm/yr	0–914	25.4–203	203–343	483–610

[a]Summarized from Table 1 in Reference 44.

TABLE 2-10. Anodically Protected Equipment in Fertilizer Solutions

Solution	Metal	Temperature (°C)	References
Ammonium nitrate Urea Ammonia Water	Steel	40	45,46
Ammonium nitrate Urea Ammonia Water	Steel	27	44
Ammonium nitrate Ammonia Water	Steel	27	43
N–P–K fertilizer solutions— ~10% KCl	OKh23N28M3DT	Ambient	47

(290.6-m^3) carbon steel vessel containing an ammonia–ammonium nitrate solution.

Steel has an interesting corrosion behavior in aqueous ammonia, which has attracted the attention of various investigators.[48–51] In many instances corrosion has not been severe under ambient temperatures. However, contamination of ammonia by iron oxide resulting from corrosion can be troublesome. Among the effects noted have been the severe pitting of tank bottoms,[48] with widely varying behavior resulting from the formation of oxygen concentration cells. Higher corrosion rates are measured in areas deficient in oxygen.

Interesting and unique features have been reported about two vessels under anodic protection. One 33.4-m-i.d. × 12-m-high vessel containing 25% aqueous ammonia with a 10,000-m^3 (2,628,922 gal) capacity was anodically protected by first being filled with a dilute ammonia solution and by then turning on the control system, after which the solution concentration was increased to 25%. The control system had sufficient capacity to maintain passivity but not to establish it at the higher concentration. Corrosion decreased to 0.0033% of the unprotected rate.

In a recent paper, Szymanski[50] described an anodic protection system using a galvanic couple rather than impressed current. A 20,000-gal (75.7-m^3) vessel containing a 28–30% ammonia solution has been under protection for more than two years using scrap titanium pipe hung from the vessel's top, with another segment of pipe bonded to the bottom to act as galvanic cathodes. The author said that no contamination or discoloration of the solutions has occurred.

Other Interesting Uses for Fertilizer Solutions

There have been some very interesting applications of anodic protection to the equipment used to transport fertilizer solutions. Two patents issued in 1945 covered the application of anodic protection to steel-tank cars carrying nitrogen fertilizer solutions and the protection of steel in maleic acid.[52,53] Both were written to cover anodic protection of chemically passivated surfaces. However, the voltage applied between the anode and cathode was controlled to fall in the range necessary for sufficient current for protection, but less than that at which the solution electrochemically decomposed. For ammonium–ammonium nitrate solutions, the range was 1.25–2.2 V when using an aluminum cathode. These voltages had to be reduced when an inert or iron cathode was used, because decomposition of the solution occurred at lower voltages. This was attributed to differences in back emf between the various metals. A test described in one of the patents[52] concerned passivating a tank car containing a solution consisting of 60% ammonium nitrate, 20% ammonia, and 20% water. The 850-gal (3.22-m^3) tank was passivated by anodic current as it was being filled, the current varying throughout the test, probably owing to buildup of the passive film. The vessel was protected for ten days after the source of the current was disconnected.

It is interesting to note that these patents have been overlooked in the accounts of anodic-passivation history. It may be that they have not been cited because the potential of the vessel as an anode was not controlled. The investigators were fortunate in that it is relatively easy to anodically protect steel in nitrogen fertilizer solutions.

Banks and Hutchison[44] describe an ambitious project in 1966 involving 136 10,000-gal (37.85-m³) steel railroad tank cars in nitrogen fertilizer service which were anodically protected. In 1967 a second experiment was conducted with 40 10,000-gal steel rail tank cars. Both tests were preceded by another program in 1962–1963 with two or three cars. Problems which developed during the 1962–1963 program were solved mostly in the 1966 experiment, and a few others, like cathode leaks and sticking valves, were solved during the 1967 program.

The cars were designed for liquefied-petroleum gas (LPG) service which peaks in winter months and which, therefore, made them available for the peak period for fertilizer service in early spring. Figure 2-14 is a schematic drawing of the mounting arrangement of the anodic protection system.

Iron-content and coupon-corrosion-rate data indicated that the anodic systems provided at least 93.5% protection. Coupons exposed 150 days with protection were unchanged, while unprotected coupons lost approximately two-thirds of their initial weight.

The economies of using anodically protected steel cars compared to alu-

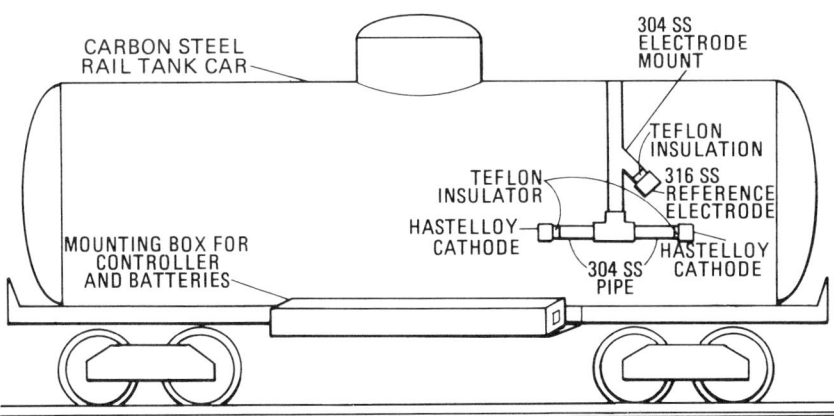

Figure 2-14. Schematic diagram—mounting arrangement for Model 5300 Anotrol system.[44]

minum cars were calculated. It was found in 1969 that use of the protected steel cars would save $900 to $1500 per car annually. Although it would be possible to use the steel cars for LPG service during the fertilizer off-season, credit for this factor was not included in the calculation.

All data from experience with the anodic protection of steel in nitrogen fertilizer service indicate it should be a prime commercial market for the technique. Surprisingly, the market is not being pursued to any great extent because companies involved in marketing anodic protection have chosen to concentrate on sulfuric acid storage and cooling applications.[31]

Pulp and Paper Industry Applications

One of the earliest large-scale applications of anodic protection was in the pulp and paper industry. A kraft digester was protected during the late 1950s, and during the 1960s several other vessels and units of process equipment were anodically protected. These installations are described below.

Mueller and Watson[54] obtained a patent entitled "Anodic Protection of Kraft Digesters," which was filed in 1958. They describe an installation on an 88-ton-capacity (79.83-tonne) steel digester. Watson[55] later discussed this installation in detail, but before reporting what he said, an introduction to digester operation processes is in order.

Paper is made from cellulose fibers recovered from a natural source like wood. The cellulose must be separated from lignins and other compounds before it can be processed into paper products. The first chemical step is to digest wood chips in a caustic-based solution at high temperatures. In the kraft process, the chemical solution is made of sodium hydroxide and sodium sulfide. A fresh makeup solution called "white liquor" and a recycled liquor called "black liquor" are used together to form the digesting chemical. The process, carried out in a batch-type operation at 350°F (177°C), results in an operating pressure of about 115 psig (792 kPa). These conditions translate into a serious corrosion problem for these large steel vessels, which can be potentially very dangerous if allowed to continue unchecked until the wall thickness of the digester decreases until it cannot sustain the operating pressure.

Watson[55] said that the digester was in an active corrosion state during the first half of each cook, after which the steel walls became naturally passive, owing to the oxidizing effects of polysulfides, thiosulfates, and other oxidizing agents in the black liquor. The goal of the anodic protection installation was to passivate the metal at the beginning of the batch and maintain it in this state during the first half of the operation. Watson's anodic protection method applied very large volumes of current (4000 A) for three minutes. A timer then actuated a switch which lowered the current to 2700 A for 12 minutes. A current of 600 A was then applied for the remainder of the process. The wall potential was monitored but not controlled. These very large currents were probably required because of the oxidation and reduction reactions within the digesting liquor, as well as from the passivation process at the steel wall.

Watson described successful reduction in corrosion rates for a digester protected for more than a year. Table 2-11 lists results of thickness measurements made on several digesters. The course designations refer to the heights at which measurements were made on the digester walls. The top of the digester is designated as the first course. Although other installations were made on digesters, they were not very successful. There are now no known anodic protection installations on batch digesters, most of which rely on welded overlays of 300 series chromium–nickel steels for protection.

Locke[56] described an experience with several other vessels and units of process equipment in pulp and paper mills. A washer drum made of AISI 316 wire was anodically protected. This drum was of the same type of construction as a rotary vacuum filter and was rotating in a pulp–water mixture containing chlorine dioxide. Anodic protection extended the life of the washer wires.

Black liquors from the kraft process are recycled and, during various stages of operation, are designated by color. Table 2-12 lists laboratory data for three such liquors versus 1020 steel which show that the green and oxidized heavy black liquors are corrosive to steel, while the weak black solutions are not. Anodic protection substantially reduced the corrosion rates of steel exposed to green and heavy black liquors.

A field installation on a steel vessel used at different times to hold all of these liquors is also described. Visual inspection after two years' operation indicated that corrosion had been reduced by anodic protection.[56]

Mueller investigated the anodic and cathodic protection of steel in white and green liquors. He found that both methods reduced the corrosion to about the same level, but that anodic protection required lower current densities.

TABLE 2-11. Corrosion Rates of Paper-Mill Digesters: Effect of Anodic Protection[a,b]

	Average losses				
	Courses 1,2,3		Cone		
Number of digesters studied	mpy	mm/yr	mpy	mm/yr	Years observed
Five not protected	37.83	0.96	65.06	1.65	9.69
One protected	2.40	0.06	10.70	0.27	1.66
Percent protection	99.90	99.90	99.80	99.80	

[a]Taken from Reference 55.
[b]All rates are averaged for each course.

TABLE 2-12. Anodic Protection of 1020 Steel in Recovery Liquors[56]

Liquor	Temperatures		Corrosion rates				Polarization potential (mV)	Current density	
			Unprotected		Protected				
	°F	°C	mpy	μm/yr	mpy	μm/yr		mA/ft²	mA/m²
Green liquor	160	71	112	2844	0.64	16	−700	11.60	124.6
Oxidized heavy black liquor	182	83	28	711	1.80	46	−600	14.40	154.7
Weak black liquor	170	76	2.3	58	1.70	43	−600	5.76	61.9

Polysulfides formed by dissolving elemental sulfur in the liquor also reduced corrosion, owing to the chemical passivation they induced.[57]

Miscellaneous Applications

There have been several other applications of anodic protection on metals exposed to a variety of corrosive solutions. This section includes descriptions taken from the literature.

Anodic protection has been used to protect stainless steels in electroless nickel plating solutions. These solutions contain a mixture of organic and inorganic salts from which metallic nickel is plated by chemical reduction on the metal parts exposed to them in a bath. The reduction step is initiated by corrosion-generated hydrogen. If the container of the solution begins to corrode, the plating reaction will begin on the walls of the container and continue until the solution is exhausted, so it is important to prevent corrosion of the container walls. Anodic protection systems to accomplish this have been described,[58-60] indicating successful prevention of the plating of nickel on the container walls and associated tubing. However, the cathode does become covered with nickel, because hydrogen generated at the cathode in turn reduces nickel from the solution. The electrodes must be cleaned periodically and the nickel recovered. One author[59] reported on a system in which the cathode was placed in a porous ceramic beaker filled with a 10% solution of sodium hydroxide which successfully eliminated the nickel-deposition problem.

Perrigo[61] reported on laboratory and pilot-scale tests of anodic protection in a 0.1–0.7 molar oxalic acid solution, in which he found that anodic protection provided 40–50% protection to carbon steel over a 22–50°C range.

Application of anodic protection to stainless steel (1KH18N10T) in a "chromeammonium alum" solution was described in the Russian literature.[62] Small electrodes and a 14,161-cm^3 (3.7-gal) vessel were utilized in the study. This application was intended to reduce accumulations of iron and other metal impurities. These laboratory and pilot-scale tests demonstrated that it is possible to protect stainless steels in this environment, to 70°C.

Robinson and Golant[63] described anodic passivation of 18Cr–18Ni–3.5Mo–2.5Cu in nitric acid concentrations of less than 10%. Ammonia increased passivation, but formic acid decreased it, while sulfur dioxide reduced the passivation range, but decreased the average passive current density. They also reported that titanium can be protected in nitric acid, but suffers catastrophic attack when ammonium fluoride is added.

Kuzub and Kachanov also studied the passivation of several stainless steels in 1.5 N nitric acid containing 0.1–3.0-M/liter NaCl.[64] They tested 1Kh18N9T, Kh17, and Kh27 stainless steels in these environments. It was possible to anodically protect 1Kh18N9T in 1.5 nitric acid with 1 M/liter NaCl and reduce the corrosion rate by a factor greater than 2000.

Anodic protection of stainless steels in hydroxylamine sulfate was investigated and applied in plant applications.[65-67] Hydroxylamine sulfate is a precurser to caprolactam, which is used in Nylon 6 production. It was found in the laboratory work that 1Kh18N10T, Kh21N6M2T, OKh21N5T, and Kh18N12M3T stainless steels could be protected in this corrosive. At 40° and 100°C the 1Kh18N10T alloy had a 200–733% reduction in corrosion with protection. A pilot-scale apparatus consisting of two 650-liter-capacity vessels was constructed. One vessel was made of OKh21N5T and the second was made of OKh21N6M2T steel. The results of this study indicated that it was possible to protect these vessels up to 60°C.

These data were used to scale up to a plant application.[66] A brief description has been given concerning the protection of carbon steel tanks containing hydroxylamine sulfate.[67] This application reduced the iron content of the product caprolactam by a factor of 7500 times. This improved product quality and color.

There are several miscellaneous systems in which anodic protection has been proven suitable listed in the literature. Table 2-13 lists several of these. Some of the systems listed in Table 2-13 were studied in the laboratory only, but pointed toward proving the feasibility of field applications.

A particularly interesting laboratory study indicated that anodic protection could be used to prevent corrosion fatigue.[74] This work was done with carbon steel, 13% Cr steel, AISI 304, and AISI 316 stainless steels. These metals were exposed to ammonium nitrate and 10% H_2SO_4 at room temperature. The stress required to cause corrosion-fatigue failure was higher for the metals

TABLE 2-13. Miscellaneous Applications of Anodic
Protection

Corrosive solution	Metal	References
NH_4HCO_3	OKh23N28M3D3T	68
$NH_2OH—SO_4^{2-}$	OKh23N28M3D3T	69
H_3BO_3 with NH_3, KOH, LiOH	ST20	70[a]
NaCNS	OKh17N13M3T OKh21N6M2T	71[a]
Formic acid reaction mix	Cr–Ni–Mo alloy	72
$CaCO_3—NH_4NO_3$	Carbon steel	73[a]

[a]Laboratory investigation only.

under anodic protection than for the unprotected specimens. In a few cases the protected samples had higher stress levels for failure than the samples fatigued in air.

Summary

It seems to be probable that present-day imperatives with respect to materials and energy conservation, avoidance of safety hazards, and ecological contamination may induce many to consider using anodic protection who have not heretofore given it favorable attention. Although practical and successful applications became a reality more than 15 years ago, the method is not well known among persons concerned with corrosion control. Even among those who have heard of it, the method is only vaguely understood and is often confused with cathodic protection.

It is the expectation of the authors of this book that more extensive use of anodic protection will follow a better understanding of its principles, its application parameters, and economic benefits.

References

1. M.G. Fontana, *Ind. Eng. Chem.* **43**(8), 652 (1951).
2. C.E. Locke, W.P. Banks, and E.C. French, *Mater. Prot.* **3**(6), 50 (1964).
3. Z.A. Foroulis, *Ind. Eng. Chem. Process Des. Dev.* **4**(1), 20 (1965).
4. A.O. Fisher and J.F. Brady, *Corrosion* **19**, 37t (1963).
5. D. Fyfe, R. Vanderland, and J. Rodda, *Chem. Eng. Prog.* **73**(3), 65 (1977).
6. *Handbook of Chemistry and Physics,* R.C. Weast, ed., Chemical Rubber Co., Cleveland, Ohio 1969.

7. R.L. Schuyler, III, *Mater. Perform.* **18**(8), 9–16 (1979).
8. O.L. Riggs, Jr., M. Hutchinson, and N.L. Conger, *Corrosion* **16**(2), 47t (1960).
9. J.D. Sudbury, C.E. Locke, and D. Coldiron, *Chem. Process.* 23, February 1963.
10. Ya.M. Kolotyrkin, *et al., Zashch. Met.* **7**(6), 722 (1971).
11. J.D. Sudbury, and C.E. Locke, *Oil Gas J.* **61**(43), 111 (1963).
12. C.E. Locke, M. Hutchinson, and N.L. Conger, *Chem. Eng. Prog.* **56**(11), 50 (1960).
13. C.E. Locke, *Tappi* **49**(1), 61A (1966).
14. J.C. Redden, *Mater. Prot.* **5**(2), 51 (1966).
15. P. Neufeld, and R.C. Williamson, *Corros. Sci.* **5**(9), 605 (1965).
16. V.S. Kuzub, *et al., Koks. Khim.,* No. 10, 48 (1973); *Ca* **80**, 9733 (1974).
17. Staff Feature, *Mater. Prot.* **2**(9), 69 (1963).
18. A. Andrezej, *Ochr. Koroz.* **19**(1), 19 (1976); *Ca* **85**, 53593g (1976).
19. Magna Corp., private communication.
20. J.F. Delahunt, and R.A. Haisch, U.S. Patent No. 3,483,101, December 9 1969.
21. V.S. Kuzub, *et al., Khim. Tekhnol. (Kiev),* No. 1, 63 (1974); *Ca* **81**, 57504 (1974).
22. R.B. Perry, and F.F. Lyle, *Mater. Perform.* **15**(8), 38 (1976).
23. N.A. Supronov, Kh. Freid, N.S. Baburina, and Tr. Ivanov, *Khim. Tecknol. Inst.* **12**, 180 (1970); Ca, **79**, 142218g (1973).
24. J.C. Redden, *Mater. Perform.* **5**(2), 51 (1966).
25. J.C. Redden, *Chem. Process.* October 1964.
26. L.R. Hays, *Mater. Perform.* **5**(9), 46 (1966).
27. Ya.M. Kolotyrkin, *et al., Zashch. Met.* **1**(5), 598 (1965).
28. American Heat Reclaiming Co., private communication.
29. D. Fyfe, *et al., Corrosion*/75, NACE, Houston, TX.
30. A. Antoniuk, *Ochr. Koroz.* **18**(4), (1975); *Ca* **83**, 101641d (1975).
31. Chemetics International., Ltd., Willowdale Ontario, Canada.
32. W. Michalik, M. Klinek, and J. Lizkowski, *Ochr. Koroz.* **21**(3), 69 (1978).
33. M.S. Boiton, *Aciers Spec.* **27**, 17 (1974); *Ca* **84**, 113382 (1975).
34. J. Montuelle, *Rev. Metall.* **75**(11), 641 (1978); *Ca* **90**, 78320g (1979).
35. P.E. Morgan, and L.S. Evans, *Mater. Prot.,* **4**(1), 60 (1965).
36. L.S. Evans, P.C.S. Hayfield, and M.C. Morris, Proceedings of the 4th International Congress on Metal Corrosion, NACE, Houston, TX (1972), p. 625.
37. V.A. Makarov, *et al., Zashch. Met.* **13**(2), 181 (1977).
38. W.A. Ashby, L.S. Evans, and W. Shephard, *Br. Corros. J.* **13**(2), 85 (1978).
39. E. Maahn, *Br. Corros. J.* **1**, 350 (1966).
40. C.E. Locke, *Mater. Prot.* **4**(3), 59 (1965).
41. S.P. Napreenko, V.A. Timonin, E.V. Uvarov, A.G. Tokarenks, and V.S. Kuzub, *Issled. Zashch. Met. Koprozii Khim. Promst.,* 55 (1978); *Ca* **91**, 29485 (1979).
42. F.W.S. Jones, *Anticorrosion,* December 1976.
43. J.D. Sudbury, W.P. Banks, and C.E. Locke, *Mater. Prot.* **4**(6), 81 (1965); *Extended Abstracts of the 2nd International Congress on Metal Corrosion,* NACE, Houston, TX (1966).
44. W.P. Banks, and M. Hutchison, *Mater. Prot.* **8**(2), 31 (1969).
45. V.S. Kuzub, *et al., Zashch. Met.* **7**(3), 361 (1971).
46. L.G. Kuzub, V.I. Gnezdilova, and V.S. Kuzub, *Zashch. Met.* **4**(5), 564 (1968).
47. V.S. Kuzub, *et al., Khimi. Promst. (Moscow)* **8**, 609 (1974); *Ca* **82**, 65939a (1975).
48. L.M. Dvoracek and L.L. Neff, *Corrosion* **18**, 85t (1962).
49. A.I. Tsinman, L.A. Danielyan, and V.S. Kozub, *Zashch. Met.* **9**(2), 156 (1973).
50. W.A. Szymanski, *Mater. Perform.* **16**(11), 16 (1977).
51. L.A. Danielyan, *et al., Zashch. Met.* **9**(4), 492 (1973).

52. C.K. Lawrence, and R.F. Engle, U.S. Patent No. 2,366,796, January 9 1945.
53. C.K. Lawrence, and R.F. Engle, U.S. Patent No. 2,377,792, June 5 1945.
54. W.A. Mueller, and T.R.B. Watson, U.S. Patent No. 3,009,865, November 21 1961.
55. T.R.B. Watson, *Mater. Prot.* **3**(6), 54 (1964).
56. C.E. Locke, *Tappi* **49**(1), 61A (1966).
57. W.A. Mueller, *Pulp and Paper Industry Corrosion Problems, Vol. 2,* NACE, Houston, TX (1977).
58. W.I. Clark, B. Griggs, D.D. Hays, and G.F. Jacky, U.S. Patent No. 3,347,768, October 17 1967.
59. A.V. Rybshenkov, *et al., Zashch. Met.* **7**(6), 718 (1971).
60. C.E. Locke, U.S. Patent No. 3,375,178, March 26 1963.
61. L.D. Perrigo, *Mater. Prot.* **5**(5), 73 (1966).
62. E.S. Letskikh, and A.G. Komornikova, *Zashch. Met.* **5**(3), 300 (1969).
63. F.P.A. Robinson, and L. Golant, *Proceedings of the 2nd International Congress of Metallic Corrosion, New York, 1963.* NACE, pp. 290–299, Houston, TX (1963).
64. V.S. Kuzub, and V.A. Kachanov, *Zashch. Met.* **2**(3), 358 (1966).
65. V.S. Kuzub, *et al., Zashch. Met.* **4**(2), 199 (1968).
66. V.S. Kuzub, *et al., Zashch. Met.* **4**(4), 362 (1968).
67. V.S. Kuzub, *et al., Khim. Prom. (Kiev)* **49**(4), 267 (1973); *Ca* **79**, 26408v (1973).
68. C-C. Chen, Hua Hsueh Tung Pao, No. 2, 117 (1974); *Ca* **83**, 30415v (1975).
69. V.S. Kuzub, *Tr. Ukr. Resp. Konf. Elektrokhim.* **2**, 88 (1973); *Ca* **81**, 85146 (1974).
70. Zh.L. Georgiev, *et al., Dokl. Bolg. Akad. Nauk.* **26**(12), 1657 (1973); *Ca* **80**, 140544 (1974).
71. V.S. Kuzub, *et al., Khim. Promst.,* **2**, 129 (1977); *Ca* **86**, 112906g (1977).
72. P. Novak, *et al., Chem. Prum.* **28**(9), 461 (1978); *Ca* **90**, 158947g (1979).
73. S. Mladinovic, *Hem. Ind.* **33**(6), 223 (1979), *Ca* **91**, 219238 (1979).
74. W.E. Cowley, F.P.A. Robinson, and J.E. Kerrich, *Br. Corros. J.* **3**(9), 223 (1968).

for estimating the resistance and design parameters for electrodes are discussed in Chapter 4.

Cathodes that have been discussed in the literature are listed in Table 3-1. Materials used range from platinum to cast iron and all are discussed in this chapter.

Platinum-Clad Cathodes

Early installations of anodic protection used platinum clad on brass as a cathode. Although this is very expensive, it is the best from a stability standpoint. An early installation used a cathode and the mounting shown schematically in Figure 3-1. Several installations were made using the cathode shown schematically in Figure 3-2.[2-8] This electrode has a high contact resistance

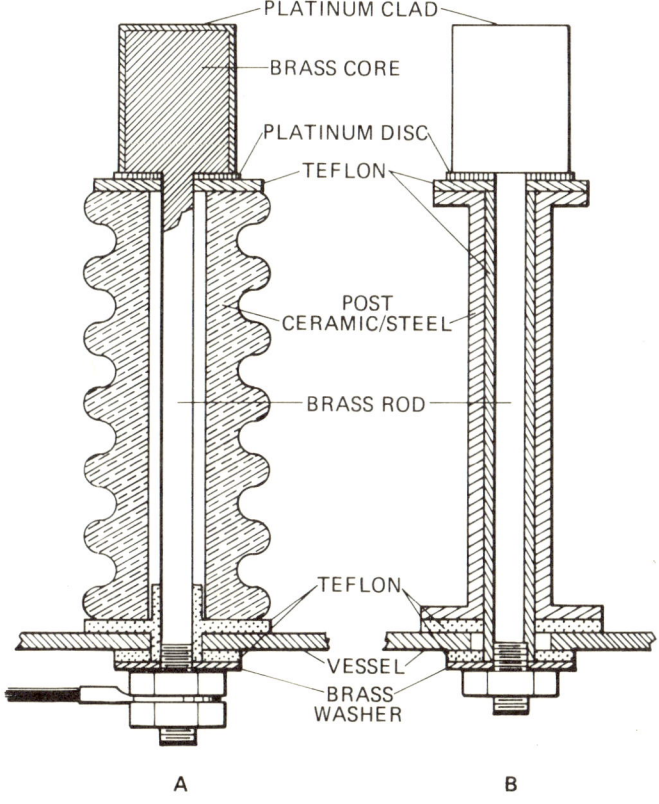

Figure 3-1. Diagram of cathode used in first field installation of anodic protection.[1] A, Earlier version; B, improved later version.

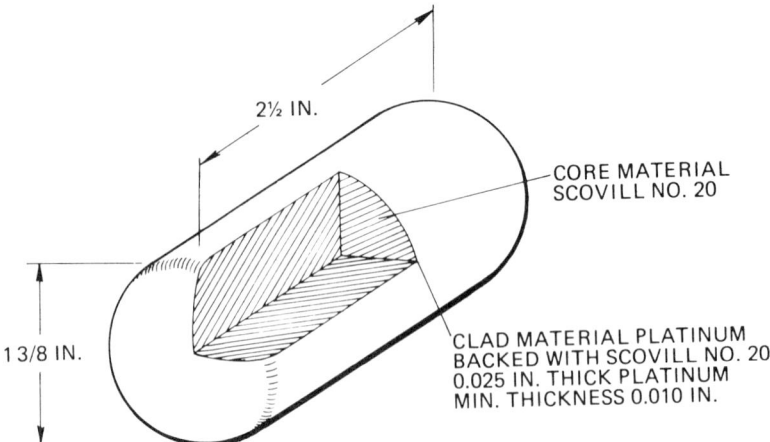

Figure 3-2. Platinum-clad electrode.[8]

owing to its relatively small surface area. The size was controlled by the price of the platinum cladding. Larger electrodes using this configuration were economically unfeasible.

Details of a mounting method for this electrode are shown in Figure 3-3. The method involved penetration of the vessel wall, which could result in prob-

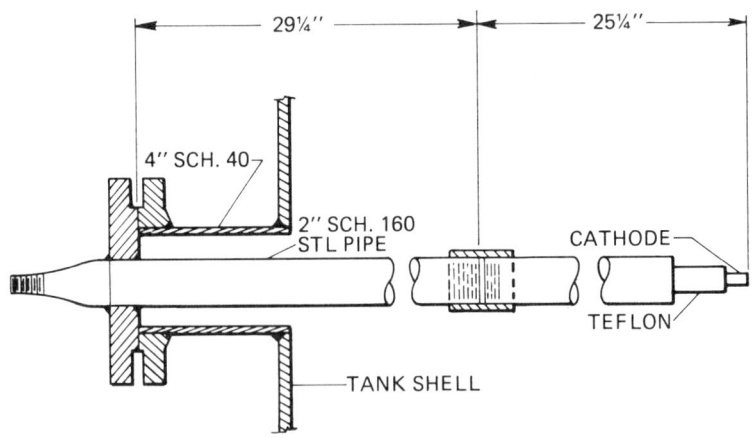

Figure 3-3. Typical electrode assembly showing the cathode connection for anodic protection in a sulfuric acid tank. The electrode extends into the tank.[2]

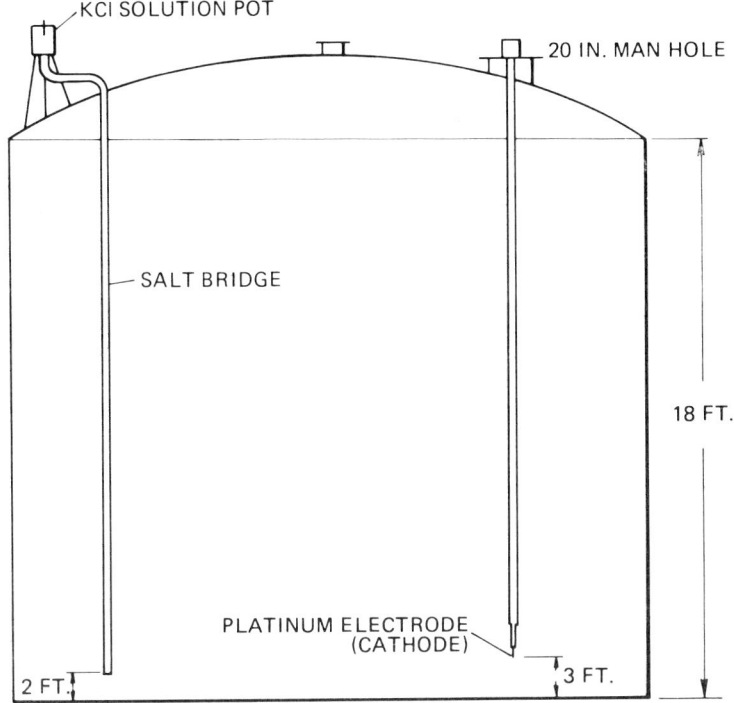

KCl SOLUTION POT

20 IN. MAN HOLE

SALT BRIDGE

18 FT.

PLATINUM ELECTRODE
(CATHODE)

2 FT.

3 FT.

Figure 3-4. Schematic of electrode installation for anodic protection.[8]

lems if a leak developed. An alternate mounting method through a tank roof is shown in Figure 3-4. It permits insertion and removal without emptying the vessel or causing a leak hazard.

Platinum-clad electrodes are excellent electrochemically. In sulfuric acid hydrogen reduction is the only reaction that occurs. Although they are permanent, they are usually not fabricated in large sizes because of cost.

Hastelloy C Cathodes

This chromium–nickel alloy (16Cr–16Mo–5Fe–4W–bal Ni) has been used in a few applications. Presently the most common use of these electrodes is in the shell and tube heat exchangers described in Chapter 2.[9] Mounting methods for these tubular electrodes are shown in Figure 3-5. Figure 3-6 shows an example of these electrode mountings extending through a water box.

Hastelloy C was also used in tank cars carrying nitrogen fertilizer solu-

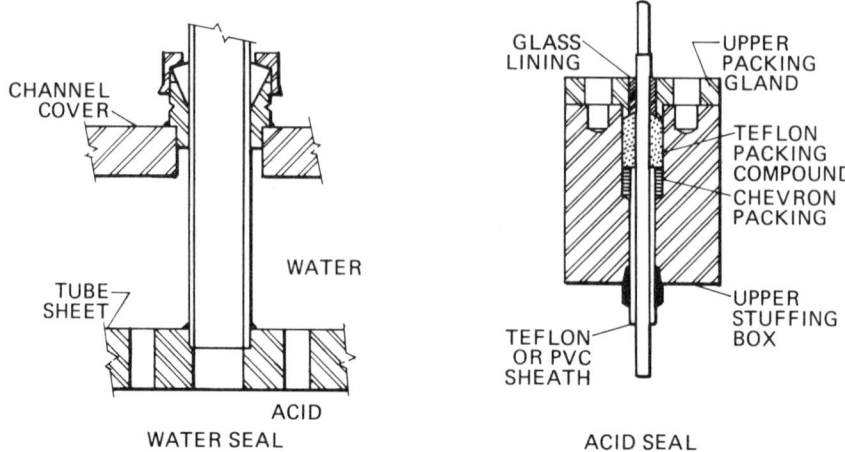

Figure 3-5. Acid (right) and water (left) seals for cathode assembly in shell and tube coolers.[9]

Figure 3-6. Electrode mounts for shell and tube heat exchangers.[50]

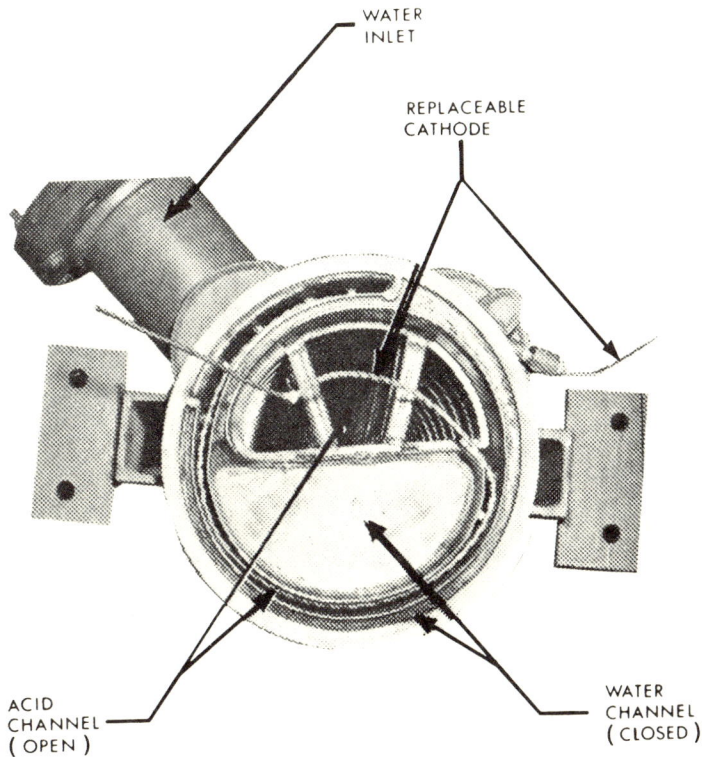

WATER
INLET

REPLACEABLE
CATHODE

ACID
CHANNEL
(OPEN)

WATER
CHANNEL
(CLOSED)

Figure 3-7. Bottom view, with cover removed, of cathode installation in heat exchanger.[51]

tions. Figure 2-13 is a sketch of the latter installation showing some of the mounting details. This alloy has a limited life in sulfuric acid heat exchangers. It is estimated that they must be changed every three or four years.[10]

Hastelloy C has been used in the form of a wire as a cathode for spiral-type heat exchangers. Figure 3-7 shows this type of cathode installation. This configuration has been patented.[11]

Stainless (Cr–Ni) Steel Cathodes

Various chromium–nickel steel alloys have been used to fabricate cathodes used in nitrogen-fertilizer and ammonium hydroxide solutions in the

USSR[12-14] in the form of tubes (usually 18Cr steel). Figure 3-8 is a diagram of a large ammonium hydroxide storage vessel which uses six 38-mm-diameter by 11.7-m-long 18Cr steel tubes. The interconnecting supports for the cathodes are noteworthy.

Stainless-steel AISI Type 304 steel is used by a commercial supplier of anodic protection systems.[10] Figure 3-9 is a diagram of a typical installation using AISI 304 pipes which are sealed by welding and have a positive pressure inside. If a hole develops in the pipe, a drop in the pressure signals it. This design is reported to be an improvement over others because "there is no acid–air seal which could leak to cause the cathode to short circuit." The lifetimes of these anodes have been about two years.[10]

Chromium–nickel steel was also used in a study of anodic protection of a stainless steel heat exchanger in sulfuric acid service.[15] Size and mounting were not described.

A high-silicon cast iron (S-15 Ferrosilide, USSR designation) was used in

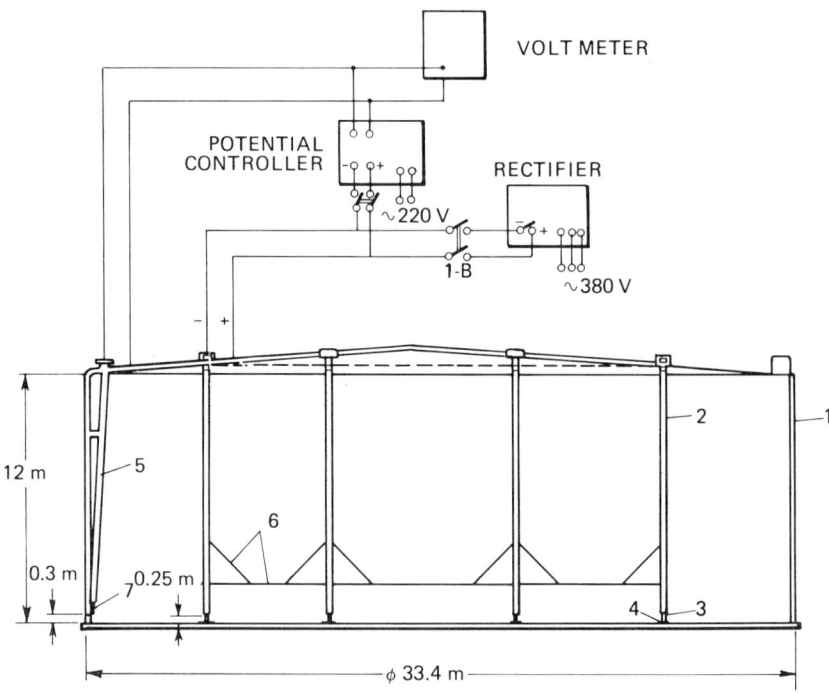

Figure 3-8. Schematic diagram of anodic protection: (1) storage vessel (steel 3); (2) cathodes (Kh18N10T); (3) sleeve; (4) rod (polytetrafluoroethylene); (5) tube for comparison electrode (1Kh18N10T); (6) supports for cathodes (1Kh18N10T); (7) pocket for comparison anode.[12]

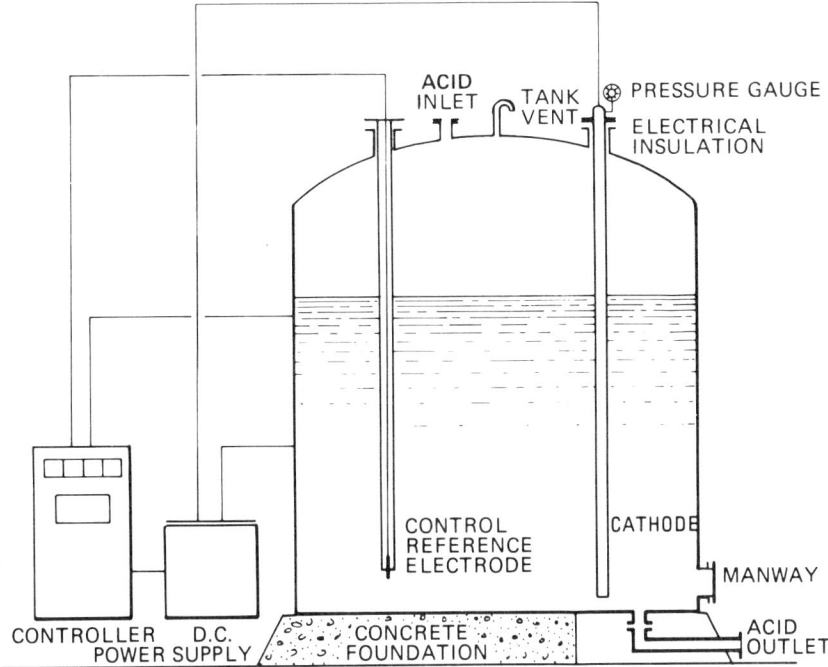

ACID INLET — TANK VENT — PRESSURE GAUGE — ELECTRICAL INSULATION — CONTROL REFERENCE ELECTRODE — CATHODE — MANWAY — CONTROLLER — D.C. POWER SUPPLY — CONCRETE FOUNDATION — ACID OUTLET

Figure 3-9. Installation of anodic protection cathode is a stainless steel pipe that is pressured to detect leaks.[52]

a system to protect a sulfuric acid storage vessel.[16] Mounting details are given in Figure 3-10. Performance was not discussed.

Steel has also been used in systems to protect kraft digesters; Figure 3-11 is a diagram of a suggested cathode installation made of steel pipe. Later installations in this environment used steel cables suspended from the top of the digester.[17] One of the authors participated in the preparation of a design in which the cables were secured at their ends by clamping mechanisms. Other designs were used in which the cables were mounted only at their top end and were swinging free at the bottom. Tips were insulated to prevent shorting against the digester walls.

Copper and molybdenum electrodes have also been mentioned in the literature.[16,18] A mounting arrangement for a copper electrode is shown in Figure 3-12.

A very interesting cathode design called the "air" cathode is shown in Figure 3-13.[19] This design is similar to that used for fuel-cell electrodes. Platinum black was deposited by filtration on porous graphite and the reverse side of the electrode was impregnated with a fluorocarbon suspension. The electrode

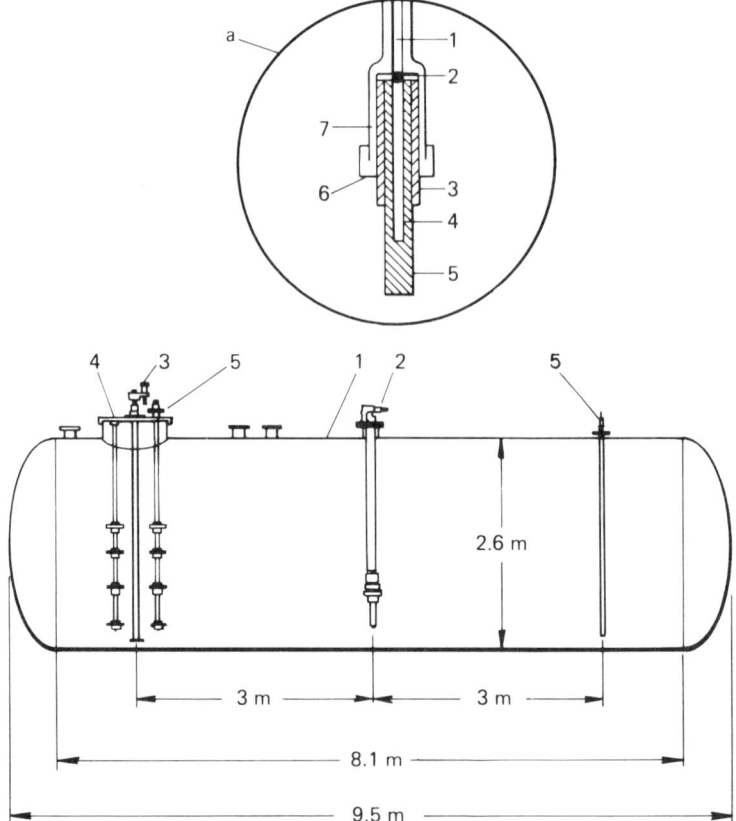

Figure 3-10. Longitudinal section of sulfuric acid storage vessel: (1) storage vessel; (2) cathode; (3) sensing element of comparison electrode; (4) and (5) rods for specimens without current and with current, respectively. (a) ferrosilide cathode: (1) electric cable; (2) terminal; (3) insulator (polytetrafluoroethylene); (4) rod (Kh18N10T); (5) casting (Ferrosilide S 15); (6) nut; (7) bush (St. 3).[16]

was then calcined for one minute at 380°C and mounted in titanium. The investigators say, "The water repellant (fluorocarbon) protects the gas side of the porous electrode from flooding when the electrode is immersed to a depth of one meter."

The problem of nickel plating on cathodes exists in the application of anodic protection to electroless nickel plating baths, as described in Chapter 2. Ryabchenkov et al.[20] suggested a novel solution to this problem. They put the cathode in a porous ceramic beaker filled with 10% NaOH. They also maintained that nickel-plated cathodes would be satisfactory and would not accumulate excessive amounts of nickel from the bath because of their small size.

This summary of the various types of cathodes that have been conceived and used indicates that a large number of designs and materials are available for effective cathodes. It is reasonable to expect that additional useful designs will be developed. Some are expendable and some are so-called "permanent" types. Which design and type are selected depends on the parameters of the system and the economics of the installation.

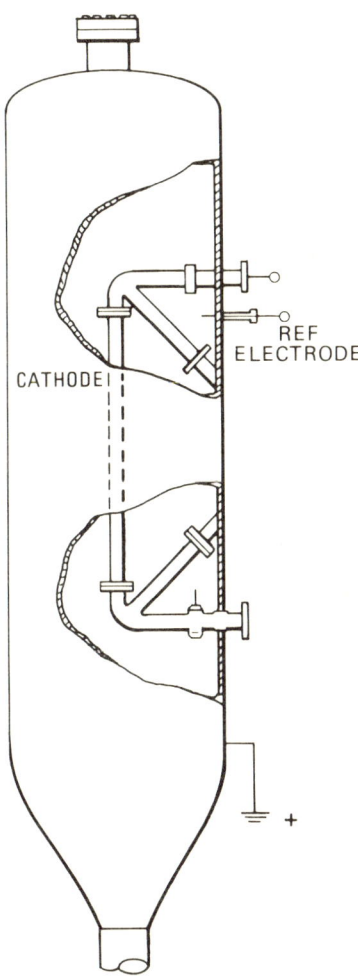

Figure 3-11. Pipe-type cathode for kraft digester.[24]

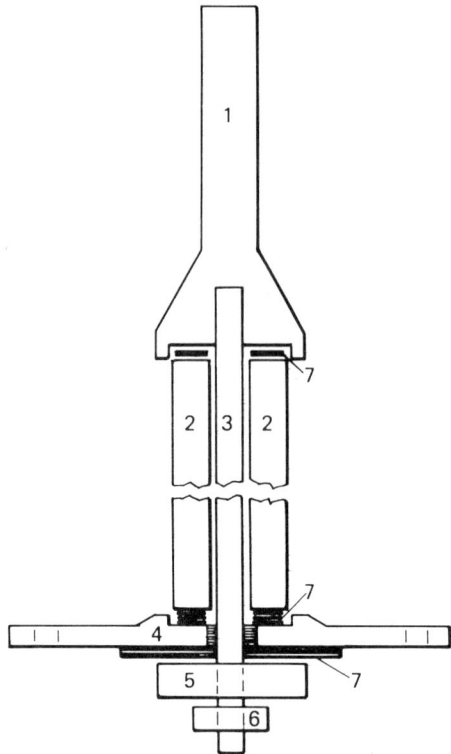

Figure 3-12. Schematic diagram of a cathode assembly for anodic protection: (1) copper cathode; (2) porcelain insulator; (3) steel rod; (4) flange; (5) nut for tightening the cathode assembly; (6) nut for fastening the lead wire; (7) insulating washers.[18]

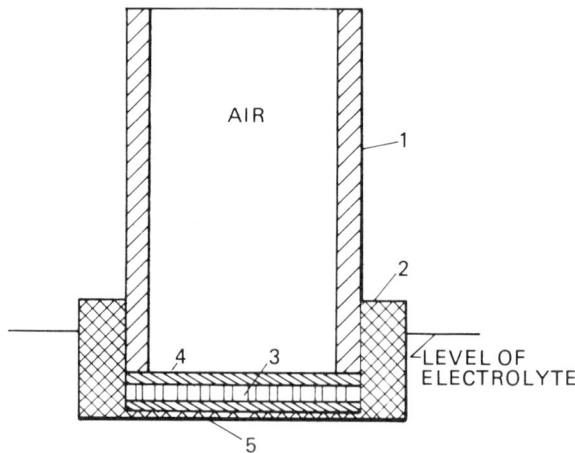

Figure 3-13. Construction of an air electrode: (1) current-conducting titanium tube; (2) polyethylene mounting; (3) graphite substrate; (4) water repellent; (5) platinum.[19]

Reference-Electrode Designs

A reference electrode is necessary because the potential of the vessel wall must be measured and controlled. The vessel-wall potential is obtained by comparison to the reference electrode. The theory of reference electrodes has been discussed adequately by Ives and Janz[21] and others.

This section will discuss the characteristics of reference electrodes to be used in the application of anodic protection.

It is impossible to accurately determine the electrochemical interaction of a metal with a solution without using a reference electrode. This is a fundamental concept in electrochemistry. The relative activities of metals have been ranked on a scale called the electromotive-force (emf) series, in which their behavior is compared to that of a hydrogen electrode which is commonly accepted as a standard base. The potential of the standard hydrogen electrode has been arbitrarily set at zero. A hydrogen electrode consists of a metal (usually platinum) immersed in a solution containing hydrogen ions of unit activity. Hydrogen gas at 1-atm (101.35-kPa) pressure is bubbled through the solution and over the metal. This standard is not particularly convenient for many studies and is especially hard to use in an anodic protection system.

Because of this, several other electrodes are often used as a reference in anodic protection systems. Many are electrochemical half-cells in which the electrode potential is reversible with respect to some anion. In a few instances, a metal is used as a reference, but when this is the case the stability of the potential depends on the particular metal and environment.

Numbers representing the magnitude and polarity of the potential of a corroding metal are related to the potential of the reference electrode used. Consequently, when a metal potential is given, the kind of reference electrode should also be identified. As an example, the potential of a calomel half-cell with a saturated potassium chloride salt bridge is $+0.24$ V with respect to the potential of the hydrogen electrode, so steel that has a potential of -0.500 V with respect to the calomel cell will have a potential of -0.26 V with respect to the hydrogen electrode. This is a particularly important factor when results of a laboratory investigation are used in the design of a field installation of anodic protection. Frequently the reference electrode used in the field installation will be different from that used in the laboratory.

A reference electrode used for anodic protection should be physically rugged, relatively insoluble in the corrosive fluid, and have a potential that is stable with respect to time and environmental changes. It has been impossible to construct a single electrode that can meet all these requirements in all the environments that have been encountered in anodic protection. As a result, many reference electrodes have been used. Table 3-2 lists electrodes described in the literature. Each type is discussed in greater detail in the following sections.

TABLE 3-2. Reference Electrodes Used for Anodic Protection Installations

Electrodes	Systems and electrolytes	References
Calomel	Sulfuric acid, miscellaneous concentrations	1,8,24
	Kraft digester	25,26
Ag/AgCl	Sulfuric acid, fresh or spent	2,4,5,6,8
	Urea–ammonium nitrate	13,14
	Sulfonation plant	1
Hg/HgSO$_4$	Sulfuric acid	15,16,28
	Hydroxylamine sulfate	29
	Design, data	27,30
Pt/PtO	Sulfuric acid	3,6
Au/AuO	Alcohol solution	6
Mo/MoO$_3$	Kraft digester	17
	Green or black liquors	34
Platinum	Sulfuric acid	10
Bismuth	Ammonia solution	12
AISI 316	Nitrogen fertilizer solution	35
Nickel	Nitrogen fertilizer solution	36
	Nickel plating solution	39,40
Silicon	Nitrogen fertilizer solution	36

Calomel Half-Cell

The calomel half-cell is the reference electrode based on the following reaction:

$$HgCl_2 + 2e^- \rightleftarrows Hg + 2Cl^-$$

Commercial electrodes are constructed with the $HgCl_2$ (calomel) made into a paste with mercury. The paste is encased in glass. Electrical contact is made with the corrosive environment by means of a salt bridge, usually made of saturated potassium chloride. In several early installations of anodic protection, the calomel electrode was placed on top of the vessel and the salt bridge extended through the roof.[8] Figure 3-4 illustrates such a system in a storage tank. This system is also patented.[22,23] Although there were many problems in using this configuration, owing to the fragility of the salt bridge, it was used because the solid reference electrodes available at the time were unsuitable. This system was abandoned when some of the other electrodes discussed hereafter were developed.

A "permanent" version of the calomel electrode is mentioned for use in a kraft digester.[24-26] Unfortunately there are no details concerning the construction of this electrode which did not require a salt bridge. The design was probably similar to the $Hg/HgSO_4$ electrode described in another place herein.

Silver/Silver Chloride Half-Cell

The silver/silver chloride half-cell was the first solid reference electrode used in anodic protection systems. It is also widely used in laboratory investigations and in marine applications of cathodic protection. Silver/silver chloride electrodes are based on the reaction

$$AgCl + e^- \rightarrow Ag + Cl^-$$

Ives and Janz report that they can be formed by electrolytic, thermal, or precipitation methods.[21] They have also been formed using melt-casting techniques. Electrodes used in anodic protection systems have been formed by one or more variations of the melt-casting technique in which molten silver chloride is poured around silver metal or the metal is dipped in the melt.

There are difficulties in manufacturing the melt-cast electrodes. If the casting is made at a temperature much above the melting point, the electrode develops a high internal impedance. This impedance is so high that it is impossible to measure the potential with an electrometer or other high-input voltmeter. Melt-cast electrodes of this type are rugged and can be easily mounted for entry into a vessel. Figure 3-14 is a sketch of one such electrode, a design used in sulfuric acid storage vessels during the developmental phases.

Silver/silver chloride half-cells have been used in storage vessels containing sulfuric acid at concentrations up to 100%. One of the authors discovered that silver chloride is soluble in oleum when an electrode dissolved in a 5000-gal (19-m^3) storage tank. This probability should have been investigated beforehand in the laboratory. The calomel electrode with salt bridge was used in oleum storage tanks until oxide-type electrodes were developed.

The silver/silver chloride electrode is used in a few miscellaneous applications as listed in Table 3-2. Interestingly, it is used without a salt bridge in these applications. Laboratory versions of the electrode are built with a salt bridge which contains potassium chloride and a silver salt, and because of this, the potential of silver/silver chloride half-cells in the storage vessels was

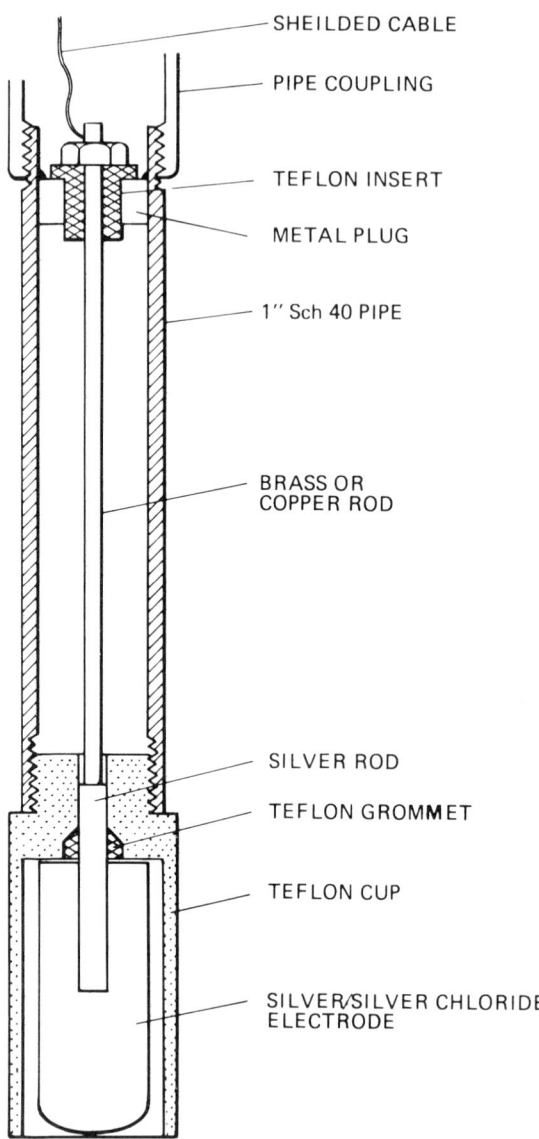

Figure 3-14. Ag/AgCl reference electrode used in anodic protection systems. Design used in development stage.

slightly different from that calculated from a known salt-bridge solution. Experience has shown that the potential of this solid electrode is only ± 20 mV different from that of the saturated calomel electrode. Theoretically, the standard potentials differ by -20 mV (Ag/AgCl to saturated calomel electrode). Because this difference was so small, laboratory data obtained with the calomel cell was used directly in the field.

Experience with this electrode is good. It is rugged and stable. The only difficulty is in the manufacturing step.

Mercury/Mercury Sulfate Half-Cell

Investigators in the USSR have discussed the use of a mercury/mercury sulfate electrode in sulfuric acid applications.[27] Installations of anodic protection for sulfuric acid storage and a reactor containing hydroxylamine sulfate have been described.[15,16,28,29] This electrode is based on the reaction

$$HgSO_4 + 2e^- \rightleftharpoons Hg + SO_4^{2-}$$

It is manufactured in much the same way as a calomel cell. A paste of mercury and $HgSO_4$ is made and mounted in a polytetrafluoroethylene case. Connection to the electrode is through an asbestos plug soaked with sulfuric acid. The electrode is immersed directly in the corrosive liquid. Diagrams showing details of this electrode are given in Figures 3-15 and 3-16.

The potential of the mercury/mercury sulfate electrode with respect to the standard hydrogen electrode was found to be $+0.674$ V in 1 M sulfuric acid,[30] which is also $+0.430$ V with respect to a saturated calomel half-cell in the same solution. Therefore, if a mercury/mercury sulfate electrode is used in a field application, laboratory data taken with the calomel cell would have to be corrected by this value.

Metal Oxide Electrodes

Several metal/metal oxide electrodes have been used in various environments as listed in Table 3-2. All these electrodes have similar half-cell reactions, according to Ives and Janz[21]:

$$M(OH)_2 + 2e^- \rightleftharpoons M + H_2O$$

Low pH: $\quad M(OH)_2 \quad\quad \rightarrow M^{2+} + 2OH^-$

High pH: $\quad M(OH)_2 \quad\quad \rightarrow MO_2^{2-} + 2H^+$

While few metals exhibit both reactions, different metals may have one of these half-cell reactions. The emf of these cells is dependent on pH, so they should

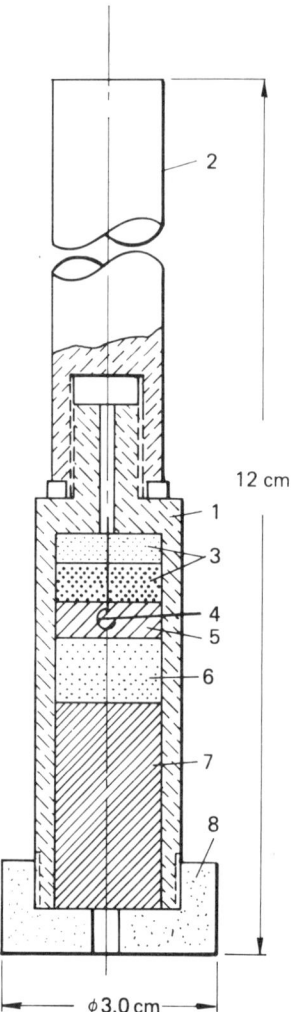

12 cm

φ 3.0 cm

Figure 3-15. Immersible mercurous sulfate reference electrode: (1) polytetrafluoroethylene; (2) mounting pipe; (3) rubber washers; (4) platinum wire; (5) mercury; (6) Hg/Hg$_2$SO$_4$ paste; (7) asbestos; (8) plastic nut.[27]

be used in systems in which the pH is constant if they are to maintain the necessary stability. Manufacturing methods for these electrodes are described by Every.[31] The metals may be oxidized chemically or electrochemically. Their emf has been reported for sodium hydroxide and phosphoric acid.[32] As predicted by the half-cell reactions, their potentials are highly pH dependent. Table 3-3 lists the potentials of these electrodes for sodium hydroxide and phosphoric acid solutions.

There is a wide shift in the potentials between base and acid solutions for the electrodes. These values must be known if laboratory data are to be successfully applied in the field. Some of the electrodes have been patented.[33]

The platinum/platinum oxide electrode has been used in oleum and concentrated sulfuric acid storage vessels.[2,5] Current commercial practice is to use a platinum electrode as a reference in sulfuric acid storage tanks.[10] The Mo/MoO$_3$ electrode is used in highly basic solutions, like those in kraft digesters and in storage tanks for recovery liquids in pulp mills.[17,34] A gold/gold oxide

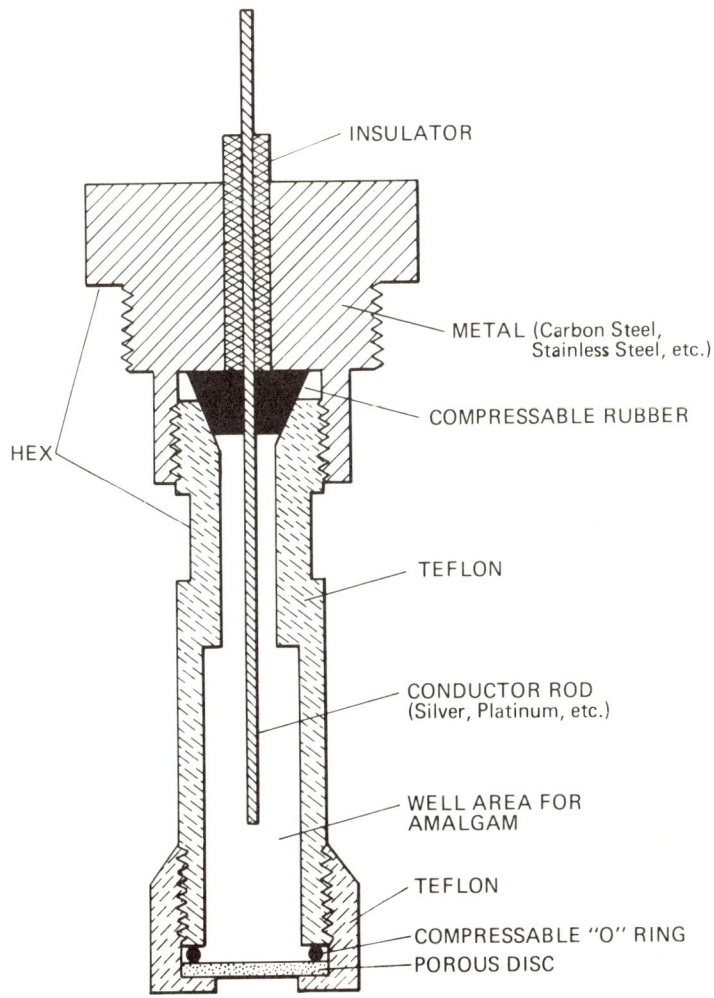

Figure 3-16. Immersible amalgam-type reference electrode (Riggs and Locke version, 1959).

TABLE 3-3. Potentials of Metal/Metal Oxide Electrodes (mV)[a]

Electrodes	Sodium Hydroxide, wt. %			Phosphoric acid, wt. %	
	10	20	40	85	95
Pt/PtO	−105	− 95	−200	+370	+340
Mo/MoO₃	−550	−600	−710	+170	+200
Au/AuO₂	−105	− 95	−200	+ 90	+160

[a]Adapted from Reference 32

electrode was used in one installation for a contacter that was hydrolizing unsaturated hydrocarbons to alcohols.[6]

Metals As Reference Electrodes

There have been a few installations in which a pure metal or alloy was used as a reference electrode. Metals are satisfactory if their potentials in the environment are stable. There are a few instances when this condition prevails.

The platinum electrode is used for commercial anodic protection systems in sulfuric acid storage tanks.[10] A photograph of one such electrode is shown in Figure 3-17. It is reasonable to expect that the potential of platinum would be very stable because it is such an inert metal. However, there is some danger in using the pure metal alone as a reference electrode. Many reduction reactions can easily occur on platinum surfaces, which can change the electrode potential. Platinum/platinum oxide is a better choice as a reference electrode.

Bismuth was used in an installation to protect a large ammonium hydroxide storage vessel.[12] The bismuth electrode has a potential of −0.45 V with respect to the saturated calomel electrode at a pH of 12. This potential is reported to be independent of pH in the range 11 to 12.8. However, in one instance it was found that the potential shifted from −0.45 to −0.30 V during the first 15 days in ammonium hydroxide. The electrode was found to be covered with dark-brown oxides after this period. The authors recommended that the bismuth should first be anodically oxidized for best results and that probably what is required is a bismuth/bismuth oxide electrode. Figure 3-18 gives mounting details of this electrode.

AISI Type 316 steel has been used successfully in nitrogen fertilizer solutions as a reference electrode.[35] This steel is passive in these solutions, so its potential should usually be stable. There is the possibility that the stainless

Figure 3-17. Platinum reference electrode with protective shield for commercial anodic protection system (Rohrback Instruments).

steel could become active, owing to contaminants. If it did so, the potential of the reference would be shifted by several hundred millivolts. This could cause serious disruption of the anodic protection system. Therefore, in the writer's opinion, this type of reference electrode should be avoided. Mounting of this electrode is illustrated in Figure 2-13. No data were given in the paper con-

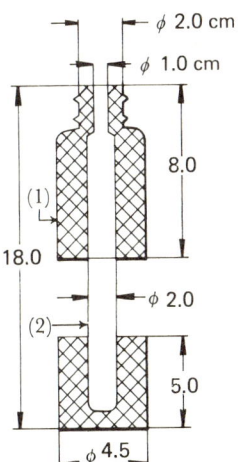

Figure 3-18. Bismuth reference electrode: (1) polytetrafluoroe-thylene body; (2) bismuth.[12]

TABLE 3-4. Potentials of Silicon
Reference Electrode at 25°C vs pH[36]

pH	mV vs SCE[a]
1	−580
5	−610
7	−480
10	−500

[a]Saturated calomel electrode

cerning the potential of the 316 stainless steel with respect to a standard reference electrode.

Several pure metals have been studied for use as reference electrodes in "complex" fertilizer slurries[36] which contain nitrates, phosphates, sulfates, and chlorides. Investigators found that nickel and silicon were suitable as electrodes for field trials. The potential of the Si electrode as a function of pH is given in Table 3-4. The potential of the nickel electrode was not given in the report.

Both electrodes were tried in anodic protection of reactors producing the "complex" fertilizers and exhibited good reliability. The nickel and bismuth electrodes were mounted as shown in Figure 3-18. The silicon electrode was mounted as shown in Figure 3-19.

Cast iron was used as a reference electrode in a laboratory evaluation of cast iron in KOH melt.[37] This does not seem to be a good selection for the reference electrode for a field application because, as discussed previously, the alloy potential could shift with time. Possibly one of the metal/metal oxide electrodes would be better.

A Swedish patent was granted on the use of molybdenum as a reference electrode for use in strong acids.[38] Sulfuric and phosphoric acids were mentioned. This is interesting in that the Mo/MoO_3 electrode has previously been found suitable for strong bases, but not for acids.

Patents were granted to a group from the USSR for nickel as a reference electrode to be used in a bath containing electroless nickel plating solution.[39,40] One was granted by Japan and the other by West Germany. The nickel-rod electrode was coated in a warm nickel chemical plating bath before being mounted. The potential was -0.40 ± 0.01 V relative to the standard hydrogen electrode.

Electronic Control and Power Supplies

Figures 2-9 and 2-10 illustrate typical anodic protection systems. Notice that each has equipment for potential sensing and a control coupled to a dc

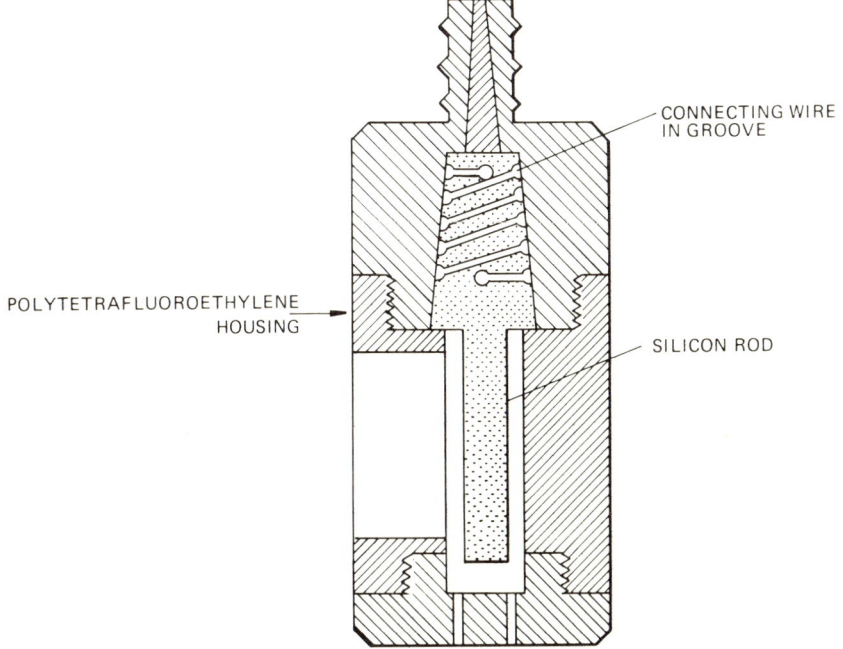

Figure 3-19. Silicon reference electrode.[36]

power supply. These electronic and electrical circuits sense the potential between the reference electrode and the vessel, compare that potential to a preset value, and control the output of a power source that supplies dc current between the cathode and the vessel wall. This discussion covers some of the basic requirements for this equipment and surveys some of that which is described in the literature.

Potential Controller

The potential controller can be viewed as a limited-function potentiostat. The circuitry and requirements of a potentiostat are described in Chapter 7. The discussion here involves only those factors unique and important in a plant or field version of the instrument.

Also, Figure 2-9 shows the two functions of the potential controller: sensing and control. The sensing circuit is basically a high-impedance voltmeter. It must have an impedance at least 10^6 ohms because the reference electrodes also have high impedance. Usually a readout device is included for visual monitoring of the operation of the system potential. As discussed in Chapter 7, this signal is compared to a preset potential value and the resulting difference is used to control the power-supply circuit. This control can be varied in an on–off manner or the level can be varied proportionally to the magnitude of the difference signal.

The precision of the control potential in field instruments need not be as high as that of a laboratory potentiostat. As an example, most field controllers would provide satisfactory control if the potential were maintained within ± 5 mV of the desired value. Laboratory potentiostats control to within 0.1 mV or better. It is possible to control corrosion of some systems under anodic protection when the potential varies by as much as 100 mV. This permits the use of on–off controls for satisfactory anodic protection. Presently, in the United States and Canada, commercial practice is to use proportional-type controls only.[10] Owing to the vast improvement in solid-state circuitry and dramatic cost reductions in the past few years, the incentive to use on–off controls has diminished. However, because on–off controls have been used in many installations over the years, their application will be discussed briefly.

The literature has many references to the use of on–off control for anodic-protection installations.[2,4,6,14,16,25,27,35,41,42] In addition, several patents concern on–off systems. The incentive to use on–off controls was that the cost of the circuit to control in this manner was much less and much simpler than that of proportional-control systems when vacuum tubes were used. Many on–off units used vacuum-tube circuits which contained a meter relay. When the potential-meter needle touched a lower limit, current was turned on and when the indicator touched an upper limit, current was turned off.

This type of control is practical in systems in which current-on time is shorter than current-off time. In a steel storage tank containing 100% sulfuric acid at 30–40°C, current may be on about 10% of the time, except when the vessel is being filled.[4] Figure 3-20 is a plot of "on" time as a function of elapsed time after acid has been added to the tank. Notice that on time increased to 85% in one hour, immediately after filling, then slowly decreased.

The fraction of on time is affected by acid concentration and temperature. A control system for 98% sulfuric acid was reported to be on for only 2% of the time.[6] In a 93% sulfuric acid vessel, the fraction of on time was 1% at 80°F (27°C) and 63% at 66°C.[6] Systems have been observed on oleum storage vessels in which there would be a few seconds of on time, while off time lasted as long as an hour. Lower-concentration acids usually have a shorter off-time interval than those of higher-concentration solutions.

Satisfactory protection was achieved with these systems, which were used principally in sulfuric acid storage. They were also used on nitrate fertilizer storage and transportation tanks.

The proportional type of control system was first used only in installations of a critical nature in which the potentials would instantly shift to a corrosive value when the current was off.[1,6,16,26] There have been many patents for pro-

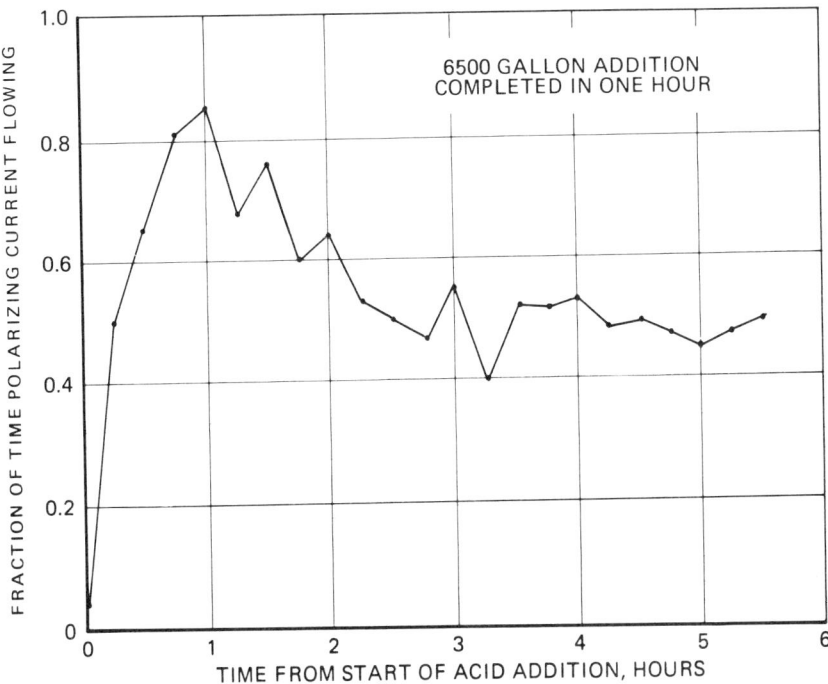

Figure 3-20. Typical effect of acid addition to storage tank on passivating-current requirement.[4]

portional-control circuitry for anodic protection systems.[43,44,45,46] A few circuits have been described in the literature.[47,48,49]

Figures 3-21 and 22 are photographs of potential controllers manufactured by Rohrback Instruments. The unit in Figure 21 is mounted in a weatherproof box and was used until about 1975. The unit shown in Figure 22 is a controller built for rack mounting used by Rohrback for installations since 1975.

Figure 3-23 illustrates a rack containing potential controllers for nine stainless-steel heat exchangers. These controllers were manufactured by Magna (now Rohrback) for Chemetics International, Ltd. in Canada.

Power Supplies

The function of a power supply in anodic protection is to be a source of dc current for the cathode–anode circuit. Batteries have been used,[35,41,46] but most installations have ac as a primary source. Equipment described herein uses ac as a primary power source.

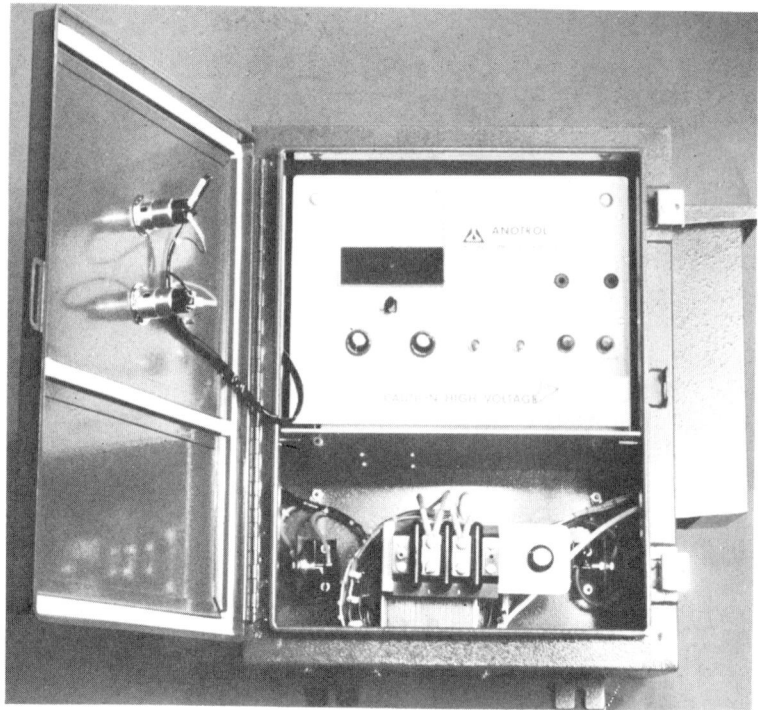

Figure 3-21. Potential controller in weatherproof box.[53]

A power supply has a circuit such as that shown in the Figure 3-24 block diagram. The transformer steps down the ac voltage to a lower, useable value. The voltage used depends on overall circuit resistance, which is highly dependent on cathode area. Voltages may range from 10 to 50 V and power supplies currently used have an output of 10 V. The current control is used to vary the

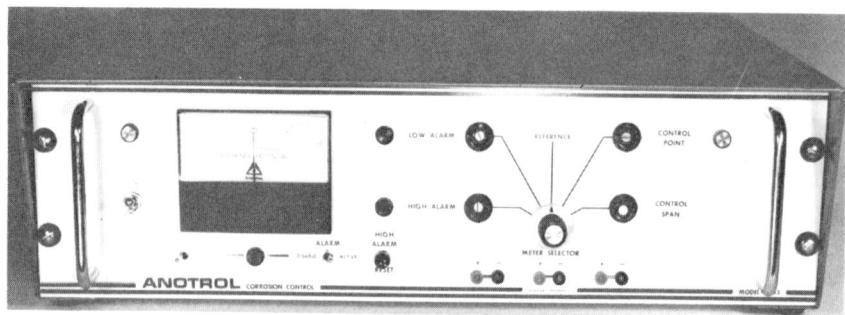

Figure 3-22. Potential controller built for rack mounting, in use since 1975.[53]

Figure 3-23. Rack containing potential controllers for nine stainless steel heat exchangers (Chemetics International, Inc.).

output of the power supply in response to a signal from the potential controller. Current control in some early on–off systems was accomplished by use of a relay. Because the entire current must be mechanically interrupted when using this method, relay-contact life was short, owing to arcing between contacts. This problem was solved by using the relay to interrupt the control signal to a current-control device. This signal current was a few milliamperes and thus did not greatly affect relay contacts.

Saturable core reactors have been used as current controllers. These devices conduct current when dc windings are "saturated" by the control signal and, theoretically, do not conduct when the control signal is removed. However, they are not perfect current interrupters and some current will flow over when the control signal drops to zero. In some installations, this "bleed-through"

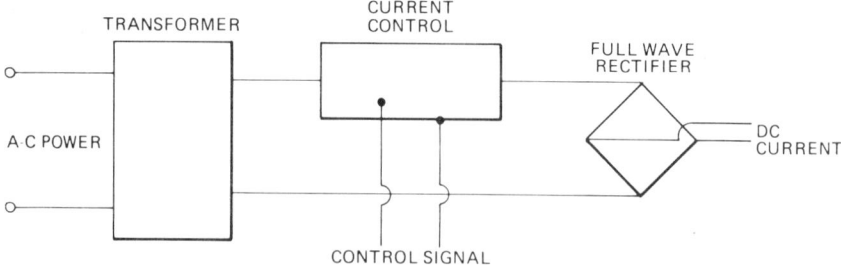

Figure 3-24. Transformer power-supply circuit.

current was sufficiently high to shift the vessel potential into a value higher than desired.[2] "Turn-down" circuits were employed to lower this "bleed through" to an acceptable value. Units now used have such a circuit.[10]

Silicon control rectifiers have also been used for current control. A silicon control rectifier will operate in such a way that a portion of the current cycle

Figure 3-25. Potential controllers with power supplies in background. Power supplies have 200-A capacity.[52]

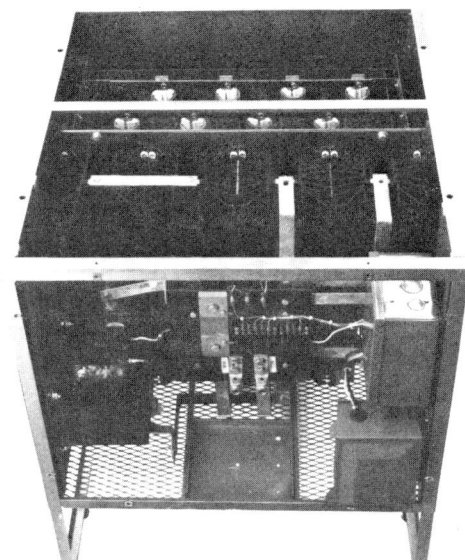

Figure 3-26. Interior view of power supply similar to those shown in Figure 3-23.[53]

is conducted through it, the amount being proportional to the control signal's magnitude. An additional "triggering" circuit is required to operate these devices. Because it is possible to completely stop all current flow from silicon rectifiers, this eliminates the "bleed-through" problem. Silicon control rectifiers are not as durable as saturable core reactors and are more easily damaged by current surges and transients.

Although rectifier circuits have been used as full-wave bridges, and both silicon and selenium rectifiers have been used, it is beyond the scope of this discussion to compare them. The reader is referred to discussions of their relative merits in electrical engineering or cathodic protection literature.

Power supplies are usually located close to the equipment that is protected to reduce the length of high-current-carrying wire to the anode and cathode. Because heat generated by the power supply must be removed, air-cooled enclosures are used, except in areas of explosion danger or in highly corrosive atmospheres in which oil-cooled containers are used.

Figure 3-25 illustrates a controller board for two anodic protection systems with power supplies in the boxes in the background. These power supplies have a current capacity of 200 A each and are used for sulfuric acid storage tanks. Figure 3-26 is a photograph of the internals of a power supply similar to those shown in Figure 3-23.

Summary

This discussion of the characteristics and application of equipment in anodic protection systems handling a wide range of corrosive solutions under typical industrial conditions indicates that sufficient experience has been accumulated to permit reliable designs. Some of the details concerning the selection of components for specific types of exposures and conditions have been given. Operational reactions have been observed and the performance of critical components reviewed in such a way that design and materials and equipment selection have been greatly simplified for the corrosives discussed.

It is apparent that extensive technological expertise has been developed and applied and that results can be anticipated with considerable precision. In any case, analogies are implicit for corrosive-materials combinations for which anodic protection has so far not been considered.

References

1. O. L. Riggs, M. Hutchison, and N. L. Conger, *Corrosion* **16**(2), 58t (1960).
2. L. R. Hayes, *Mater. Prot.* **5**(9), 46 (1966).
3. C. E. Locke, *Mater. Prot.* **4**(3), 59 (1965).
4. A. D. Fisher, and J. F. Brady, *Corrosion* **19**(2), 37t (1963).
5. J. C. Redden, *Mater. Prot.* **5**(2), 51, (1966); *Chem. Process.* **10** (1974).
6. J. D. Sudbury, C. E. Locke, and D. Coldiron, *Chem. Process.* February 11 1963.
7. Staff Report, *Mater. Prot.* **2**(9), 69 (1963).
8. C. E. Locke, M. Hutchison, and N. L. Conger, *Chem. Eng. Prog.* **56** (11), 50 (1960).
9. D. Fyfe, D. Sanz, F. W. S. Jones, and G. M. Camerdy, *Corrosion* 75/63, NACE, Houston, TX.
10. Chemetics International, Ltd., 47 Sheppard Ave. E., Willowdale (Toronto) Ontario, Canada M2 N 3A1.
11. C. E. Locke, and G. D. Harrel, U.S. Patent No. 3,409,530, November 5 1968.
12. L. A. Danielyan, A. I. Tsinman, V. S. Kozub, V. G. Moise, and N. N. Statsenko, *Zashch. Met.* **9**(4), 492 (1973).
13. L. G. Kuzub, *et al.*, *Zashch. Met.* **4**(5), 564 (1968).
14. V. S. Kuzub, *et al.*, *Zashch. Met.* **7**(3), 361 (1971).
15. V. A. Markov, *et al.*, *Zashch. Met.* **13**(2), 181 (1977).
16. Ya. M. Kolotyrkin, *et al.*, *Zashch. Met.* **1**(5), 598 (1965).
17. W. P. Banks, M. Hutchison, and R. M. Hurd, *Tappi* **50**(2), 49 (1967).
18. N. D. Tomashov, and G. P. Chernova, *Passivity and Protection of Metals Against Corrosion*, Plenum, New York, 1967.
19. G. N. Trusov, and Ya. Kryuchkova, *Zashch. Met.* **10**(4), 440 (1974).
20. A. V. Ryabchenkov, *et al.*, *Zashch. Met.* **6**, 718 (1971).
21. D. J. G. Ives, and G. J. Janz, *Reference Electrodes—Theory and Practice*, Academic, New York (1961).
22. M. Hutchison, O. O. Riggs, and J. D. Sudbury, U.S. Patent No. 3,126,328, March 24 1961.
23. M. Hutchison, O. L. Riggs, and J. D. Sudbury, U.S. Patent No. 3,152,058, October 6 1964.
24. W. A. Mueller, and T. R. B. Watson, U.S. Patent No. 3,009, 865, November 21 1961.

25. W. A. Mueller, Tappi **42**(3), 179 (1959).
26. T. R. B. Watson, *Mater. Prot.* **3**(6), 54 (1964).
27. V. S. Kuzub, *et al., Zashch. Met.* **5**(1), 56 (1969).
28. Ya. M. Kolotyrkin, *et al., Zashch. Met.* **7**(6), 722 (1971).
29. V. S. Kuzub, *et al., Zashch. Met.* **4**(4), 362 (1968).
30. T. M. Alkhazishvili, *et al., Zashch. Met.* **8**(6), 742 (1972).
31. R. L. Every, *J. Electrochem. Soc.* **112**, 524 (1965).
32. R. L. Every, and W. P. Banks, *Corrosion* **23**, 153 (1967).
33. R. L. Every, and W. P. Banks, U.S. Patent No. 3,462,353, August 19 1969.
34. C. E. Locke, *Tappi* **49**(1), 61A (1966).
35. W. P. Banks, and M. Hutchison, *Mater. Prot.* **8**(2), 31 (1969).
36. V. S. Kuzub, and G. L. Makovei, *Zashch. Met.* **10**(4), 437 (1974).
37. B. Stypula, and A. Piotrowski, *Zesz. Nauk. Akad. Gorn. Hutn. Cracow Zesc. Spec.,* **45**, 26 (1973); *Ca* **83**, 87098 (1975).
38. B. Wallen, and H. E. Anderson, Swedish Patent No. 393,404, May 9 1977; *Ca* **87**, 191232f (1977).
39. Yu. Y. Nikhaenku, Japan Kokai 7,864,623, June 9 1978; *Ca* **89**, 15475 (1978).
40. Yu. Y. Nikhaenuku, B. A. Gru, and V. S. Kuzub, Ger Offen. 2,655,679, June 15 1978; *Ca* **89**, 82116 (1978).
41. C. E. Locke, *Mater. Prot.* **4**(3), 59 (1965).
42. V. G. Moisa, and V. S. Kuzub, German Democratic Republic Patent No. 126,330, July 13 1977; *Ca* **88**, 67244t (1978).
43. M. Hutchison, O. L. Riggs, and J. D. Sudbury, U.S. Patent No. 3,208,925, September 28 1965.
44. N. L. Conger, U.S. Patent No. 3,280,020, October 18 1966.
45. G. R. Hoey, U.S. Patent No. 3,442,729, May 6 1969.
46. Statsanko, N. N., *et al.,* Ger Off. 2,642,163, March 23 1978; *Ca* **88**, 199983T (1978).
47. V. G. Moisa, *et al., Zasch. Met.* **8**(6), 745 (1972).
48. V. G. Moisa, and V. S. Kuzub, *Zashch. Met.* **7**(2), 203 (1971).
49. V. G. Moisa, and V. S. Kuzub, *Zashch. Met.* **5**(6), 718 (1969).
50. *Sulfuric Acid Coolers,* Chemetics International, Ltd., 47 Sheppard Ave. E., Willowdale (Toronto) Ontario, Canada M2 N 3A1.
51. American Heat Reclaiming Corp., 1270 Sixth Avenue., New York, N.Y. 10020.
52. *Anodic Protection System for Sulphuric Acid Storage Tanks,* Chemetics International, Ltd., 47 Sheppard Ave. E., Willowdale (Toronto) Ontario, Canada M2 N 3A1.
53. Rohrback Instruments, 11861 E. Telegraph Rd., Santa Fe Springs, California 90670.

Design, Operation, and Maintenance of Anodic Protection Systems

Anodic protection installations must be planned carefully. As discussed in Chapters 2 and 6, this method of corrosion control is successful only in systems that exhibit an active–passive anodic-polarization behavior. Protection by anodic polarization also has some inherent dangers in that it is possible to accelerate corrosion by incorrect application. However, if correctly designed, operated, and maintained, this technique of corrosion control can be a very powerful tool in the hands of the corrosion engineer.

This chapter describes what needs to be considered in designing anodic protection systems. This includes electrochemistry and electrical and mechanical factors. Both installation and start-up of anodic protection systems require special attention which will also be discussed. Operation and maintenance of systems will be covered.

Design Requirements

Electrochemistry

The first essential question that must be answered regarding any proposed anodic protection installation is: Will anodic protection be feasible for this particular system? This question can be easily answered by the results of suitable experiments in the laboratory, in which anodic-polarization behavior must be determined using the techniques described in Chapter 8.

The metal or alloy used in the laboratory tests must be the same as that selected for use in the field installation. Usually this presents no particular problem, because a sample with the same alloying constituents and percentages will be satisfactory. It is not necessary to cut a test coupon from the vessel itself unless the alloy used is unique or rare.

The test solution, however, presents a completely different problem. If possible, a sample of the actual solution to be encountered in the installation should be used. Such a sample will insure that incidental contaminants present in the process stream will be included in the evaluation. If an actual sample of the process solution is not available, then a simulated solution should be prepared incorporating the elements whose presence can be anticipated. Extreme difficulties and false data can result if the sample is not representative. Furthermore, because the characteristics of the process stream are so important to successful operation, it is essential to anticipate any major changes in process-stream contents during operation of the system so that the necessary adjustments can be made.

To illustrate the importance of this factor, the following case history involving use of a nonrepresentative sample is relevant. One of the authors had a perplexing experience while protecting a tank holding spent sulfuric acid. Laboratory investigation had indicated that the system could be easily protected, so protection was designed on data from the acid sample from the vessel.

Extreme difficulty was experienced with the field installation. Current requirements were much higher than indicated by laboratory tests and it was very difficult to maintain protection when a tank-car load of spent acid from a treating process was pumped into the tank. Coupons installed in the vessel were removed and found to be covered with a thick layer of soft corrosion product. This was analyzed as lead sulfate which had been picked up by the acid during use in the treating process. When polarization curves were made using coupons covered with this layer, it was found that current requirements were extremely high and that anodic protection was a marginally economical method of corrosion control.

This series of events is related to reinforce the prerequisite that a reasonably representative sample of the solution must be obtained. Also, it is necessary to be sure that the operating conditions are simulated in laboratory tests. For example, pressure differences between laboratory-test conditions and field installations may result in a changed solution composition due to differences in dissolved gas content.

Temperatures should be duplicated and particular attention paid to metal temperature, especially if heat transfer is involved. The solution may be at a temperature lower than that of the vessel wall if heat transfer is occurring. This is important because the wall temperature is a factor in the corrosion rate and because heat flux is a known accelerator of corrosion of most alloys.

Establishing Electrochemical Parameters

The next important step is the determination of two basic parameters: the protected potential range and current requirements. The protected potential range can usually be represented graphically as shown in Figure 4-1. Within

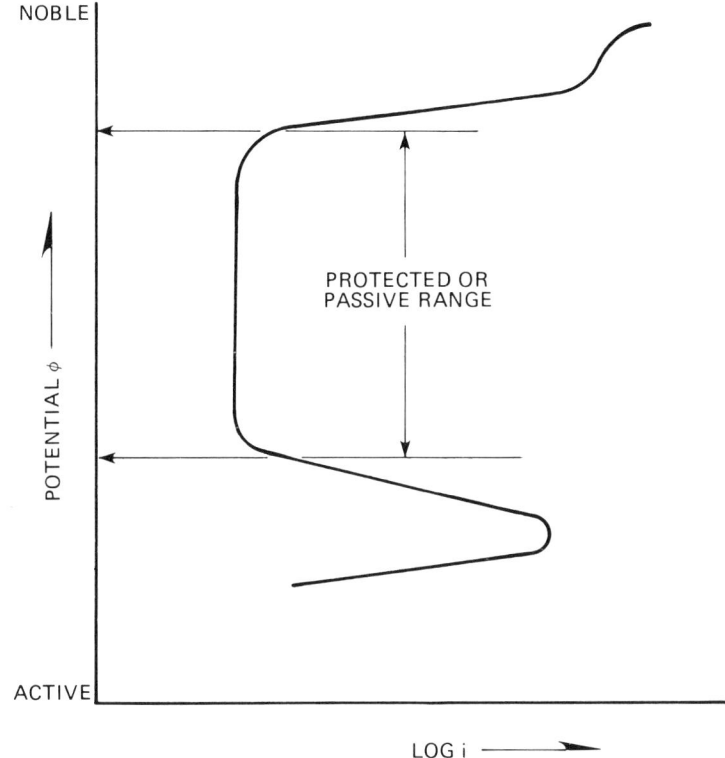

Figure 4-1. Potentiostatic scan showing passive or protected zones which is used to locate optimum criteria for electrical factors.

the protected range there may be optimum values or a preferred range of potentials.

The optimum potential for the vessel will be the point on the polarization curve within the passive region at which current demand is lowest. This can be read easily from the curve obtained during initial polarization experiments. Fortunately, it is possible to use laboratory data directly in field installations. The only modification necessary, usually, is an adjustment if the reference electrode used in the laboratory and the field is not the same. Potential differences among reference electrodes are described in Chapter 3.

Next to be considered are the values of current necessary to achieve and maintain protection. While values for these currents can be calculated from polarization curves, caution must be observed in using them directly. The currents are highly time dependent. In most cases, currents indicated by laboratory tests will be higher than those necessary for field installations. Although the laboratory values tend to provide a useful safety factor, they may cause excessive power costs. Actual current demands for field installations may be smaller.

The power supply should be sized to deliver sufficient current to achieve passivity. However, in many early installations, a portable power supply was used for initial passivation and the system power supply sized to maintain it.

Time Effects on Passivation Currents

A few studies have been published illustrating time effects on the current required to achieve passivation. Figure 4-2 is a summary of one combination of field and laboratory studies[1]. The figure shows that the current required to passivate decreases with time up to 30–40 minutes. These extended times to

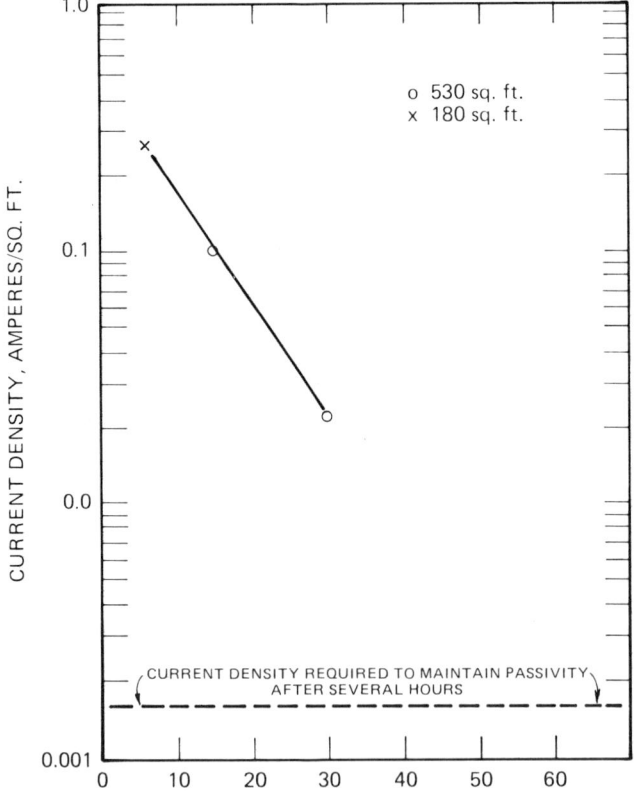

Figure 4-2. Effects of time on current density required to obtain passivity of carbon steel, sulfuric acid storage tanks containing 93% acid at 80°F (27°C).[1]

achieve passivation can be used if the installation has an unprotected corrosion rate of less than 20–30 mpy (0.51–0.76 mm/yr). A sulfuric acid storage vessel is an example of such an installation.

If the system has a high corrosion rate, the passivation current value should be selected so the vessel can be passivated in a relatively short time. The value obtained in the laboratory corrosion curve is good to use for this purpose.

To summarize, the electrochemical design parameters (control potential and current requirements) are obtained from laboratory experiments using the metals and solution to be encountered in the field installation. The values of potential can be used directly with appropriate corrections for reference-electrode differences. Current densities obtained in the laboratory tests will probably be higher than those required for field installations. However, if they are used directly, they provide a safety factor.

Electrodes

Factors involved in the design of electrodes for a system include materials to be used for the reference electrode and cathode and determining the geometry and areas of cathodes. Both factors are discussed here.

Reference Electrodes. Limiting factors in reference-electrode selection include compatibility with the corrosive environment and electrochemical stability in it. An electrode must be substantially insoluble in the environment and, in addition, should be electrochemically stable with respect to the solution and temperature changes anticipated. Reference electrodes suitable for a wide range of solutions are listed in Chapter 3.

It should be possible to use one or more of the listed electrodes in most systems in which anodic protection is feasible. If it is necessary to use a reference electrode whose characteristics have not been established, factors discussed in that chapter and in the book by Ives and Janz[2] should be considered. The size of the reference electrode is not critically important, as a rule. Mounting and location details are discussed below.

Cathodes. Cathode-materials selection should be based on stability in the environment. They should be inert or cathodically protected by the impressed currents used. The cost of the anode material is also an important factor. Power costs of the system will depend to a large extent on the area of the cathode, because resistance in the current-carrying circuit is largely dependent on contact resistance between the cathode and the solution. This latter consideration is the controlling factor in this resistance, so it is desirable to use as large a cathode surface area as is economically possible to reduce power consumption. In early installations of anodic protection, platinum-clad electrodes were used to insure stability. Power consumption in designs using platinum-clad electrodes was usually high because it was economically prohibitive to use platinum cathodes with a large surface area. As discussed in Chapter 3, many other metals are now being used to overcome this problem. The size and number of electrodes are determined after the material of construction is selected.

Size and Number of Cathodes. The size and number of cathodes are based on circuit resistance and current distribution. Current-distribution considerations and mounting details are discussed below. Contact resistance of cathodes can be predicted by using an equation for cylindrical electrodes published by Stammen and Townsend[3]:

$$R_c = \rho \left(\frac{\log (r_2/r_1)}{1.97\pi h} \right)$$

where R_c = resistance of cathode to solution (ohm), ρ = resistivity of solution (ohm cm), r_1 = radius of cathode (cm), r_2 = distance from centerline of cathode to tank wall (cm), and h = length of cathode (cm). Stammen and Townsend published comparisons of calculated and measured values with this equation and found reasonable agreement, as shown in Figure 4-3. However, these data show there is a departure from calculated values for longer electrodes and that the resistance approaches an asymptotic value. This indicates that there may be a limiting resistance with long electrodes. Present practice (discussed in Chapter 3) is to use long cylindrical electrodes to achieve low resistance. As an example, a circuit resistance of 0.04 Ω has been reported in concentrated sulfuric acid with long cathodes.[4]

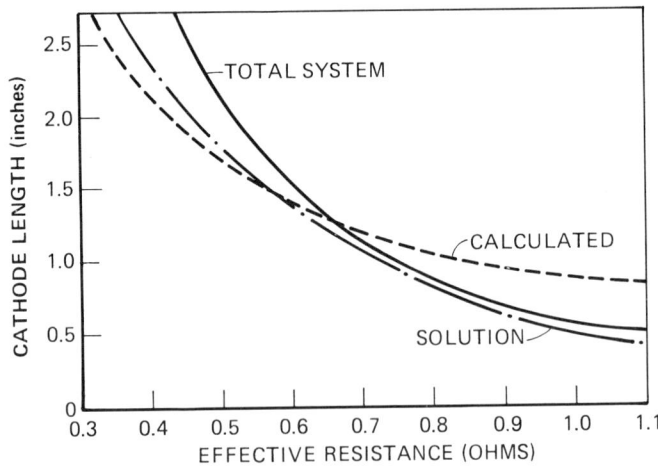

Figure 4-3. Comparisons of calculated and measured values of cathode length versus resistance.[3]

Electronic-Hardware Selection

The design of an anodic protection installation is dependent on electro-chemical factors, electrode selection, and, finally, on electronic hardware. A potential controller is necessary because the potential of the protected surface must be maintained at a value predetermined in the laboratory. The potential controller regulates current output of a dc power supply to accomplish this control. Each of these factors is discussed in this section.

As explained in Chapter 3, on–off potential and proportional controls have been used. However, because of the development of vastly improved and rela-tively inexpensive solid-state electronics, proportional controls are preferred.

Therefore, among considerations in the design of an installation, selection of a controller may be the simplest task of all. One of the instruments described in Chapter 3 may be selected for the intended application.

Power-Supply Criteria

Power-supply decisions involve determination of adequate current and power capacity. The current capacity is established by the anticipated current required to passivate the equipment. As explained elsewhere, it is possible to use a portable or auxiliary power supply to establish initial passivity and then to select an on-line system with sufficient capacity to maintain this passivity.

One of the authors participated in the design of several systems on storage tanks in which a dc welding-machine generator was used to provide power to passivate the vessels after they were full. The power supply for permanent on-line use had sufficient capacity to maintain protection and possibly to passivate the tanks when the fluid levels were low. This power supply was also capable of maintaining protection as the vessels were filled.

An interesting and innovative method of initially passivating a vessel is described in Chapter 2.[5] It concerned a vessel containing ammonium hydroxide which was passivated at a low current-density level, using a dilute solution. After this, a concentrated solution was put into the vessel to fill it. The power supply was adequate for that procedure, but not large enough to passivate the vessel when it was full of concentrated solution.

In most cases, extraordinary procedures of this kind are not desirable. A power supply should be sized to passivate the vessel in a reasonable time. Because the current required to passivate is highly time dependent, it can be decreased by an order of magnitude by increasing the time to passivate from 1 s to 30–60 min. Because corrosion rates are accelerated during passivation, long passivation times should be used only on installations in which repassiva-tion occurs infrequently and the unprotected corrosion rate is acceptably low.

Consequently, systems that must be repassivated frequently, or in which the unprotected corrosion rate is high, should be designed for passivation times as short as possible.

Table 4-1 lists some typical values of current required for passivation and maintenance of protection. These values, in most cases, are based on laboratory data, so they can be considered high. Values for longer times will be somewhat lower.

The second value to consider in power-supply design is output voltage. This voltage is based on the following factors in the aggregate circuit resistance: wire, vessel wall, solution, cathode, and cathode-solution contact. Cathode-solution contact resistance is the critical factor among these, as has been pointed out. Power supplies presently popular use 10–20 V to deliver the required current, which can range up to 750 A.

There can be a problem at the other end of the current spectrum if a saturable core reactor is used to control current. As mentioned in Chapter 3, a "turn-down" circuit may be necessary to reduce the minimum or "bleed-through" current. There have been cases in which the bleed-through current was greater than the current required to maintain protection, so that the potential was shifted well above the control point by it. In these cases, however, no damage was done, because the current was insufficient to shift the potential into the transpassive range.

TABLE 4-1. Anodic Protection Current Requirements

Solution	Concentration, %	Temperature, °C	Metal	Passivate		Maintain	
				mA/ft²	mA/cm²	mA/ft²	mA/cm²
Oleum		25	Carbon steel	120	2.640	0.15	0.00380
Sulfuric acid	93–98	25	Carbon steel	120	2.640	1.80	0.03960
Sulfuric acid	93–98	20	316 SS	2	0.044		
Sulfuric acid	90–94	25	Carbon steel			0.90	0.00198
Sulfuric acid	78	25	Carbon steel	140	3.080	2.50	0.05500
Ammonium hydroxide	25	25	Carbon steel	1850	40.070	0.22	0.00484
Oxalic acid	0.1 M	20	Carbon steel			5.80	0.12760
Fertilizer solutions (NH₄)₂CO,CO(NH₂)₂		20	Carbon steel			0.46	0.001012
Sulfuric acid	93	60	316 L			0.40	0.00880
Sulfuric acid	67	25	317	164	3.608	0.09	0.00198

Installation and Start-up

Techniques to be used for an installation on a vessel already in service differ from those employed on pre-packaged systems. As described in Chapter 2, heat exchangers are now marketed with factory-installed anodic protection systems. These will not be discussed here.

Electrode-Installation Modes

Electrodes usually are installed through the roofs of vessels, as shown in Figure 3-9. In a few early installations electrodes were inserted through vessel sidewalls, as shown in Figure 3-3. Roof entry is preferred because it is possible to make installations, inspections, or repairs on roof-installed electrodes without emptying the vessel or removing it from service and because the possibility of leaks is diminished.

Figure 4-4 illustrates a method for mounting a cathode through a roof. It is necessary to weld a flanged nozzle to the roof to allow for mounting of the cathode. The cathode is electrically insulated from the vessel wall by appropriate dielectric gaskets which are commercially available. Insulation should be resistant to vapors from tank solutions and to condensation inside the tank.

It is not necessary to locate the cathode exactly in the center of a vessel when only one cathode is used, because there is usually sufficient throwing power for adequate current distribution in spite of the geometrical imbalance. However, when multiple cathodes are used, they should be evenly distributed in a pattern related to the vessel wall. For example, when three are used, they should be spaced 120 degrees apart. Location of the cathodes with respect to the distance from the sidewalls is not critical. However, for efficient current distribution, it is best that they be located at a distance from the wall equal to one-half the radius of the tank.

Cathodes should extend to within a foot or so of tank bottoms. Although this distance is not critical, spacing should be at least one foot (30 cm) to avoid a locally transpassive region beneath the electrode. In addition, the lower ends of the cathodes should be at levels such that they are covered by solution during normal operation of the vessel.

Reference electrodes should be spaced radially as far from the cathodes as possible to adequately sense the average potentials of the vessel walls. If only one cathode is used, the reference electrode should be located 180 degrees from the cathode. If three cathodes are used with 120-degree spacing, the reference should be located 60 degrees from any cathode. Figure 4-5 illustrates this arrangement. Usually, the reference electrode is located in the same radius as the cathode, so it will be at approximately the same distance from the vessel wall. Although it can be nearer to the vessel wall than the cathode, if it is, the potential it senses may not be the average for the vessel wall.

The reference electrode should extend close to the bottom of the vessel. If the reference electrode is uncovered by a reduction of the fluid level, potential controllers now available will automatically interrupt the current. The end of the reference electrode should be at least six inches (15 cm) above the end of the cathode to avoid the possibility of a spark being generated as the liquid level falls. These spacings are diagrammed in Figure 4-6.

Present commercial practice is to install one cathode in small [less than 20 feet (6.1 m) in diameter] storage tanks and three in all others. Three reference electrodes are used: one for sensing, one spare, and one as a level sensor.

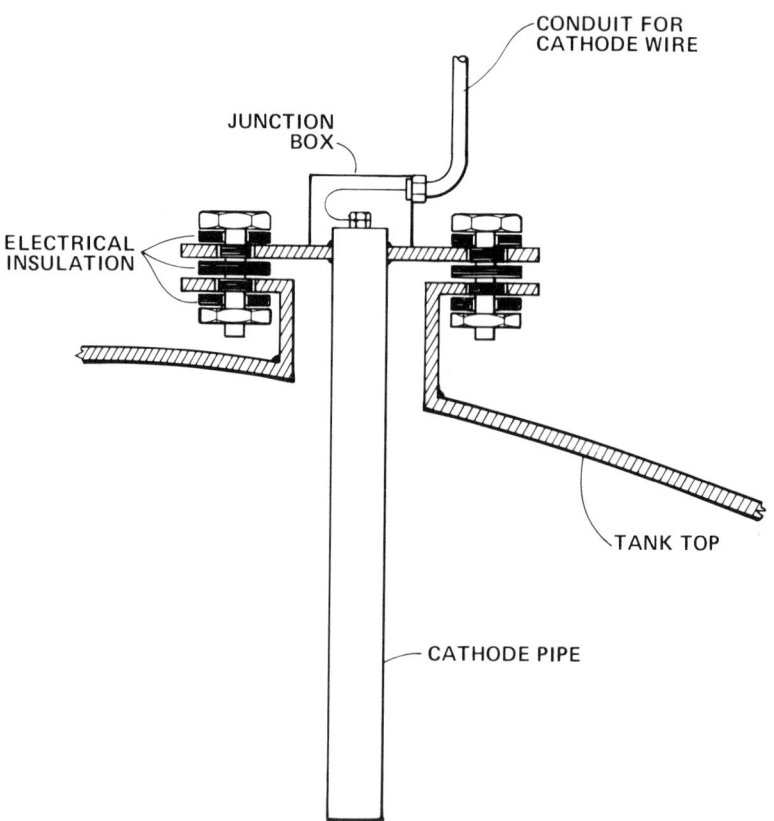

Figure 4-4. A suggested method for mounting a cathode through a roof (Chemetics International, Inc.).

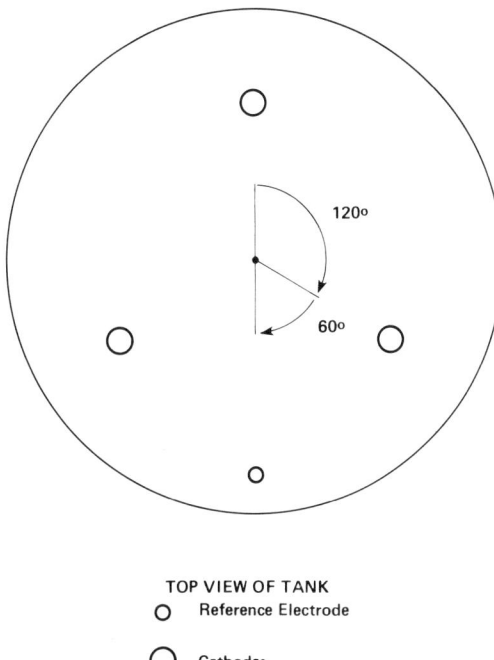

TOP VIEW OF TANK
○   Reference Electrode

○   Cathodes

Figure 4-5. Typical arrangement of three cathodes and one reference electrode in a circular tank.

If horizontally instead of vertically mounted vessels are used (discussed above), some differences in the spacing are necessary. Vertical-spacing relations between reference electrodes and cathodes are the same. If one cathode is used, it should be located in the center (end to end). If multiple cathodes are used, they should be spaced evenly along the length of the vessel.

If the vessel to be protected has a complex geometry, electrode placement is more difficult to determine. Cathodes must be located so that there is adequate current distribution to all areas, while avoiding local areas of transpassivity. Reference electrodes must be located to sense a meaningful average potential. It may be necessary to construct a pilot-scale model to accurately determine electrode placement.

Electrical-Wiring Criteria

Wiring between the cathode and the power supply should conform to pertinent electrical codes. Wire sizes should be based on the maximum expected current to each cathode. A separate wire should be run from the power supply

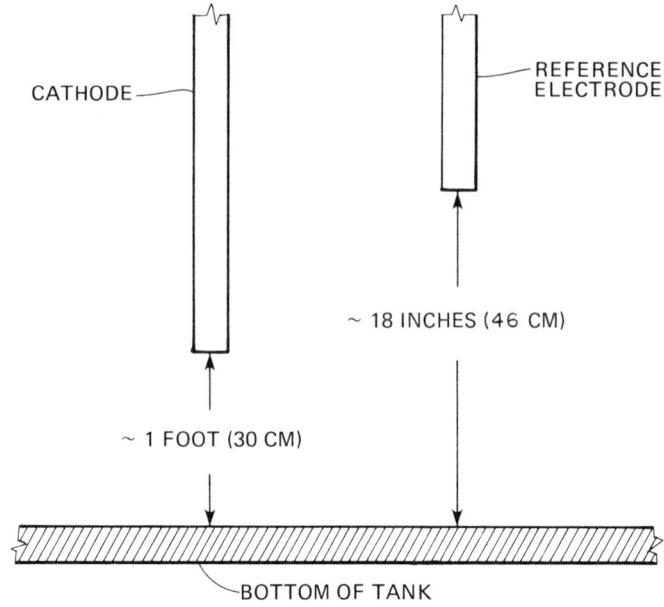

Figure 4-6. Distances of cathode and reference electrode from tank bottom.

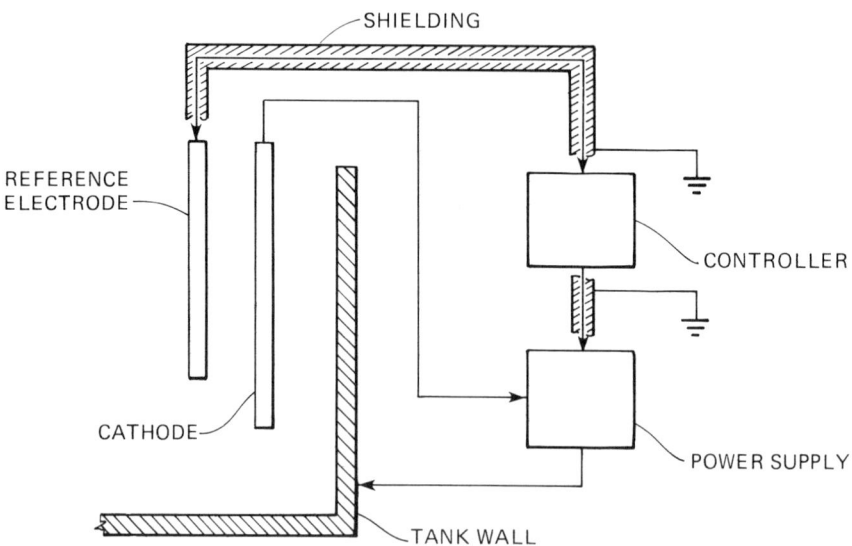

Figure 4-7. Diagram of power and control circuits identifying shielded leads. Power supply should be close to protected vessels.

to each cathode. The wire to the vessel wall should be sized to carry the total current applied between the vessel wall and the cathode and can be connected to a clip welded to the tank wall.

Reference Electrode. The lead from the reference electrode to the potential controller carries a low-level signal. It is vulnerable to stray-current pickup, so a shielded cable is recommended with the shield grounded at one end only to avoid stray-current loops, as shown in Figure 4-7. It is also best to connect the controller potential-sensing circuit directly to the vessel wall by a separate conductor. This will avoid the potential drop which will occur if the current-carrying circuit to the vessel is also used for the potential-sensing circuit.

Leads between the controller and power supply should also be shielded to prevent stray-current pickup.

All leads should be in conduits and appropriate junction boxes should be used. Code restrictions for all electrical work should be observed.

Location of Power Supply and Controller

The power supply is usually placed close to the vessel being protected to minimize the length and resistance of the current-carrying wire. In some installations it has been placed in a control room close to the controller. Present commercial practice is to locate the power supply on a concrete pad close to the protected vessel. Electrical components are immersed in an oil-filled container. The oil will quench a spark which might cause an explosion and is also an exchange medium to remove heat generated during operation.

The potential controller is located in a control room where it can be monitored periodically. In some locations where atmospheres are very corrosive, or near marine exposures, it may be necessary to maintain a positive pressure of filtered, dry air to prevent introduction of destructive air into the electronic package. Commercially available instruments are packaged so that they can be integrated into a control panel.

Start-up Procedures

The startup of a system involves passivating the equipment for the first time. It is preferable that the installed system have sufficient current capacity to passivate the surfaces of interest in a reasonable time. If sufficient capacity to do this is lacking, then use of a supplementary power supply will be necessary.

Initial passivation is made easier if one or more of the following conditions can be met:

1. A small area of wetted surface.
2. Lowered temperature.
3. Reduced corrosivity of solution.

Passivation of a storage tank, for instance, is easier when the liquid level is low and the area exposed to the corrosive solution is at a minimum. Then, after the initial wetted area is passivated, additional liquid is pumped in, gradually increasing the wetted area. The system will passivate this newly wetted area and maintain the initial wetted areas in a passive state. The current demand is much less than would be required if the entire wetted surface must be passivated silmultaneously. In any case, current demand is a function of temperature and the area to be passivated.

Similar benefits in reduced current demand can be achieved if temperatures are increased to working levels *after* the system is under protection. However, in high-temperature applications, such as exist in acid coolers, there should be an electrical capacity sufficient to passivate the equipment at normal operating temperatures.

Operation and Maintenance Parameters

An anodic protection system ordinarily does not require an excessive amount of attention from operating personnel. The traditional maxim that implies that satisfactory operation of a plant is evident when operators are relaxed in the control room with their feet up applies to the operation of an anodic protection system. It may be desirable to keep a record of current demand as a function of time and operating conditions. If so, the current can be registered on a strip chart recorder, manually or by a computer.

Control units are equipped with potential-sensing limits that can be connected to an alarm system. When the potential equals one of these limits, the alarm is activated, the unit turned off, and the operator alerted to a problem with the system.

Corrosion-Rate Measurements

Also, it will be useful to monitor the effectiveness of the system in controlling corrosion. This can be accomplished through the use of one or more of the following techniques:

1. Weight-loss coupons.
2. Analysis of solution metal content.
3. Use of resistance-type rate meters.

Preparation of coupons to measure weight loss, pitting, or other effects should follow established practices. Procedures for coupon tests are adequately discussed in the recommended practices published by the National Association of Corrosion Engineers.[6,7]

Weight-loss coupons must be connected to the vessel wall so that they will participate in the protection conferred on the wall by the system. As a means of measuring the unprotected rate, it is desirable also to electrically expose insulated coupons in the solution. Evaluation of anodic protection effectiveness in this manner has been described elsewhere.[8–10]

Because purity of product is the criterion of protection in many installations, monitoring the metal content of the solution is the usual way of determining success. Analytical procedures involved may be found in books on analytical chemistry. However, sampling procedures must be described.

How to Take Solution Samples

Solution samples for analysis must be taken in such a manner that they truly represent the process stream of interest. Care must be taken that the samples

1. Are not taken from a stagnant zone or bypass line. Bypass lines must be flushed before sampling.
2. Care must be taken not to allow introduction of additional oxygen that may affect analysis values.
3. The sampling container must be resistant to the solution.

Special instrumental techniques such as ac impedance and, possibly, linear polarization may be utilized. The use of these for anodic protection is presently only suitable for laboratory investigations.

Commercial Units Reliable

Commercial anodic protection systems now available have a high degree of reliability. They incorporate materials and electronic and electrical components which represent the culmination of decades of experience making and installing thousands of units. Units from experienced manufacturers should require minimum maintenance. They also provide the latest safety factors and warning circuits to permit continuous observation of performance.

Summary

It is obvious that technical considerations and engineering decisions that must be made when contemplating the installation of anodic protection are not trivial. In this they do not differ significantly from equivalent factors that must be taken into account in designing and protecting any unit designed to handle corrosive solutions.

Sufficient information has been provided to guide those who use corrosive solutions in making decisions about the desirability of anodic protection. An understanding of the installation and monitoring criteria will help those who install systems to give them proper supervision.

References

1. W. P. Banks and J. D. Sudbury, *Corrosion* **19**, 300 (1963).
2. O. I. G. Ives and G. J. Janz, *Reference Electrodes—Theory and Practice,* Academic Press, New York (1961).
3. J. M. Stammen and C. R. Townsend, *Corrosion* **23**, 343 (1967).
4. Chemetics International, Ltd., Willowdale, Ontario, Canada, private communication.
5. L. A. Danielyan, A. I. Tsinman, *et al., Zashch. Met.* **9** (4), 492 (1973).
6. TM-01-69, *Laboratory Testing of Metals for the Process Industries* (1976 revision), NACE, Houston, TX.
7. RP-07-75, *Preparation and Installation of Corrosion Coupons and Interpretation of Test Data in Oil Production Practices,* NACE, Houston, TX.
8. A. O. Fisher and J. D. Brady, *Corrosion* **19**, 372 (1963).
9. W. P. Banks and M. Hutchinson, *Mater. Prot.* **8** (2), 31 (1969).
10. D. Fyfe, R. Vanderland, and J. Rodde, *Chem. Eng. Prog.* **73** (3), 65 (1977).

Economic Evaluation of Anodic Protection

The task of making an economic study to determine the desirability of using anodic protection does not differ from that in evaluating any other engineering decision. Cost factors to be considered are listed in Table 5-1. Not all of the factors listed must be considered for every evaluation. It is virtually impossible to quantify other, perhaps equally pertinent, factors such as effects of extended delivery times for certain types of equipment, availability of materials at any price, and the cost of liability insurance covering one type of installation as compared to another.

Another factor that enters into the evaluation for certain kinds and classes of product is that relating to selling price differentials between pure and impure solutions, that is, those which have been contaminated with corrosion products.

In any case, because anodic protection is usually one among several options for the solution of a corrosion problem, several analyses must be made and their respective costs compared. This procedure often results in the discovery that the least in initial cost is not the most economical. If long-term absolute availability of trouble-free equipment is essential, the economics of present importance can become a negligible consideration.

Methods for making studies of this kind can be found in almost every good engineering text, as well as in the literature of cost accounting. A useful document on the method is available which defines the procedures and shows how to make the necessary comparisons.[1] Stephens and Verink have criticized this document and suggested modifications and simplifications.[2] Because of the ready availability of this information, it is not necessary to give further details here.

Known Costs of Anodic Protection

Not much information is available in the literature concerning the costs of anodic protection. Such information as is available is reviewed here and additional information from suppliers of commercial systems is given. These

TABLE 5-1. Factors To Be Considered in Economic Analysis of Anodic Protection
Installations[a]

Capital Investment	Inflation allowance
Design and blueprints	Taxes
Materials selection	Depreciation
Equipment and accessories	Projected service life
Construction costs	Estimated maintenance costs
Supervision and overhead	Supervision and monitoring
Corrosion allowance, if any[b]	Additional safety equipment[c]
Cost of capital	Electricity

[a]It is assumed that the cost of the site is absorbed elsewhere.
[b]This is the difference in cost between a material with no allowance for corrosion attack and that with an allowance.
[c]Such as instantaneous rate meters or additional costs of warning devices to alert operators of upset of conditions.

costs were published over a number of years and cannot be compared with present costs unless a correction is made. A convenient method of correction is provided by the CE Plant Cost Index as published in the magazine *Chemical Engineering*.[3] This index is based on chemical-plant costs and includes a special index for process instruments and controls. It is published in every issue of the magazine. Table 5-2 lists the cost index for process instruments and controls for the years 1960–1978.[3]

The index may be used to calculate present values by multiplying the cost of an item in a particular year by the ratio of the cost index for the present year to the cost index for the year of the known cost. As an example, an item costing $1000 in 1963 would cost $1000(223.5/105.7) = $2144 in 1978.

Costs of Small Tanks

It will be useful as a frame of reference to consider the costs of small tanks made of different metal and nonmetal materials. Table 5-3 lists some costs as of 1977.[4] It is important when considering the various options in this table to take into account some of the factors pertinent to the properties of the materials, as well as the amount of capital invested. For example, a rubber lining for a carbon steel tank adds about 50% to the original cost. This is a 44% advantage in initial cost over an AISI 316 tank of the same capacity. This is a poor decision, however, if the lining requires replacement after five years at a cost of about $3000 and a second replacement at the end of a second five years at $4000 (allowing for 33% inflation), bringing the cost of the carbon steel tank at the end of ten years to about the same as the original cost of the AISI 316 tank. This analysis does not consider the cost of money, taxes, and other factors which can be significant and assumes that the 316 tank will not be corroded during the decade. Also not considered is the value of lost production during down-time for relining the carbon steel tank, or as an alternative to this, the cost of redundant tankage to replace the tank under repair.

Under circumstances similar to these, one of the major factors in favor of anodic protection for a tank made of expensive alloys is that, in many environments, corrosion can be reduced to negligible or essentially zero rates. This, in effect, extends the useful life of the tank to its ultimate obsolesence and insures a substantial recovery from the sale of scrap.

Data on Larger Tanks

Walker[5] provided a study of two 25-kgal (94.6-m³) tanks, one carbon steel bare (lined with polyvinylchloride) and the other made of CrNi steel. He also provided a comparison between a carbon steel tank with anodic protection and an aluminum tank, each of 1-million-gallon (3785-m³) capacity. These comparisons will be found in Table 5-4, where it is indicated that anodic protection is the most economical method of control for both sizes. However, because the method of obtaining the "annual costs" was not given, it is difficult to be sure of their accuracy. It is impossible either to know what lifetime estimates were made for the tanks. The data used were obtained from an advertising brochure published by an anodic protection engineering company which compiled the

TABLE 5-2. C.E. Plant Cost Index for
Process Instruments and Controls[3]
(1957–1959 = 100)

Year	Index	Year	Index
1960	105.4	1970	132.1
1961	105.9	1971	139.9
1962	105.9	1972	143.8
1963	105.7	1973	147.1
1964	105.8	1974	164.7
1965	106.5	1975	181.4
1966	110.0	1976	193.1
1967	115.2	1977	203.3
1968	120.9	1978	223.5
1969	126.1		

TABLE 5-3. Relative Costs of 7000-gal
(26.5-m³) Tanks[4]

Materials	$ U.S. 1977
Carbon steel, bare	6324
AISI 304	14210
AISI 316	16985
Carbon steel, rubber lined	9455
Fiber-reinforced polymer	
Bis-phenol A Polyester	5850
Furan	7753

TABLE 5-4. Comparative Annual Cost of Tanks With and Without Anodic Protection[5] ($U.S. 1970)[a]

Parameters	Tank cost	Anodic protection system	Lining	Electricity	Maintenance	Totals
25,000 U.S. gallons (94.6 m³)[b]						
mild steel, ft² (m²)						
anodically protected[b]	0.38 (0.035)	0.21 (0.019)	—(—)	0.01 (0.0009)	0.12 (0.0112)	0.72 (0.0725)
lined[c]	0.38 (0.035)	—(—)	0.90 (0.084)	—(—)	—(—)	1.46 (0.136)
AISI 304 steel[b]	1.02 (0.095)	—(—)	—(—)	—(—)	—(—)	1.02 (0.095)
1,000,000 U.S. gallons (3785 m³)						
mild steel, ft² (m²)[b]						
anodically protected[b]	0.38 (0.035)	0.01 (0.0009)	—(—)	0.01(0.0009)	0.12(0.0112)	0.52 (0.0048)
aluminum[b]	0.78(0.0725)	—(—)	—(—)	—(—)	—(—)	0.78 (0.0725)

[a] Costs are in dollars/square foot/year.
[b] Ten-year tank and equipment life.
[c] Three-year lining life.

costs in 1967. Furthermore, substantial changes in materials and power costs in the intervening decade influence the validity of the estimates.

Banks and Hutchison[6] compared the costs of used steel, anodically protected, railroad tank cars, with aluminum cars transporting fertilizer solutions. The steel cars were designed for and used to transport liquefied petroleum gas. The study indicated that substantial savings could be made by using anodically protected steel. Details of this study are given in Table 5-5.

The method of cost calculation for this system was not rigorous. However, it does provide a comparison of anodically protected, steel, railroad tank cars, with cars made of AISI 304 steels and aluminum.

Current Costs of Protecting Steel Tanks Against Sulfuric Acid

Some information on the cost of anodic protection systems has been obtained from commercial suppliers in North America. These include the design, specification of installation details, electrodes, potential controller, power supply, and start-up of the system, but not for mechanical and electrical work.

Storage tanks for concentrated (over 78%) sulfuric acid are usually made of carbon steel. Many of these tanks have been in use for periods up to 40 years

TABLE 5-5. Cost of Anodically Protected 10 kgal (37.85 m³) of Steel and of Aluminum Railroad Tank Cars to Transport Nitrogen Fertilizer Solutions[b]

Components	Annual costs/Car ($ U.S. 1969)
Aluminum: rent at $165–$220/month	1960–2640
Steel: Anodically protected cost	600[a]
Anodic protection	
Instruments	355[b]
Labor	20[c]
Operation	80[d]
Removal	20[e]
Total	1075

[a]Based on $3000 cost of used steel car (assumed 5-yr life).
[b]Based on costs: potential controller, electrodes, and mounts, $1150; controller mounting box and conduit, $250; batteries (3) $150 (system life = 5 yr; batteries, 2 yr).
[c]Installation and removal.
[d]Actual during 1966.
[e]No credit for salvage.

and, as has been discussed in Chapter 2, a critical factor in this extended life has been the incidental copper content of the steels from which they were constructed.[7] Experience with tanks made with steel with a low copper content indicates that they will probably not have the same resistance to sulfuric acid as the older tanks. This fact reinforces the importance of anodic protection as an economical way to extend their useful lives and avoid leaks or catastrophic failure.

Why Tanks Require Protection

At one time the sole economic justification for anodic protection (or any other protection) for sulfuric acid storage vessels was the reduction of product contamination by corrosion products. Now, however, an added incentive is increasing the life of the tanks.

It has been customary to add a "corrosion allowance" to the thickness of the plates of at least one-eighth of an inch (0.32 cm). This adds substantially to the cost of the plates, a factor which is incrementally significant when engineering cost analyses are prepared. This expense can be avoided if anodic protection is used, so many recently constructed tanks do not have the added plate thickness and this saving can be included when economic studies are made.

Costs of anodic protection systems vary in relation to tank size. These costs are not linear, because some of the components and installation costs will be the same for tanks of any size. Added costs are entailed for the power supply and electrodes (usually to a maximum of three) and sometimes for other components, depending on the tank geometry. No added costs related to size usually accrue for potential controllers, reference electrodes, and certain other components, such as the instrumentation or warning signals.

Cost of Internally Lining Tanks

Because there is a limited number of options for corrosion control of tanks and process vessels, somewhat fewer for such equipment as heat exchangers, comparisons need to be made between two, or at most three, procedures. Some of the principal control measures are: coatings, cathodic protection, and anodic protection. Cathodic protection may not be feasible for certain combinations of materials and corrosive solutions, or more expensive than the other options. There are definite limits to the choices for coatings that are adequate to resist attack by many acids and alkalis, as well as mechanical or economic limitations that may make them undesirable. These options prevail after a decision has been reached on the material to be used. When certain highly resistant materials are selected, then the necessity for additional protection may be reduced or eliminated entirely.

It is useful to consider the internal-coatings option as applied to concentrated sulfuric acid storage tanks. Almost the sole organic coating material with sufficient resistance is baked phenolic, which must be applied in a rigid schedule of application procedures and then baked to "cure" the resin. Multiple coats are usually applied, each of which is cured by heating. Then, after application of the final coat, a concluding heat treatment is carried out. There must be a fairly close control of the heating cycle, which is accomplished in large tanks by timed application of 205°C air into an externally insulated tank to crosslink the phenol–formaldehyde polymer.

Obviously, these procedures are not only costly and time consuming, but also have limits as a function of tank size. The cost of coatings of this kind range from $2.50 to $6.00 U.S. per ft² ($26.86 to $64.46 per m²).[8] Costs in this range are believed to be adequate for tanks in sizes from 500 tons (diameter × height in ft = 25 × 18 or 7.62 × 5.49 m) to about 6500 tons (60 × 40 ft = 18.29 × 12.9 m). Table 5-6 lists the costs of lining these vessels.

Estimated lifetimes of coatings of this type are from 2 to 5 years[7] in acid concentrations up to 100%. Lifetimes are shorter in oleum.[9] Other sources say phenolic linings are not suitable for concentrations above 90%.[7] Because of the difficulties with heat curing of phenolic linings, the upper limit in size for a tank is cited to be 60 × 40 ft.

It is of interest to compare the cost of phenolic lining to the cost of anodic protection. Data in Table 5-7 show the cost of anodic protection to diminish in a ratio 30:4 as a function of increased tank size. The cost of the baked phenolic lining per ton of stored acid is not changed very much for the larger tank sizes. Anodic protection is less expensive than the lining for the larger tank sizes

TABLE 5-6. Costs of Baked Phenolic Lining for Concentrated Sulfuric Acid Storage Vessels[8]

Capacity, tons (metric tons)	500 (453.6)	6500 (5897)
Dimensions, ft (m)	25 (7.62)	60 (18.29)
Diameter	25 (7.62)	60 (18.29)
Height	18 (5.49)	40 (12.19)
Wetted area, ft² (m²)	1414 (131.4)	7540 (700.5)
Costs, $ U.S.[a]		
per ft² (m²)	2.50 (8.20)	6.00 (19.67)
per ton (metric ton)	7.07 (7.79)	6.96 (7.67)
totals	3535	45247

[a]Costs as of April 1979.

TABLE 5-7. Capital Costs of Anodic Protection for Sulfuric
Acid Storage[9] (100% Sulfuric Acid)

Tank capacity (ton)		$ U.S./ton[a]		
English	Metric	English	Metric	Total ($ U.S.)
1000	907.2	30	27.22	30,000
20000	18144.0	4	4.40	80,000

[a]Costs as of April 1979.

when compared on the cost per ton of acid stored. Anodic protection costs $4.00 per ton of acid, while a lining costs $6.96 per ton of acid. However, the cost of the anodic protection system is much more expensive per ton of stored acid in the smaller-sized vessels. The anodic protection system costs $30.00 per ton of acid and the baked phenolic lining cost $7.70 per ton of acid.

This comparison is not completely adequate for a true engineering estimate. The costs of money, future worth, taxes, operating expense, and other economic factors were not considered. However, it does indicate that anodic protection does become economically attractive as the vessel size is increased. This is generally true when comparing the economics of using higher-grade alloys to anodic protection of a lower-grade alloy.

The operating expense of an anodic protection system is primarily the cost of electrical power supplied to passivate and protect the vessel. In most applications this cost is not of any major significance. The power consumed on a continous basis is usually only a few watts, even though the power supplies discussed in Chapter 3 can be as large as 7.5 kV A. The currents required to maintain a system are orders of magnitude smaller than those to establish protection so that these power supplies operate well below the maximum capacity.

Sulfuric Acid Coolers

Heat exchangers for sulfuric acid cooling discussed in Chapter 2 present a good opportunity for the application of anodic protection in many plants. Coolers of corrosion-resistant CrNi alloys designed to replace the traditional "trombone" cast-iron cascade coolers can be protected efficiently.

The cast-iron coolers have been a large item of maintenance costs, because most plants using them had to keep a large stock of replacement units and usually had a large pile of discarded units damaged by corrosion. Jones[10] itemized losses attributed to the cooling system in 400–600-ton sulfuric acid plants using cast-iron coolers as $0.38 Canadian per ton of acid. This amount equalled 10% of the net profit. Table 5-8 lists the factors in this estimate. He also estimated that costs would escalate to about $1.25/ton for a 1000-ton/day plant.

TABLE 5-8. Cost of Losses Using Cast-Iron Coolers in
Sulfuric Acid Manufacture (Canadian $ 1976)[a,b]

Item	Costs[c]	
	English ton	Metric ton
Scheduled repairs	0.12	0.110
Unscheduled repairs	0.05	0.045
Acid lost in spills	0.01	0.009
Lost production due to unscheduled shutdown	0.18	0.163
Miscellaneous (spill neutralization, coatings maintenance, etc.)	0.02	0.018
Totals	0.38	0.345

[a]F.W.S. Jones, *Anticorrosion,* December 12 1976.
[b]Based on 600 English ton (544 metric ton) and 400 ton (362.9 metric ton) per
day plant operations.
[c]Canadian $ = 0.8667 U.S., 18 May 79.

These anodically protected stainless steel shell and tube exchangers are
described in Chapter 2. Maintenance costs for operation of these coolers have
been $0.01–$0.02 per ton and there was no lost production owing to cooling
equipment.

The capital cost of stainless steel exchangers is substantially more than
that for cast-iron coolers, but the operating-cost reduction should make them
attractive investments. Costs of anodic protection systems only are shown in
Table 5-9, where an increase in area by a factor of six increases costs by a

TABLE 5-9. Cost of Anodic Protection
Systems for Heat Exchangers in
Sulfuric Acid Service[a,b]

Heat-exchanger area		Costs (U.S. $)		
ft²	m²	ft²	m²	Total
1000	92.9	25	2.32	25,000
6000	557.4	7	0.65	42,000

[a]Chemetics International, Ltd., Willowdale, Ontario,
Canada.
[b]Costs as of April 1979.

factor of 1.68. The economic calculations for the stainless steel exchangers must also include the increased value of acid with lower iron content.

Summary

Economic studies have shown that in many corrosive solutions, and with a relatively wide range of materials, anodic protection is a desirable economic alternative to other methods of reducing or controlling corrosive attack. An important advantage of anodic protection is that, in contrast to other options, laboratory tests will define with considerable precision whether or not it is a viable alternative.

Economic analysis will permit a reliable decision to be reached when comparing the merits of alternative systems. A major advantage is the high order of protection that can often be achieved, usually considerably better than that provided by other systems.

There is little or no debate about the necessity for conserving scarce or expensive metals and alloys and there is considerable economic incentive to use corrosion-control systems that assure uninterrupted operation of equipment, particularly when processes involve handling hot, toxic, or highly corrosive solutions. Accepted methods of economic analysis permit advance determination of which option is best.

Finally, when considering the merits of alloys with high specific resistance to corrosion attack, such as Hastelloy or titanium, engineers may find anodic protection a useful adjunct to their decision to use explosion-clad materials as alternatives to solid vessels. When the backing material is thicker than one inch (2.54 cm), the price per pound of explosion-clad Ti or Hastelloy on carbon steel may be as much as 50% of that for solid materials and about 70% of that for clad materials of less costly alloys or metals.[11]

Laboratory data indicate that anodic protection may be used successfully to control corrosion in many combinations of aggressive solutions and metals for which its successful application has not been proved unequivocally. There is ample reason to believe, however, that many of these combinations will be subjected to practical tests and that anodic protection will prove to be a useful and economical solution to many corrosion problems for which there have not been good solutions so far.

References

1. Standard RP-02-72, *Direct Calculation of Economic Appraisals of Corrosion Control Measures*, NACE, Houston, TX.
2. W.K. Stephens and E.D. Verink, Jr., *Mater. Perform.* **17**(10), 15–18 (1978).

3. T.H. Arnold and C.H. Chilton, *Chem. Eng.,* (NY) Feb. 18, 143 (1963); C.H. Chilton, *Chem. Eng.,* Apr. 25, 184 (1966); O.R. Thorsen, *Chem. Eng.,* Nov. 13, 168 (1972); L.J. Ricci, *Chem. Eng.,* Apr. 28, 117 (1975); P.M. Kohn, *Chem. Eng.,* May 8, 189 (1978).
4. R.W. Leitheiser *et al., Corrosion* 77/162, NACE, Houston, TX.
5. R. Walker, *Metallurgia,* No. 8, 51 (1970).
6. W.P. Banks and M. Hutchison, *Mater. Prot.* **8**(2), 31 (1969).
7. D. Fyfe, R. Vanderland, and J. Rodda, *Chem. Eng. Prog.* **73**(3), 65 (1977).
8. Heresite Chemical Co. Manitowoc, WI, private communication. (1979).
9. Chemetics International, Ltd., private communication, Willowdale, Ontario, Canada.
10. F.W.S. Jones, Anti-Corros., December 12 1976.
11. E.I. Du Pont de Nemours & Co., Inc., private communication, Coatesville, Pennsylvania (1979).

Principles of Anodic Protection

An understanding of the electrochemical principles on which the application of anodic protection is based is beneficial for anyone who believes that the method may be useful in controlling corrosion of his equipment. This is particularly true when its use must be technically justified or when decisions must be made about alternative proposals, equipment, or operational modes.

Most engineers are more comfortable with systems that they understand. Fortunately, there is nothing exotic about anodic protection and nothing so complex that its principles cannot be absorbed in a few minutes. It is entirely possible to understand how anodic protection works without following through the mathematical and thermodynamic details that are presented in this chapter. Nevertheless, it is recommended that anyone who expects to be involved with the management of anodic protection be sufficiently conversant with the facts herein that he can analyze system reactions.

It is probably true, also, that understanding of anodic protection principles should precede any decision to apply it for protection of equipment, especially by those who are responsible for technical decisions. In this way the data provided by the control feedback circuits are understood in terms related to the passivity of the surfaces exposed to the corrosive solutions. The mental picture provided by polarization curves defining the electrochemical environment at the electrolyte–metal interface makes the feedback data more meaningful and undoubtedly enhances the ability to recognize upset conditions or system operational aberrations.

Finally, the system described here is a way to achieve passivation of a surface exposed to a corrosive. The objective is exactly the same as that contemplated by the application of a protective coating or the addition of an inhibitor or providing a compressive stress by mechanical methods. When the fundamental reactions are analyzed, the inevitable conclusion is that the methods achieve an electrochemical barrier between the corrosive electrolyte and the base metal. In the case of anodic protection, this barrier may be a layer of oxides, a protective configuration of electrons or ions or some other mechanism, the exact nature of which may differ from one combination of corrosive–metal to another, or which may comprise elements of all the theoretical modes. These relationships are covered in this chapter.

Thermodynamic principles imply that most metals should react with their environments. However, some metals show essentially no corrosion under conditions in which severe attack can be expected. This lack of reaction is the result of passivity. Passivity can be induced on some metals by anodic polarization.

Electrochemical Description of Passivity

It would seem that forcing a metal into more oxidizing potentials would cause increasing dissolution rates. In the case of metals which undergo an active–passive transition, more oxidizing conditions may actually lead to a much decreased dissolution rate (lower corrosion rate). This tendency of a metal to remain visibly unchanged for a prolonged period when it should be thermodynamically unstable in the given electrolyte is called *passivity*. Passivity is generally attributed to the formation of a protective layer at the metal–electrolyte interface. This protective (passive) interface (layer when referring to dimensional compositions, or film when referring to adsorptive species) prevents direct contact of the metal with the electrolyte. Measurements of a metal's tendency to passivate are important for determining its electrochemical corrosion characteristics. These measurements can be made by constructing an anodic-polarization curve.

Characteristics of Anodic-Polarization Curves

Figure 6-1 illustrates typical polarization curves for an active–passive metal under dissolution conditions. The intersection of the oxidation reaction $(M \rightarrow M^+ + e^-)$ and the reduction reaction $(M^+ + e^- \rightarrow M)$ corresponds to the reversible equilibrium potential, E_{M/M^+} and the exchange-current density, I_{OM/M^+}.

The irreversible corrosion reaction

$$Fe \rightarrow Fe^{2+} + 2e^- \qquad \text{(iron dissolution)}$$
$$2H^+ + 2e^- \rightarrow H_2 \qquad \text{(hydrogen evolution)}$$

is best discussed using the schematic drawing (Figure 6-2) for the kinetic behavior of an iron electrode in an acid solution.

The steady state for this particular system is at the intersection of the polarization curves for iron dissolution and hydrogen evolution. This intersec-

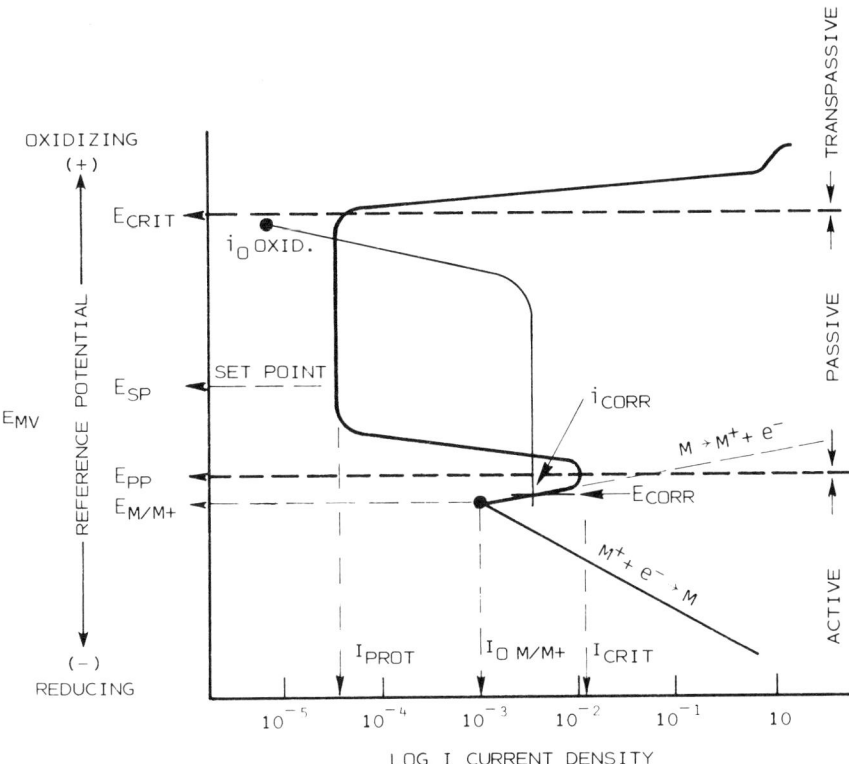

Figure 6-1. Schematic polarization curves typifying anodic dissolution behavior of an active–passive metal.

tion further describes the corrosion current, i_{corr}, and the corrosion potential, E_{corr}.

Applying this concept to the active–passive metal under typical dissolution conditions (see Figure 6-1), one can schematically impose a reduction curve for hydrogen extending from the oxidation exchange-current density such that this polarization curve intersects the oxidation-polarization curve (M $\rightarrow$ M$^+$ + e^-) of the active–passive metal. Where the two polarization curves intersect describes the steady-state corrosion current and corrosion potential of this system.

This corrosion current (in $\mu A/cm^2$) multiplied by the electrochemical equivalents of the metal provides a corrosion rate calculated in mils per year (mpy). When current is supplied such that the reference potential shifts to more negative values, the reduction processes become more active. When current is supplied such that the reference potential shifts to more positive values,

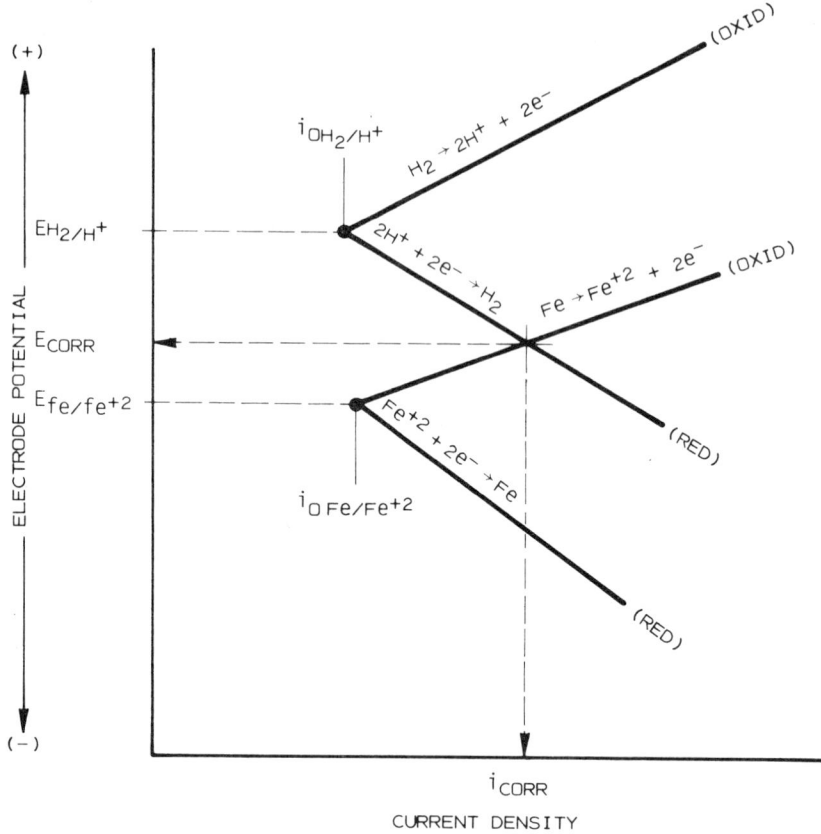

Figure 6-2. The schematic kinetic behavior of iron in an acid solution.

the oxidation processes become more dominant. The E_{mV} -vs-I curve for an active–passive metal under this condition results in an anodic-polarization curve. Generally, the anodic-polarization curve is divided into three zones: (1) active, the portion of the curve represented by potentials more negative than the primary passive potential, E_{pp}; (2) passive, the portion of the curve between the primary passive potential E_{pp} and the critical potential E_{crit}; (3) transpassive, the portion of the curve more positive than the critical potential. The maximum current required to cause the onset of passivity is the critical current, I_{crit}. The minimum current required to maintain passivity, I_{prot}, occurs at the desired potential, E_{sp}.

    To further define the electrochemical parameters of the anodic-polarization curve, a schematic potentiodynamic anodic-polarization curve can be drawn (Figure 6–3) showing the effects of both forward and reverse scans of a typical active–passive metal. The heavily shaded curve represents the typical anodic-polarization curve obtained by the forward scan (reference potentials traversed to succeedingly more positive values). The reverse scan is illustrated by the dashed line (reference potentials to succeedingly more negative values). This scanning technique is useful in determining the pitting potential E_{pit} of the metal in its environment. The closer the pitting potential is to the critical potential E_{crit}, the safer the metal is from isolated corrosion under controlled passivity (anodic protection). Also, this technique is useful in determining the most effective set-point potential E_{sp} for maintaining passivity. The preferred potential would be that value which occurs at minimum reverse-scan current within the prescribed passive zone.

The immersion of an active–passive metal into a corrosive environment can produce unusual results. Figure 6-4 illustrates three possible theoretical and respective actual cases with the active–passive metal in a corrosive acid solution. Reduction processes under activation polarization control are shown here as generalized cases (Nos. 1,2,3) and apply regardless of their curve shape

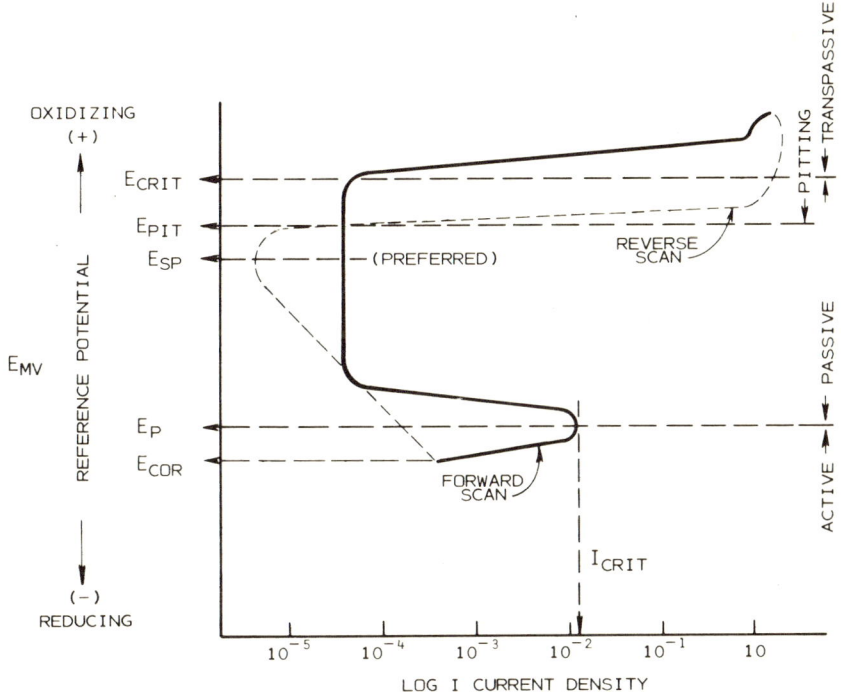

Figure 6-3. Schematic of electrochemical processes which occur during potentiodynamic anodic polarization of an active–passive metal.

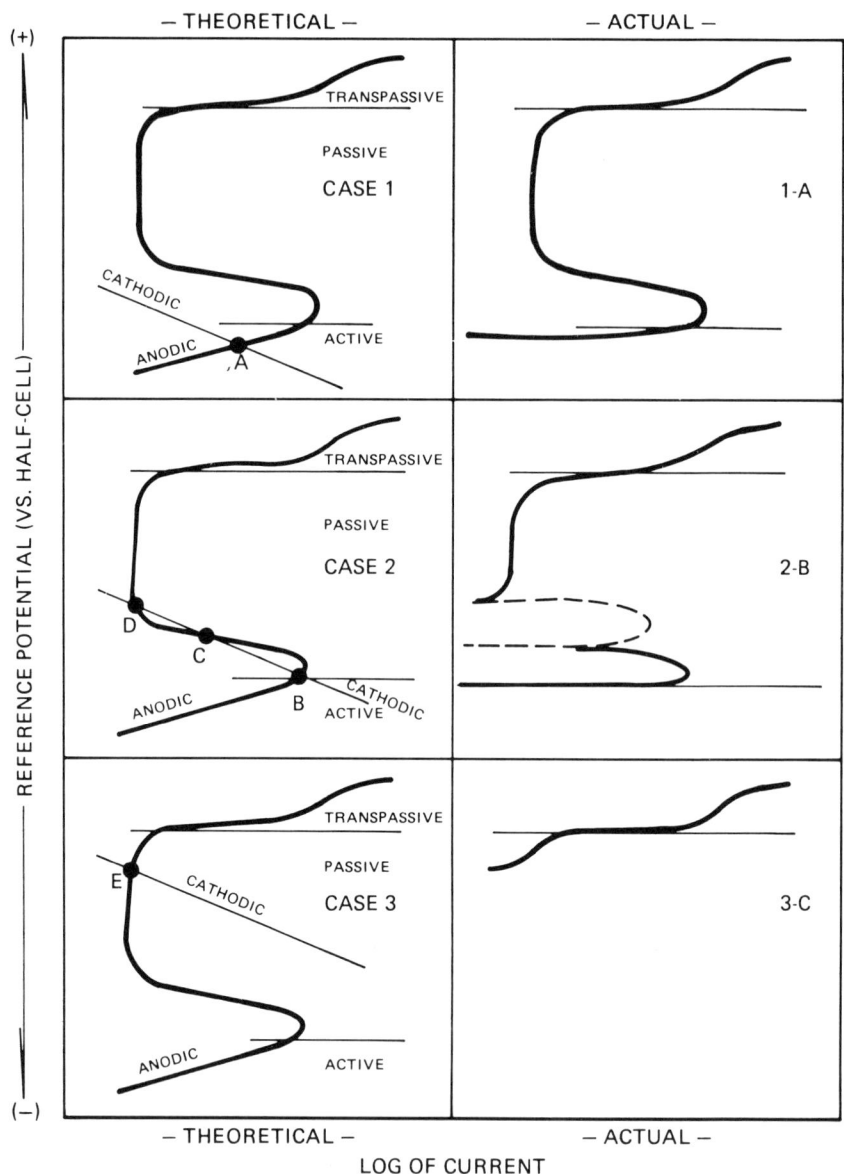

Figure 6-4. Schematic potentiodynamic polarization plots for active–passive metal.

(the process drawn is for hydrogen evolution with three different exchange current densities, i_0).

Case 1: Only one stable intersection point, A, exists which lies in the active zone and high corrosion rates are experienced. Stainless steel in sulfuric acid is characteristic of Case 1. Under these conditions, stainless steel corrodes freely, but it can be anodically protected. Case 1-A is a representation of the actual polarization curve.

Case 2: There are three possible intersection points at which the sum total oxidation rates and reduction rates are equal (points B, C, and D). Point C is electrically unstable. Point B lies in the active zone and high corrosion rates are experienced. Point D lies in the passive zone and corrosion rates are at a minimum. This system may exist in either the active or passive state. An example would be iron in nitric acid upon scratching the surface (transition from a passive to active state). The dashed line in Case 2-B (active curve) represents the "cathodic loop" (peak potential here corresponds to zero current and no corrosion). A typical system is type 304 SS in NH_4NO_3 solutions.

Case 3: Only one stable point, E, exists (in the passive zone). A metal in this system spontaneously passivates and will remain passive. The system cannot become active so the metal experiences low corrosion rates. One typical example is Type 304 SS in nitric acid. Another, is Type 304 SS in H_2SO_4 or H_2PO_4 solutions where iron and vanadium ions are present at higher oxidation states, i.e. (Fe^{3+}, V^{5+}).

From an engineering viewpoint, the most desirable system is Case 3. However, with proper understanding of the system, Cases 1 and 2 can be satisfactorily controlled through the proper installation of an anodic protection system.

Figure 6-5 shows the temperature, velocity, and noncorresponding ion-concentration effects on the anodic polarization of an active–passive metal. An increase in these variables tends to increase the critical-anodic-current density required to obtain passivity. Their effects on the primary passive potential and passive dissolution rates are relatively insignificant. Some reduction in the passive potential range generally occurs.

The magnitude of these effects (whether to slightly increase current densities, or to destroy the passive layer) can be experienced with various concentrations of other elements such as copper, magnesium, etc. (more on this later).

The Passive Metal Layer

The passivity of metals in electrolytic solutions results from the formation of an interface which prevents the direct contact of the metal with the electrolyte. Measurements of a metal's tendency to passivate, important for deter-

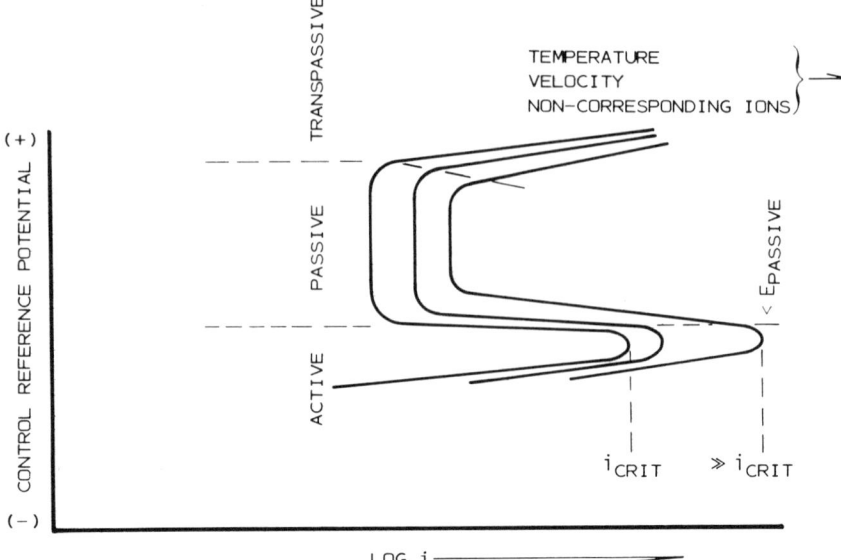

Figure 6-5. Effects of temperature, velocity, and noncorresponding ion concentration on the anodic polarization of an active–passive metal

mining its dissolution behavior, can be obtained from potentiodynamic anodic-polarization curves. The behavior of the interface depends on its thickness, conductivity, potential at which it was formed, chemical composition, corresponding rate of dissolution, electrolyte in which it was formed, and time duration.

It is disconcerting to discover the extent to which solution-environment effects have been ignored in the development of concepts related to the passivity of metals. In the development of passivity mechanisms, solution composition is usually ignored, except perhaps reporting it to be a source of protons. Almost as though no other possibility existed, investigators equate the composition of the passive interface with that of the metal oxide. This concept must be materially modified because electrolytically induced passive films, or layers, are radically different from the air-temperature-formed oxide films.

Some general discussions of the thermodynamic properties of metal electrodes with films on their surfaces and of possible kinetics of anodic passivation of metals in such systems have been given by Göhr and Lange.[1] Iron seems to be the metal most often studied in this field. The growth of the passive oxide film has been discussed by Snavely and Hackerman[2] and Vetter,[3] with Cohen[4] studying its properties.

The surface condition of passive iron in acidic and alkaline systems was investigated by Wade and Hackerman.[5] Their forced coulometric-decay data from oxygen evolution to hydrogen evolution showed that the extent of any passive film was less than one-fifth of a monolayer for any "reasonable chemical species." This low coverage precluded species other than those involved in oxygen evolution. Their data further support the assumption that hydroxyl radicals adsorb in acidic solutions and that hydroxide ions adsorb in alkaline solutions.

The formation of surface oxide films is considered by Evans[6] to cause passivity and his theory is supported by the report[7] that very thin oxide films have been stripped from passive iron surfaces. Also, γ-ferric oxide has the physical and chemical properties necessary for such passive protection. However, there is no direct evidence of the existence of γ-ferric oxide on a passive iron surface. Bonhoeffer[8] introduced cathodic pulses to an iron electrode passivated in concentrated nitric acid, and found that 200 microcoulombs/apparent square centimeter was required to destroy passivity. He believed that this was caused by the reduction of a monolayer of γ-ferric oxide.

Other theories concede that the passivity of some metals is due to visible oxides, but consider the invisible passivation layer on another basis. Uhlig's electron-configuration theory[9] is one and finds application to the metal-alloy systems. Hewmann[10] raised some serious objections and Uhlig[11] extended improved concepts emphasizing the importance of specific interaction between the passivating layer and the substrate.

Cohen[12] has presented a mechanism for passive film growth of iron in neutral solutions. He states that when an anodic potential is first applied to a film-free iron specimen, iron is oxidized to form both a film of magnetite and ferrous ions in solution. In the initial stages, the film is thin, so a high potential gradient exists across the film, so film thickening occurs rapidly. As the potential gradient decreases with thickening, the rate of migration also decreases and anodic currents decrease abruptly.

An excellent review-type compilation of work done to elucidate the composition and structure of passive films has been prepared by Brusic.[13] Electron diffractometry[14] and early ellipsometry[15-22] are exemplary of passive-film studies which may not completely define or interpret film composition (exceptions are reflectance ellipsometry[41-42]). Sato and Okamoto's study[23] on the anodic passivation of nickel in H_2SO_4 solutions discussed the role of OH^- in the formation of the passive film.

Recently, with the advent of Auger spectroscopy, the literature has been flooded with investigators' results from metal surface-layer studies[24-32]. While this technique monitors the elemental composition of the layer, it also has serious drawbacks for definitive identification of the passive layer. It is possible that hydration (water) is removed during bombardment *(in vacuo)*, hence results mislead investigators into believing that the passive layer formed is a mixture of oxides of metallic compounds in the metal-alloy substrate.

Early in 1964, research of the layer formed on carbon steel in 20 N H_2SO_4 at 25°C under potentiostatic control indicated that hydrates of iron sulfates were the cause of passivity.[33] These passive films were sufficiently oriented within 15 s after tests began that they could be identified. Minimum currents were noted at 260, 520, 810, 1060, 1320, and 1600 mV positive with respect to a saturated calomel electrode (SCE). The various chemical compositions were identified with reproducible computerized X-ray diffraction data. The passive carbon steel layer was identified as $FeSO_4 \cdot H_2O$ (260 mV), $Fe_2(SO_4)_3 \cdot H_2O$ (520 mV), $FeSO_4 \cdot 4H_2O$ (810 mV), and $FeSO_4 \cdot 7H_2O$ (1320 mV).

Proposed Mechanism of Iron Passivity

Figure 6-6 schematically presents what are considered generally to be the critical steps in the passivation of iron in sulfuric acid. The mechanism also would cover reaction of metals other than iron, whose surface affinities were not satisfied and would readily donate electrons to some electron-accepting species adsorbing to their surfaces. Such metals include actinium, chromium,

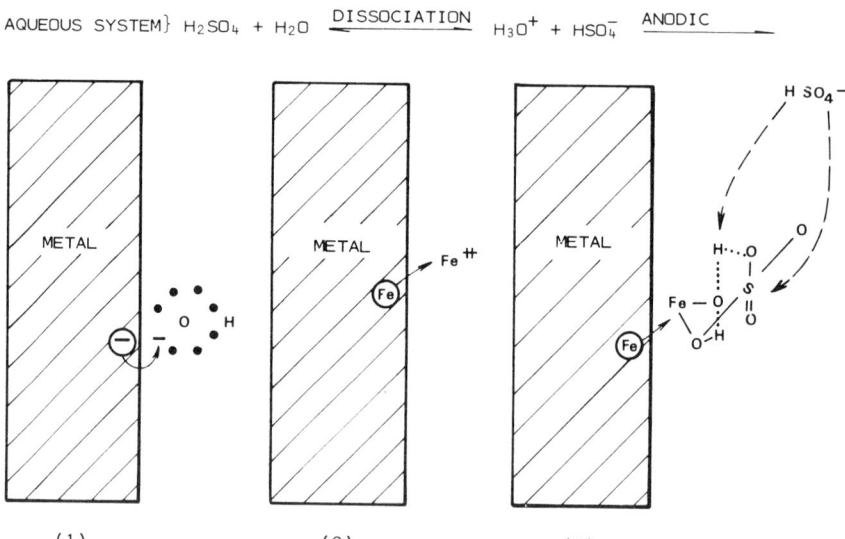

Figure 6-6. Possible mechanism for passive film formation: (1) electron donation step, (2) metal/cation dissolution, (3) passive chemical composition reaction.

cobalt, gold, hafnium, iridium, lanthanum, manganese, molybdenum, nickel, niobium, osmium, platinum, rhenium, scandium, tantalum, titanium, tungsten, vanadium, and zirconium. The first step is the open-circuit corrosion of iron (metal) when placed in an aqueous solution of sufuric acid. Step two is the adsorption step of OH-radical chemisorbing to the iron surface. This is accomplished by the acceptance of an electron by the hydroxyl radical from the metal surface. The aqueous solution of sulfuric acid can be represented to have the following species in some state of equilibrium:

$$H_2SO_4 + H_2O \rightarrow H_3O^+ + H_2SO_4^-$$

When the electrical circuit is closed, permitting the flow of anodic current, step three commences. The various constituents are readily available under the controlled potential system to form a passive chemical composition.

Ferrous ions are introduced immediately following application of the anodic current. The adsorbed hydroxyl radical readily accepts the proton from hydrogen in the system ($H_2SO_4^-$) under the influence of the anodic energies applied with the subsequent or instantaneous combination with the SO_4^{2-} of the ionic species HSO_4^-. The chemical composition of the passive layer would be $FeSO_4 \cdot H_2O$. This opinion can be given also for the process at potentials where ferric ion is introduced. Similarly then, the passive-layer chemical composition would be $Fe_2(SO_4)_3 \cdot H_2O$. If reproducible X-ray data available[32] are accepted, these then are correct chemical compositions. The compositions were identified respectively at 260 mV and 520 mV noble to SCE.

This assumption recognizes the bridging mechanism offered by the hydroxyl (as defined in Figure 6-7) radical for the passivation of iron in acid solutions (metal substrate/OH radical/bulk passive layer).[34] This mechanism also logically links the accepted portions of the three schools of thought regarding the mechanism of the passivity of metals. It is also amenable to the concept of decreasing thickness of the passive layer as the potential becomes increasingly positive within the passive potential range.

In regard to the alkaline system, evidence is insufficient to support an acceptable opinion. However, it is believed that available information points definitely towards the possibility of OH^- ions as the adsorbing, bridging step in the passive phenomena observed for iron in alkaline systems.

It is also believed that the electrostatic bonding of the hydroxide ion (defined Figure 6-7) to the metal surface does occur, and upon the application of anodic current undergoes certain undetermined surface reactions.

No one has reported on "—monolayer of oxide—" work with regard to passive films in place. Regardless of the structure of a stripped film, there is no direct and convincing proof that this film is the one responsible for the passive

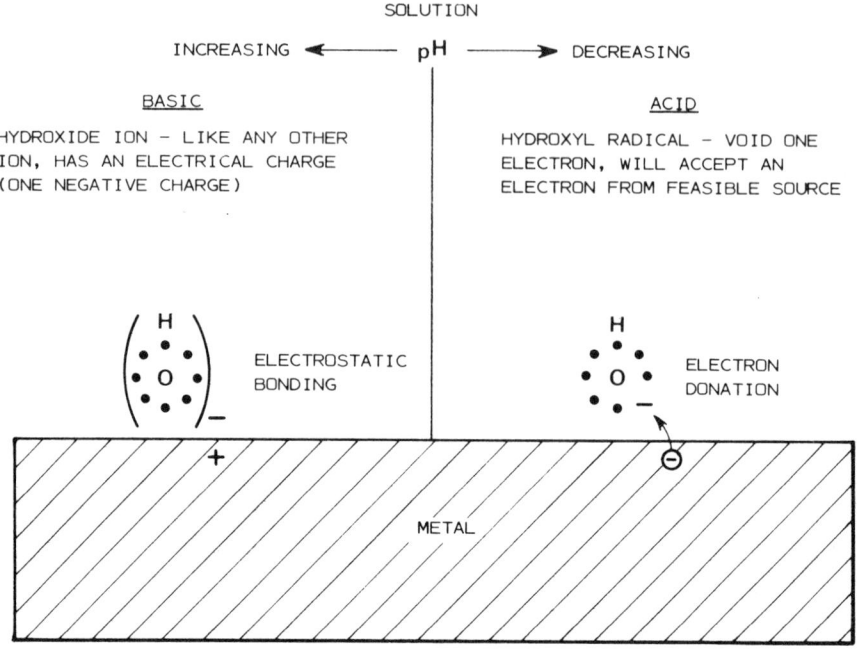

Figure 6-7. Adsorption of species dependent upon pH of aqueous system

characteristics of the metal. It is also fairly evident that the films, prior to stripping, were much thicker than that of the stripped film. It is reasonable to assume that some alteration has occurred, and that the γ-Fe_2O_3 is not the same species as that responsible for passivity. Chemisorbed oxygen is believed not to contribute to the passivity of iron in the 20 N H_2SO_4 system. A film resulting from this activity would not be stable and therefore could not exist in an acid environment.

The formation of ferrous iron prior to the time the critical current ("Flade potential") is reached is assumed to indicate the initial step, rather than the initial adsorption of oxygen. The oxygen-adsorption step has been associated repeatedly with the critical-current-density peak. However, no decrease in current density is recorded until the initial formation of ferrous sulfate monohydrate.

It is then believed that the electron-donating phase of hydroxyl-radical adsorption is the initial step in the anodic passivation of iron in acid solutions. This is consistent with all the available information, as well as the experimental knowledge available.

The anodic behavior of nickel was investigated potentiodynamically using air, nitrogen, and hydrogen-saturated 1 N H_2SO_4.[35] Depending on the sulfate ion concentration on the surface, the passivation of nickel was delayed until a potential of 500 mV (nhe) was reached. The presence of sulfide ions and NiS on the surface was demonstrated by chemical analysis and electron diffraction. Polarization of the [110] and [111] single-crystal planes shows that $H_2SO_4^-$ and HS^- were adsorbed differently on these planes.

Others[36-40] have made contributions to the adsorption theory of metal passivation and, according to Kolotyrkin,[40] passivity is a specific widespread phenomenon which demonstrates a change in the kinetics of electrodic reactions due primarily to the activated adsorption of the oxygen by water (could be the hydroxide or hydroxyl—depending on electrolytic pH); the dissolution rate can be described as:

$$i = K_1 C^1 \exp\left(\frac{\alpha_1 F}{RT} V\right)$$

where α_1 is the transfer coefficient equal to βz; when the valence of the dissolving ion, Z, is unity, α is equal to the symmetry factor β, or to the fraction of the potential available to lower the energy barrier for dissolution. The quantity C^1 is the number of atoms/cm², and with the superimposition of adsorption, C^1 changes with potential so that

$$C^1 = C_0^1 \exp\left(-\frac{\alpha_2 F}{RT} V\right)$$

and the total current is expressed as

$$i = K_1 C_0^1 \exp\left[\frac{(\alpha_1 - \alpha_2)F}{RT}\right] V$$

This equation then explains the observed current drop with potential and the potential dependence in the passive region. The basic anodic processes then are:

1. Direct dissolution of the metal
2. The competitive adsorption of anions

The above explanations and references are possibly the most important discussions about the passivation process and the cause of metallic passivity.

In recent years, the passive layer has received much attention. Very sophisticated instrumentation such as reflectance ellipsometry[41,42] and Auger electron spectroscopy[32,33,40,43,44] have begun to contribute analytical information about the passive layer. Mössbauer-type spectroscopic results were consistent with the concept of a passive film of hydrated iron oxide on iron. Okamoto *et al.*[44-46] reported on the importance of hydrated water to the corrosion protection provided by the passive layer. They suggested that the passive film is a hydrated oxide film as shown in the following model:

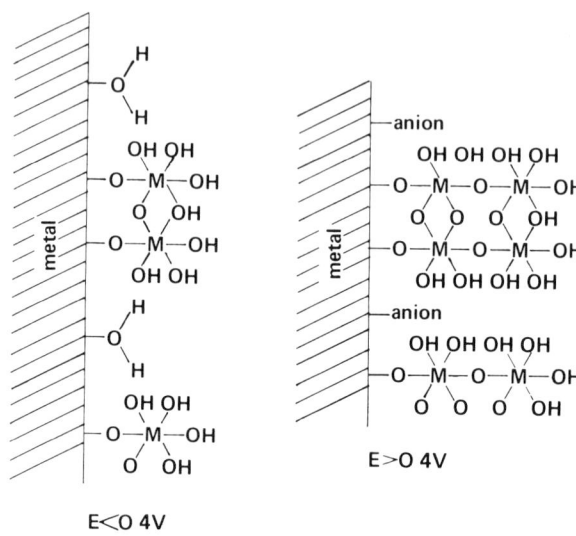

and at any stage of aging there exist three different bridges between metal ions in the passive film:

$$H_2O—M—OH_2$$
$$HO—M—OH, \text{ and}$$
$$O—M—O.$$

Some years ago during a seminar at the University of Oklahoma, Riggs discussed his theory on the passive metal layer. It is quoted as follows:

My position with regards to the passive 'film' or 'layer' began from a rather prejudiced immature opinion of what could and could not occur to a metal's surface in an acid medium. Some may say my opinion hasn't progressed from the original position. Perhaps, some may have an accurate assessment. However, my opinion becomes more convincing (to me at least) as I continue to study the process and read the reported works of others.

Some years ago [Ref. 47], I discovered [schematic concept shown in Figure 6-8] that during the course of potentiodynamically polarizing a carbon steel anode in 20 N H_2SO_4 that ferrous (Fe^{+2}) ions were forced into the electrolyte at a potential (E_{m+}) more negative than that at which

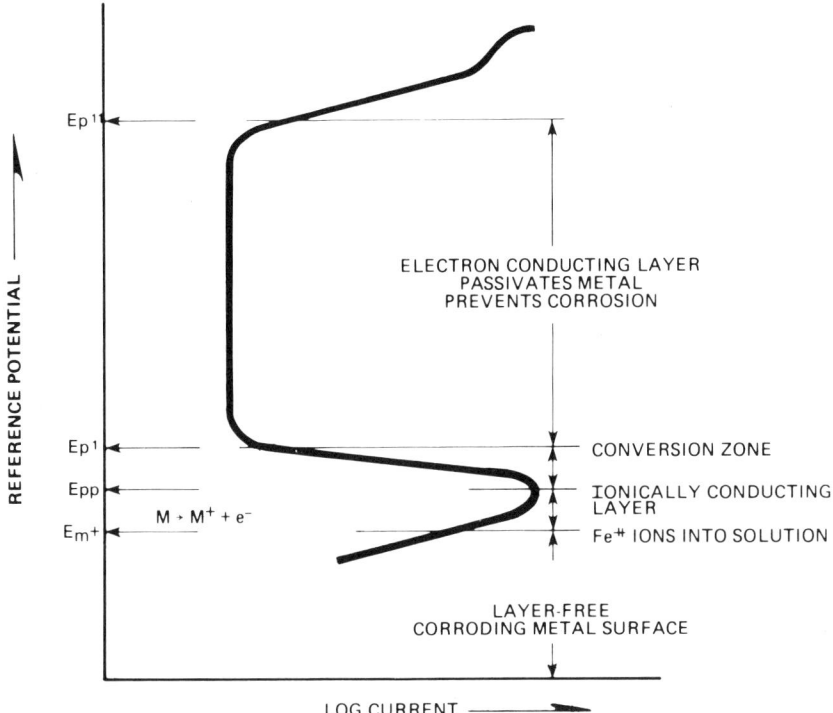

Figure 6-8. Schematic illustration of Riggs concept of the passive metal layer.

the primary passive potential (E_{pp}) occurred. At that time it was believed that this process was a precondition for the onset of passivity. In recent years, it has been referred to as the precursor film[Ref. 48]. Immediately following this event (at the instant of the primary passivation potential) the current decreases sharply. Concurrently, the visible physical surface of the metal is altered and corrosion is diminished. Although the sequence of these events is very rapid, there is some measurable time period for formation of the precursor (ionic conducting layer) and the passive 'film' or 'layer' (electron conducting layer). Also, with the metal now maintained at a potential within the passive potential range ($E_{p^I} - E_{p^{II}}$), the layer continues to be affected by time. The current continues to decrease (believed to decrease due to the continued construction of the passive layer—a preferred orientation).

The passive layer is the product of current and time, but is also the result of a respective reference potential maintained as the layer is formed. The layer is composed primarily of the -ous valence state at the more negative passive potential, and as the potential shifts to more positive values, the layer composition is primarily the -ic valence state [Ref. 41].

Further, it was interesting to learn that the thickness of the passive iron layer in 20 N H_2SO_4 varied within the passive potential range depending upon the reference potential maintained [Ref. 50]. The layer formed at the less positive (more negative) passive potential was essentially thinner than the layer formed at the more positive passive potential. The thicker passive layer was composed primarily of Fe^{+3} (ionic radius of 0.76 Å) [Ref. 51] and the less thick passive layer was composed primarily of Fe^{+2} (ionic radius of 0.64 Å) [Ref. 51].

As one then gathers the information and observations into some cohesive explanation, it seems that the most reasonable passive layer formation theory involves a property change (ionic to electronic layer) and that this layer is of some thickness (generally between 10–100 Å). Also, this layer is the direct result of a forced reaction between -ionic metal/anion of electrolyte's salt/ hydroxide. The compositional variations of the passive layer are then determined by the metal, solution medium, and passivating potential, and passivity is the result of this formation of a solid-state layer.

Metal Passivity Breakdown

An industrially useful material should probably exist primarily in the passive state. Such a material is, therefore, at a potential more electropositive than the corresponding Flade potential, and is covered with a protective film. The following discussion is by no means a definitive work, but rather is meant to make one aware of the possibilities for passivity breakdown.

Although the electrolyte is unable to attack the metal surface, it does dissolve the passive film at some rate. The passive film must remain intact if the metal is to remain in a passive (noncorroding) state. In other words, a new passive film must form electrochemically at the same rate at which the original film dissolves. Corrosion in the passive state is a property of the passive film and the boundary between this film and the solution, and in some specific cases a property of the substrate metal.

If the passive film cannot be formed at the same rate it dissolves, then metallic passivity breakdown occurs. As mentioned previously, SH^-, Cl^-, Cu^{2+}, etc. concentrations in the electrolyte are sufficiently strong depolarizers and each can destroy the original passive nature of the metal. Galvanic contact with heavy metals like copper and manganese can also destroy metallic passivity. Other conditions such as the reduction of the reversible oxygen overpotential below the penetration potential of the agressive reductive anion in the electrolyte may also exist.

The substrate metal itself may cause the breakdown of metallic passivity. Some of the conditions which contribute to this possibility are: metallic carbides and inclusions; also, the sensitization of metal in the zone adjoining the weld seam. Each of the stated conditions renders the metal more locally electroactive than its bulk surface. If care is then not exercised to shift the passive potential to a value which can protect the entire surface, the metal could experience localized corrosion. The products of this corrosion process and perhaps the process itself will cause the breakdown of passivity.

Passivation may be hindered if the Helmholtz double layer, at the potential where the reaction

$$M + H_2O \rightarrow MO_{solid} + 2H^+ + 2e^-$$

can begin, contains a considerable density of adsorbed anhydrous anions, in addition to OH^-. Readily adsorbable ions such as those formed by chlorine must be displaced by water molecules first. This displacement may be slow and difficult, and is hindered by a rise in potential,[52] so an alloy which quickly passivates in H_2SO_4 "may passivate slowly and with difficulty in a similar solution containing the chloride ion." Certain concentrations of chloride ion can prevent passivation of an alloy. As a rule of thumb, it is accepted generally that a concentration of one chloride to 4 sulfate ions is the maximum possible chloride ratio at which austenitic steels will remain passive in sulfate solutions.

Further, perhaps the most lucid description of metal passivity breakdown, was given by Dr. T. P. Hoar[53] as follows:

Three of four halide ions jointly 'adsorb' on the oxide film (passive) surface around a lattice cation—one next to a surface anion vacancy for preference. The transitional complex thus formed will be of high energy and the probability of its formation at any instant will be very small. But, once formed, the complex can readily and immediately separate from the oxide ions in the lattice, the cation dissolving in the solution very much more readily than the non- or aquo-complexed cations present in the film surface in the absence of halide ion. Under the anodic field a further cation comes up through the film to replace the dissolved cation—the field at constant anode potential increases at the 'thinned' point of the film; but arriving at the film/solution interface, it finds, not stabilizing oxide ion formed from water (nor, in de-aerated solution, oxygen molecules), but several halide ions, so that the 'catalytic' process, once begun, has a strong probability of repeating itself, and of accelerating because of the increasing electrostatic field. Thus, once localized breakdown starts with the initial transitional complex, it accelerates 'explosively'.

Evidence then is growing that the anions of the electrolyte not only participate in the formation of the passive layer on metal,[54] but also participate directly in the process of passive layer dissolution together with OH^-. Others have proposed dissolution of passive films by anions, including Acello and Green,[55] Riggs,[56] Lorenz,[57] Vetter,[58] for chloride, Heusler and Cartledge for iodide,[59] and Iofa[60] for SH^-. (See Additional Reading—following Chapter 6 references.)

References

1. H. Göhr and E. Lange, *Z. Elektrochem.* **63**, 673 (1959).
2. E. S. Snavely and N. Hackerman, *Can. J. Chem* **37**, 268 (1959).
3. K. J. Vetter, *Z. Elektrochem.* **62**, 674 (1958).
4. M. Cohen, *Can. J. Chem.* **37**, 286 (1959).
5. W. H. Wade and N. Hackerman, *Trans. Faraday Soc.* **53** (420), Part 12 (1951).
6. U. R. Evans, *Metallic Corrosion Passivity and Protection,* 2nd ed., pp. 51–56, Arnold, London (1948).

7. U. R. Evans, *Nature* **126,** 130 (1930).
8. K. F. Bonhoeffer, *Z. Elektrochem.* **57,** 157 (1941).
9. H. H. Uhlig, *Corrosion Handbook,* pp. 24–33, John Wiley and Sons, New York, (1958); *Corrosion and Corrosion Control,* 2nd ed., John Wiley and Sons, New York (1971).
10. Th. Hewmann, *Z. Elektrochem.* **55,** 287 (1951).
11. H. H. Uhlig, *Ann. N. Y. Acad. Sci.* **58,** article 6, pp. 843, 854, September 15 (1954).
12. M. Cohen, *J. Electrochem. Soc. II* **110** (6), (1963).
13. V. Brusic, *Oxides and Oxide Films* **1,** 1 (1972).
14. C. L. Foley, J. Druger, and C. J. Bechtoldt, *J. Elecktrochem. Soc.* **114,** 994 (1967).
15. A. B. Winterbottom, *J. Sci. Instrum.* **14,** 203 (1937).
16. A. B. Winterbottom, *Trans. Faraday Soc.* **42,** 487 (1946).
17. L. Tronstad, *Trans. Faraday Soc.* **31,** 1151 (1935).
18. A. K. N. Reddy and J. O'M. Bockris, Natl. Bur. Stand. U.S. *Misc. Pub.* **256,** p. 229, Washington, (1964).
19. J. Kruger, Natl. Bur. Stand. U. S. *Misc. Pub.* **256,** p. 131, Washington (1964).
20. J. O'M. Bockris, A. K. N. Reddy, and B. Rao, *J. Electrochem. Soc.* **113,** 1133 (1966).
21. B. Rao, Ph.D. thesis, University of Pennsylvania (1966).
22. W. K. Palk and J. O'M. Bockris, *Surf. Sci.* **28,** 61 (1971).
23. N. Sato and B. Okamoto, *J. Electrochem. Soc.* **110,** 605 (1963).
24. C. C. Chang, *Surf. Sci.* **25,** 53 (1971).
25. N. J. Taylor, *Techniques of Metals Research,* Vol. 7, R. F. Brunshah, ed., Interscience, New York, (1971).
26. P. W. Palmberg, *Anal. Chem.* **45,** 549A (1973).
27. A. M. Horgan and I. Dalins, *Surf. Sci.* **36,** 526 (1973).
28. F. Meyer and J. J. Vrakking, *Surf. Sci.* **33,** 271 (1972).
29. M. P. Seah, *Surf. Sci.,* **40,** 595 (1973).
30. C. C. Chang, *Characterization of Solid Surfaces,* P. F. Kane and G. B. Larrabee, eds., Chap. 20, Plenum, New York (1974).
31. J. B. Lumsden and R. W. Staehle, *Scr. Metall.* **6,** 1205 (1972).
32. M. Seo, J. B. Lumsden, and R. W. Staehle, *Surf. Sci.* **42,** 337 (1974).
33. C. E. Locke, J. H. Peavey, O. Rincon, and M. Afzal in *Characterization of Metal and Polymer Surfaces, Vol. 1, Metal Surfaces,* L. H. Lee, ed., Academic, New York, (1977).
34. O. L. Riggs, Jr., *Corrosion* **20,** 275t–381t, September (1964).
35. M. Kesten and H. G. Feller, *Electrochim. Acta* **16,** 763–778 (1971).
36. A. N. Frumicia, U. S. Bagot Skii, Z. A. Iofa, and B. V. Kabanov, *Kinetics of Electrode Processes* (English Translation), Clearinghouse for Federal Scientific and Technical Information, 1967.
37. B. N. Kabanov and D. I. Leikis, *Z. Elektrochem.* **62,** 660 (1958).
38. T. J. Popova, V. S. Bagotsky, and B. N. Kabanov, *Zh. Fiz. Khim.* **36,** 1432 (1962).
39. B. N. Kabanov, *J. Electrochem. Soc.* **113,** 1142 (1966).
40. Ya. M. Kolotyrkin, *Z. Elektrochem.* **62,** 664 (1958).
41. J. Kruger, *J. Electrochem. Soc.* **110,** June (1963); Natl. Bur. Stand. U.S. Tech. News Bull., June (1963).
42. N. Sato, K. Kudo, and R. Bishimura, *J. Electrochem. Soc.* **123,** 1419, October (1976).
43. R. W. Revie, B. G. Baker, and J. O'M. Bockris, *J. Electrochem. Soc.* **122,** 1460, November (1975).
44. W. E. O'Grady and J. O'M. Bockris, *Chem. Phys. Lett.* **5,** 249 (1943).
45. G. Okamoto and T. Shibata, *Proceedings of the Fourth International Congress on Metallurgy and Corrosion,* NACE, Houston (1972).
46. G. Okamoto, *Corros. Sci.* Houston **13,** 471 (1973).

47. O. L. Riggs, Jr., *Corrosion* **20**, 367t–369t (1964).
48. J. O'M. Bockris and A. K. N. Reddy, *Modern Electrochemistry, Vol. 2*, p. 1325, Plenum, New York, (1970).
49. O. L. Riggs, Jr. *Corrosion* **20**, 275t–281t (1964).
50. O. L. Riggs, Jr., in *An Interim Report*, Continental Oil Company (1957).
51. W. L. Bragg and J. West, *Proc. R. Soc.* London **A114**, 450 (1927).
52. T. P. Hoar, *Corros. Sci.* **7**, 341–355 (1967).
53. T. P. Hoar, *Nature* **216**, December 30 (1967).
54. E. S. Snavely and N. Hackerman, *Can. J. Chem.* **37**, 268 (1959).
55. J. J. Acello and N. D. Greene, *Corrosion* **18**, 286t, August (1962).
56. O. L. Riggs, Jr., *Corros. Vol.* 19, 180t, May (1963); **25**, 130, March (1969).
57. W. J. Lorenz, *Corros. Sci.* **5**, 121, (1965).
58. K. J. Vetter, *Angew. Chem. Int. Ed. Engl.* **1**, 583 (1963).
59. K. E. Heusler and G. H. Cartledge, *J. Electrochem. Soc.* **108**, 732 (1961).
60. Z. A. Iofa, V. V. Batrakov, and Cho-Ngok-Ba, *Electrochim. Acta* **9**, 1645 (1964).

Suggested Reading

A collection of papers which perhaps records the most up-to-date thinking with regards to passivity was published by the Corrosion Division of the Electrochemical Society. The monograph is composed of selected papers which represent the proceedings of the Fourth International Symposium on Passivity, October 1977. *Passivity of Metals,* Robert P. Frankenthal and Jerome Kruger, eds., Electrochemical Society, Inc., Princeton, 1978.

7

The Potentiostat

Modern polarographic instrumentation evolved from electrically equivalent circuits, and each essentially has within its electronic circuit a potentiostat to control cell potential. Inclusion of the third electrode permitted the maintenance of any desired control potential.

The present day analog potentiostat has several advantages. It is extremely fast. An operational amplifier potentiostat can have a rise time, while under resistive load, of less than a microsecond. It can maintain a set control potential on the order of 0.1 mV for days. The load on the reference electrode can be less than 1 poA. Speed and accuracy of potential control have become increasingly important as the modern techniques for studying electrodeposition and corrosion processes have become more sophisticated.

Potentiostat Mode

Systems which exhibit active–passive transitions require a reasonably sophisticated electronic instrument for constant potential measurements. The difficulty encountered arises from the negative slope of the current-density–voltage curve which develops during the transition from the active to the passive state. To overcome this difficulty it has become common practice to use an electronic instrument called a potentiostat. A potentiostat can be defined as an electronic instrument which can maintain the electrode potential constant at any desired value. Several such instruments exist and all consist of a voltage-error sensor connected to a power supply. When any deviation between the specified and actual electrode potential occurs, the sensor transmits a signal to the power supply to increase or decrease the applied current as is required to correct the error. This working feature is shown schematically in the block diagram of Figure 7-1. Three distinct sections, (1) potential control, (2) control amplifier, and (3) current supply, comprise the potentiostat. Three electrical leads (length no longer critical due to unique grounding circuits) are connected to three electrodes exposed in an electrolytic solution. The measuring error between the reference and test electrodes is signaled to the control amplifier.

Fundamentally, the modern solid-state basic potentiostat is essentially a three-terminal device requiring an ancillary electrochemical cell and electrodes. The type of cell used depends upon the system studied, but it usually consists of three electrodes and a reasonably conductive electrolyte.

The three electrodes are: the working electrode (the potential of which is controlled), the auxiliary electrode, and the reference electrode.

The electrical characteristics of any electrochemical cell and its electrodes are complex, but for simplicity a cell is illustrated in Figure 7-2 showing resistive components only.[1]

The series resistance (R_s) simulates the effective series resistance produced by the auxiliary electrode and the cell electrolyte, while the potential developed across the resistance (R_c), due to the flow of current between the working and auxiliary electrodes, simulates the controlled potential of the work electrode with respect to the reference electrode.

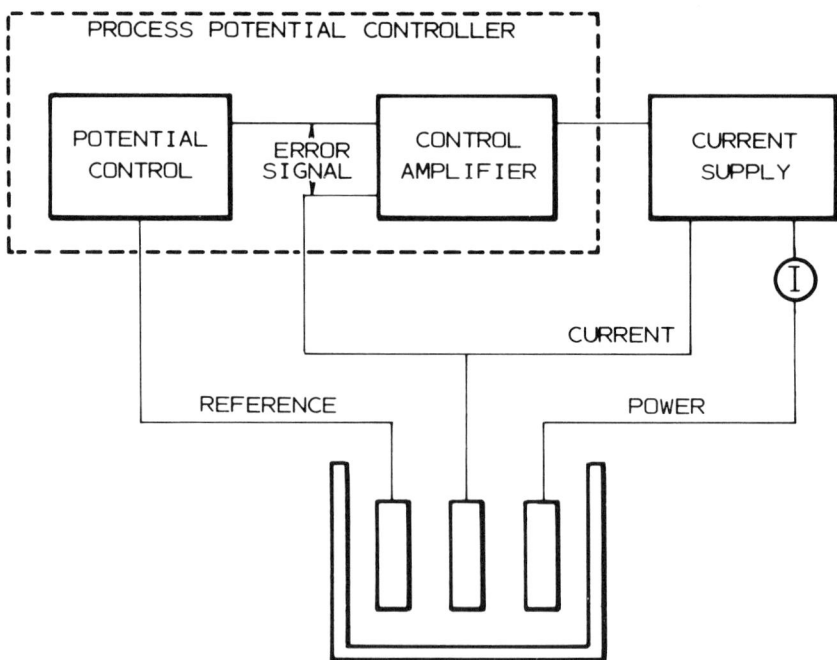

Figure 7-1. Block diagram of a potentiostat.

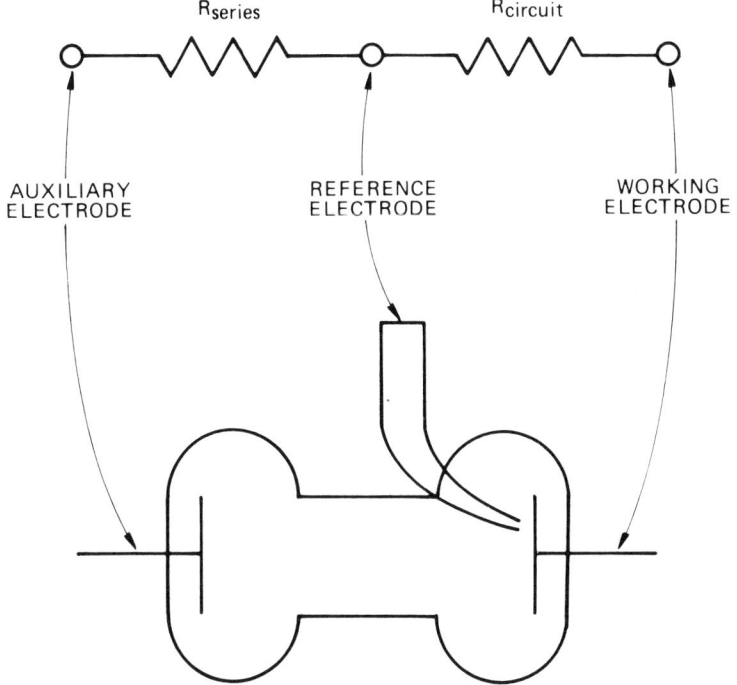

Figure 7-2. Resistive components in simple cell.

Development of the Potentiostat

Potentiostatic methods have been used for a variety of applications, i.e., metal deposition for analytical analysis, electrolytic conditioning electrodes, organic synthesis by oxidation/reduction processes, and electrometallurgical and corrosion investigations.

The first published potentiostat circuit known was reported by Hickling[2] and, following this historic instrument, other circuitry has been developed, directed primarily to application modes.

Analytical Work. This area of use required an instrument which can provide high currents and low-voltage dc output.

Organic Synthesis. Oxidation/reduction reactions (mostly less conductive systems) require a high current and high-voltage dc output.

Electrochemical Kinetics. Because of the need for high-frequency response, the potentiostat for this field must have a low current and high-voltage dc output.

Corrosion. The study of corrosion processes utilizes a potentiostat which must have a high frequency response, as well as high-current dc output. An instrument which also has high dc voltage output offers a much wider application range.

The potentiostat maintains the potential difference between the working (test) and reference electrodes at a preset value, irrespective of the change in the control-resistor value. This is equivalent to changes in the characteristics at the working electrode/electrolyte interface produced by chemical reactions. The actual control mode operating in the potentiostat is dependent then on the characteristics of the changes in the electrochemical cell.

The description and detailed electronics of the modern electronic potentiostat is truly not within the scope of this chapter. An excellent development of this is given in *Potentiostat and Its Applications,* by J. A. von Fraunhofer and C. H. Banks, Betterworth and Co., Ltd., 1972. To provide some idea of the sophisticated nature of the modern potentiostat, Figures 7-3, 7-4, and 7-5 are presented from description in the early Conger and Riggs [3] precision high-current potentiostat.

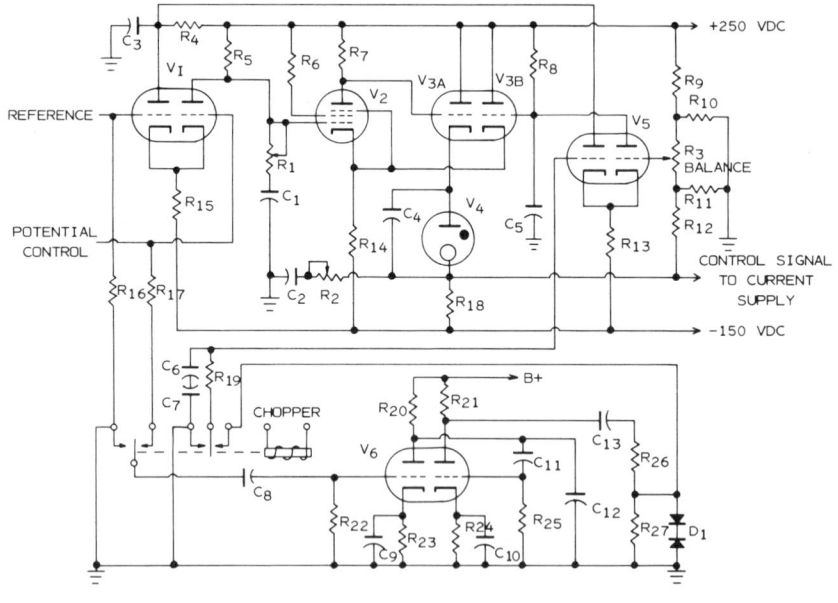

Figure 7-3. Control-amplifier circuit.

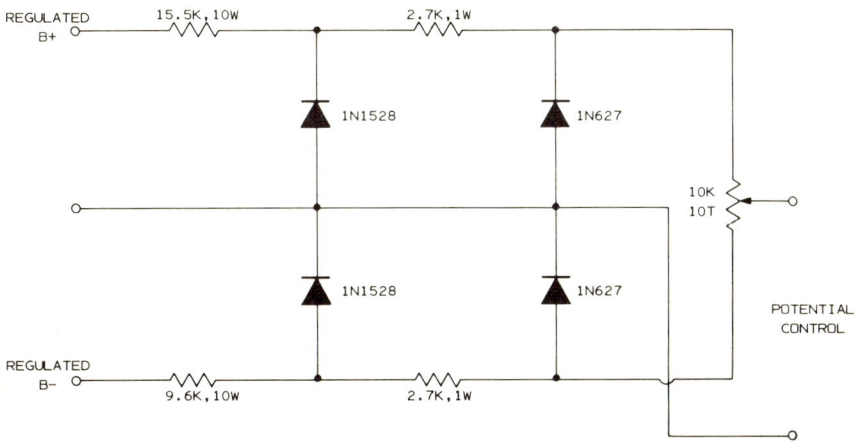

Figure 7-4. Potential-control circuit.

Potential Control

The potential-control circuit (Figure 7-3) is a highly stable yet variable source of dc voltage. Positive ($+250$) and negative (-150) voltages from the main regulated power supply are applied to zener preregulators. The zener preregulators reduce the supply voltages to ± 28 V dc and remove any fluc-

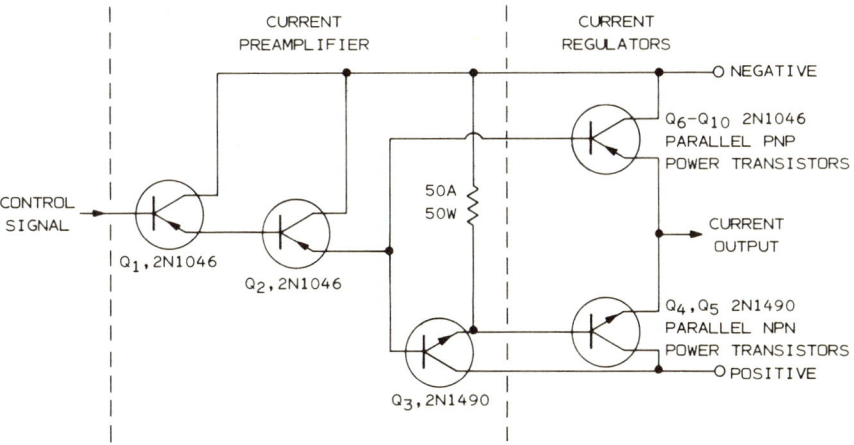

Figure 7-5. Current regulator.

tuations that remain. The final zener regulators generate ± 6.2 V dc and have a temperature coefficient of 0.001 percent. By maintaining their current constant and housing them in an airtight enclosure, the output voltage can be maintained constant to a few hundred microvolts over long periods. The 10-turn potentiometer allows one to accurately adjust the potential-control voltage over a ± 6.2-V range. When more resolution is needed at lower voltages, resistors are placed in series with each leg of the 10-turn potentiometer to reduce the voltage across it. If switching means are included to obtain several potential-control ranges, the switch is designed to connect a resistor in parallel with the output zeners on each range to maintain constant zener current.

A second 10-turn potentiometer may be connected in parallel with the first. A motor-driven potentiometer connected between the movable arms of the 10-turn potentiometers can then be used to supply potential-control voltage. This results in a very flexible potential-control scanning means, where each end of the scan is essentially independently adjustable.

Control Amplifier

The control amplifier consists of a three-stage fast response, low-noise dc amplifier plus a three-stage "potential-lock" circuit (Figure 7-4). The dc amplifier makes an instantaneous correction to the control signal, but because it has finite "gain," a small difference between potential control and reference signals remains. The potential-lock circuit nulls out this difference and locks the reference-electrode signal to the potential-control signal.

The dc amplifier consists of tubes V1, V2, V3A, and V4. The first tube compares the signal from the reference electrode with the potential-control voltage and amplifies the difference. Tube V2 provides additional amplification and V3A is a cathode follower to provide a low-output impedance. Tube V4 is a constant-voltage tube used to change the level of the dc control signal without attenuating it.

The resistor–capacitor networks $R_1 C_1$ and $R_2 C_2$ adjust the frequency response of the circuit to control oscillation. Oscillation can occur because the wide-band, high-gain potentiostat and the electrochemical cell (where phase shifts can occur) form a feedback loop in which feedback can approach 100 percent.

The potential-lock circuit consists of a mechanical chopper, tubes V5, V6, and V3B. Any differences between reference-electrode voltage and potential-control voltage is converted to a pulsating dc signal by the chopper. This signal is passed through a capacitor, converting it to an ac signal, which is then amplified by a very stable two-stage ac amplifier V6. The ac signal is then reconverted to dc by a second set of contacts on the chopper. Thus the chopper and V6 form a stable dc amplifier having essentially no drift. The resulting signal is amplified further by V5 and a cathode coupled to V2 by V3B. Thus, due to the symmetry of V1 and V5, the voltage difference between the reference-electrode and potential-control signals receives the product of the amplification of the dc amplifier and the potential-lock amplifier. This amplification is approximately 10^3. Balance potentiometer R3 adjusts the potential-lock circuit to control exactly at the potential-control voltage.

Current Regulator

The current regulator in Figure 7-5 is basically a current amplifier. The control amplifier has developed the voltage needed to control the electrochemical cell, the current regulator develops the current required. Current preamplifiers Q1, Q2, and Q3 provide the current capacity needed by the control signal to operate the current regulators. The *pnp* transistors regulate the so-called negative current (electron flow from the working electrode to the power electrode) and the *npn* transistors regulate the so-called positive current (electron flow from the power electrode to the working electrode). The *pnp* regulators can develop up to 10 A and the *npn* regulators supply up to 5 A. The regulator will supply ± 10 V at $+5$ and -10 A, with a rise time of 4 μs or less and ± 20 V at ± 5 A, with a rise time of 7 μs.

The polarity of the current is adjusted automatically according to the needs of the electrochemical cell. The crossover from one polarity to the other is barely discernible when viewed with an oscilloscope under pulse conditions.

The dc power for the current regulators is a back-to-back full-wave rectifier with a capacitor filter. Two voltage ranges are available, one supplying ± 20 V dc unloaded for output voltage to ± 10 V dc, the other supplies ± 30 V unloaded for output voltages to ± 20 V dc.

Recent [1,4-16] developments in electronic potentiostats not only provide improved accuracy and precision, but greatly increased power capabilities.

Solid-State Potentiostats

The development of solid-state electronic devices has been very important to the potentiostat user, just as it has been to all the other electronic device users. Operational amplifiers made from these solid-state devices are widely used in homemade and commercially produced electrochemical instrumentation. The term operational amplifiers was originally used to describe amplifiers used in analog computers to add, subtract, integrate, differentiate, and other mathematical operations.

A full description of the fundamentals of operational amplifiers is given in several basic textbooks. A concise description that is particularly useful is

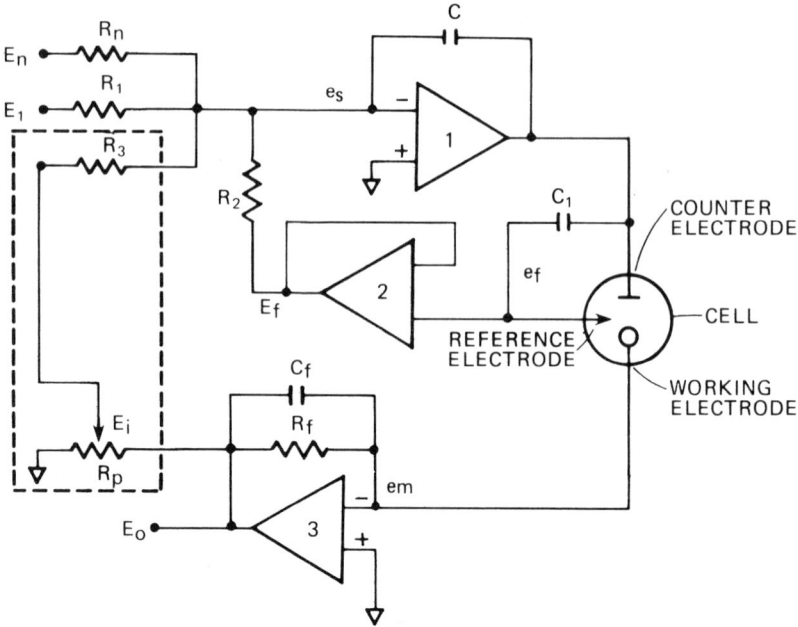

Figure 7-6. "Adder"-type operational-amplifier potentiostat. Components in the dashed rectangle provide positive feedback IR compensation. Amplifier 1, control amplifier; Amplifier 2, voltage follower; Amplifier 3, current follower; C, C_1, C_f, network stabilizing capacitors.

given by Sawyer and Roberts.[17] The reader is referred to one of these standard references for the background on operational-amplifier principles and performance characteristics.

The following description of potentiostatic circuits from operational amplifiers follows the description given by Sawyer and Roberts.[17] Figure 7-6 is a popular potentiostatic circuit. The elements of this circuit include three operational amplifiers, which are used to control the electrode potential and provide a voltage signal proportional to the electrode current.

Amplifier 1 provides a current to the counterelectrode which flows through the solution to the working electrode. The level of this current is proportional to the value of e_s. This "control signal," or "error signal," is the difference between the signal from the voltage-follower circuit configuration used with amplifier 2 (discussed below) and the voltage inputs E_n, E_1. The signals E_n, E_1 can be fixed-level signals or time-variant signals. The current flow is directed in the cell to reduce e_s to zero.

The reference-electrode signal e_f is directed to the voltage follower 2. The signal E_f is equal to e_f. The operational amplifier 2 has the high-input impedance necessary so current is not drawn from the reference electrode.

The working electrode is held at virtual ground by operational amplifier 3. The reference signal e_f is compared to ground and is thus the negative of the potential of the working electrode compared to the reference electrode. The output voltage of 3, E_0, is proportional to the value of the current through R_f, which is the current flowing through the working electrode,

$$E_0 = -iR_f$$

The signal E_0 can be monitored with a voltmeter or recorded.

The circuit in the dotted-line box is used to compensate for voltage drops in the test solution. If there are voltage drops in the cell or on the electrode surface, the voltage E_f is given by

$$E_f = -E_{cell} + iR_u$$

E_{cell} is the potential of the working electrode compared to the reference electrode. R_u is the resistance of the solution and electrode combination. The value of E_i is varied so it equals $-iR_u$. Then when inputted to the summing point, E_f is corrected so it equals $-E_{cell}$. All the capacitors have been ignored in this discussion. They are included in the circuit to stabilize the frequency response.

A very similar circuit that contains some interesting modifications is shown in Figure 7-7.[18] The output of operation amplifier 1 is directed to the base of two power transistors. One is a *pnp* and one is a *npn*. The polarity of the current forced through the counterelectrode can be changed automatically, depending on which of these transistors are excited so they conduct current. The polarity of the signal from the operational amplifier controls which transistor conducts. The same feature is used at the output of operational amplifier 3 so that the compensation signal is of the correct polarity. This circuit is shown with digital meters to monitor current and potential.

Three input control signals are also shown. The applied potential signal would be time invariant and changed manually. The Ramp. Gen. signal would be a voltage which is varied linearly with respect to time. The Stepping CRT would be a signal which was varied in small step functions with respect to time. Other suitable signals could also be inputted.

Commercial Electronic Potentiostats

Several precision potential-measuring and potential-controlling potentiostats are available on the market today. A few of these potentiostats with listed specifications are in Table 7-1. A very versatile line of precision potentiostats

TABLE 7-1. Specification Data on Modern Electronic Potentiostats

	dc output					Stability (mV/
Potentiostat	Current (A)	Voltage (V)	Rise time (μs)	Slew rate (V/μs)	Noise and ripple	24 h)
Anotrol TRW 200A	± 10	± 10	<1 at 10 A	<2, at rated loads	$<50\ \mu$V rms	± 0.5
	± 5	± 20				
PAR-173	± 1	± 100	2 at 1 A	10	$<50\ \mu$V rms	± 0.1 (after
						warm-up)
Tacussel PIT 20-24	± 2	± 20	1	N.D.[a]	$<50\ \mu$V rms	± 0.5
Wenking 70 HP 10	± 10	± 20	<10	3	$20\ \mu$V rms	± 0.5
Aardvark Model V2LR	± 1	± 20	1 at 1 A	10	$20\ \mu$V rms	± 0.2
Amel Model 555-B	± 10	± 30	$10\ \mu$s/V		$100\ \mu$V rms	± 0.2
Model 555-L	± 20	± 15	$50\ \mu$s/V		$300\ \mu$V rms	± 0.2

[a]N.D. = not determined.

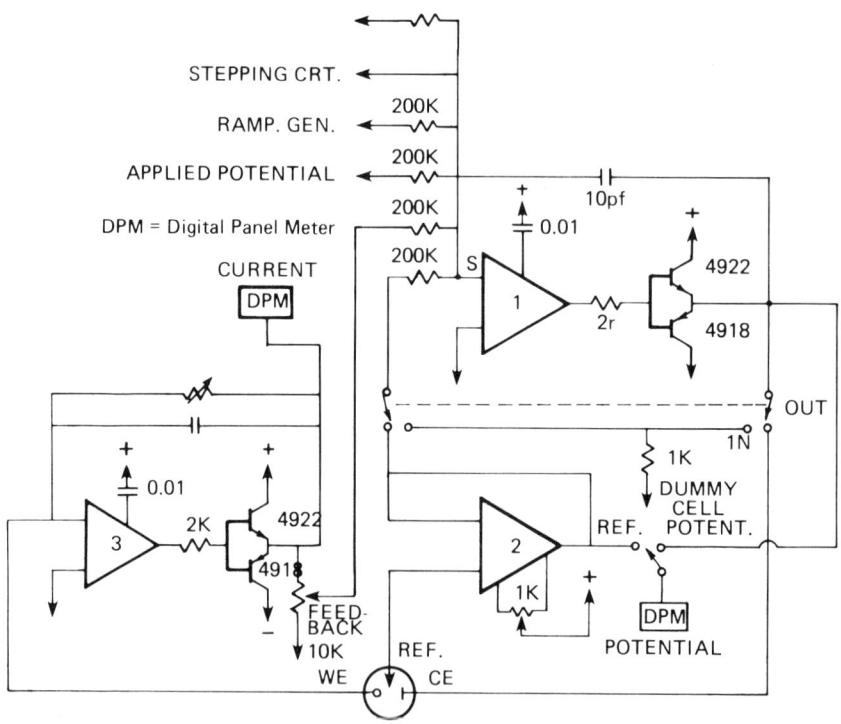

Figure 7-7. Basic potentiostatic circuit.

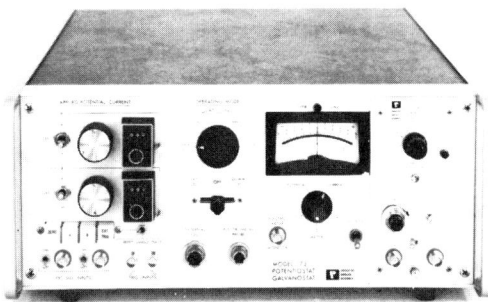

Figure 7-8. Photograph of PAR Model 173 Potentiostat/Galvanostat.

are manufactured by the Princeton Applied Research Corporation. The PAR Model 173 Potentiostat/Galvanostat is shown in Figure 7-8.

The Model 173 Potentiostat/Galvanostat offers complete flexibility in potential or current control for electrochemical applications. It features a current capability of 1 A, with compliance voltages as high as 100 V in either polarity and a slew rate of 10 V/μs. The Model 173 incorporates two independent built-in potential/current sources, each adjustable between ± 4.999 V, as well as complete logic and switching circuitry for front-panel or remote switching between these two sources. Two additional external potential/current programming signals may be added to the potentials provided by the instrument, and a wide variety of triggering and switching waveforms may be employed to control the overall potential-or current-application programs.

For maximum flexibility the Model 173 Potentiostat/Galvanostat is designed to operate with plug-in readout devices. Plug-ins include the Model 176 chopper-stabilized current follower and the Model 179 Digital Coulometer.

For those applications requiring the utmost in convenience, or where computer control and interfacing is desired, the Model 173D provides front-panel digital display of either the measured current or the applied potential. Addition of a Model 176D or a Model 179 plug-in provides binary-coded decimal (BCD) output of current or potential, including range information.

Also, the unique Model 178 Electrometer Probe is included, which puts the reference-electrode buffer amplifier right at the electrode, thus eliminating shielding problems normally associated with long reference-electrode leads and providing high-speed response, even with high-impedance electrodes.

As mentioned above, electrochemical kinetics require high voltage/low-current dc outputs. One such potentiostat which can also be used for corrosion studies is the AMEL Model 552. Figure 7-9 is a photograph of the instrument.

The operating principle of the Model 552 is based on very sophisticated

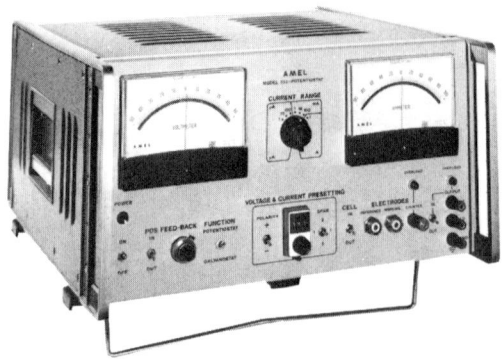

Figure 7-9. Photograph of AMEL Model 552 Potentiostat.

electronics so that it can make measurements which would otherwise require many ancillary units, as would be the case when using competitive potentiostats based on a single operational amplifier.

The potentiostatic operation, as shown in Figure 7-10, employs five operational amplifiers and a logic circuit indicated as "sequencer." The heart of the instrument is the power operational amplifier Q1 which drives current into the cell via the counterelectrode and receives the negative feedback via the reference electrode. The operational amplifier Q3 acts as current-to-voltage converter (i/E converter) providing an output with a grounded terminal. The positive feedback picked up by a potentiometer at this output is summed with eventual external modulation and dc polarization by means of the operational amplifier Q4. The electrometer Q5 has a dual function: first, to measure the working-electrode "static potential" prior to switching on the cell; second, to monitor the eventual voltage shift when current flows.

The purpose of the sequencer is to preset a logic sequence when switching the cell on and off. In such a way the electrodes are protected against voltage shocks arising from an incorrect insertion sequence of the counterelectrode. Another logic circuit protects the instrument against short circuits and line transients.

The potentiostatic operation includes two overload monitors, Q6 and Q7, respectively, for the out-range of the counterelectrode and for saturation of the output amplifier Q3.

The Aardvark Model V-2LR Potentiostat (Figure 7-11) employs very sophisticated, advanced signal processing. Extensive buffering and level-restoration techniques permit simple, low-noise, high-accuracy connections to instruments intended for experimental as well as other uses. None of its electrodes are connected internally to ground. Three large meters are provided, which allow the simultaneous display of all experimental values. Other important features include (see Figure 7–12):

(1) A guarded reference electrode. The shield of the electrode in this system is driven to the measured potential by a special circuit included for that purpose. As the input is completely surrounded by the equipotential guard/shield, no leakage can occur to the case or related circuit elements. In this way, humidity and contamination cannot compromise the instrument's input impedance.

(2) A non-standard Working Sense electrode. This input adds a four-terminal measurement capability which eliminates *IR* drops in the working-electrode connections. Typically, an improvement of 25 mV or more may be observed in the accuracy of the potential controlled at the cell during high-current operation.

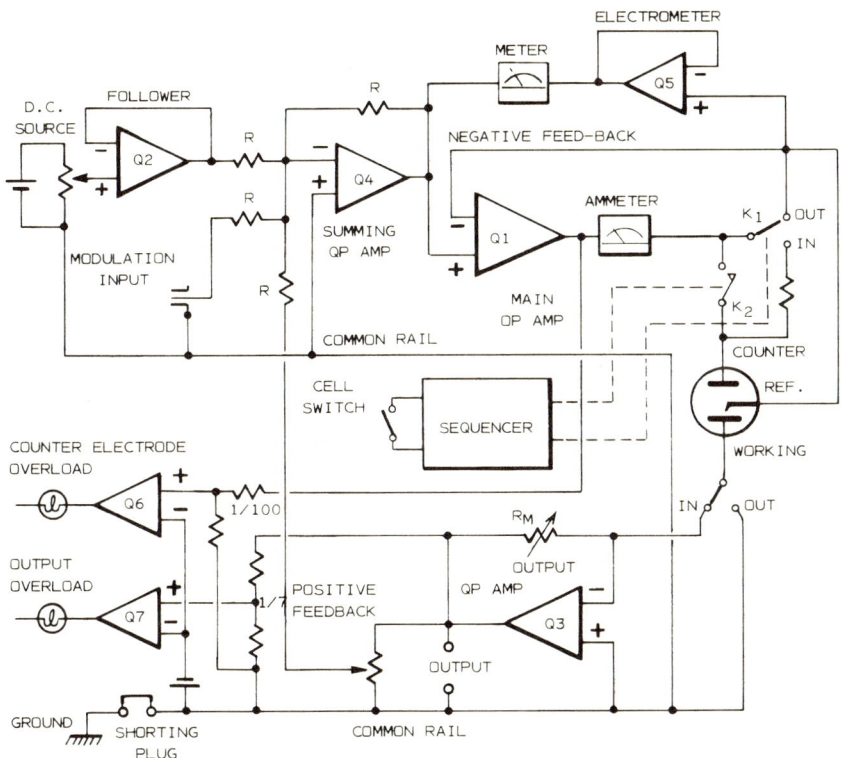

Figure 7-10. Diagram of the AMEL Model 552 potentiostat.

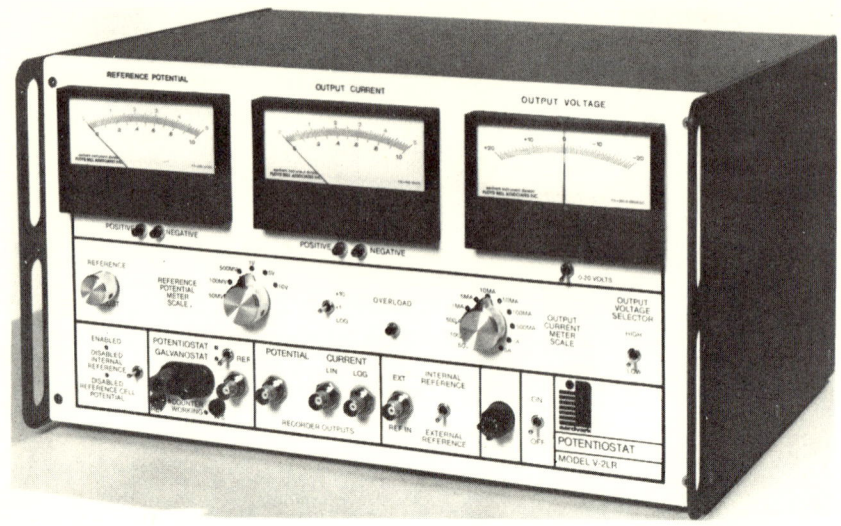

Figure 7-11. Photograph of the Aardvark Model V-2LR Potentiostat.

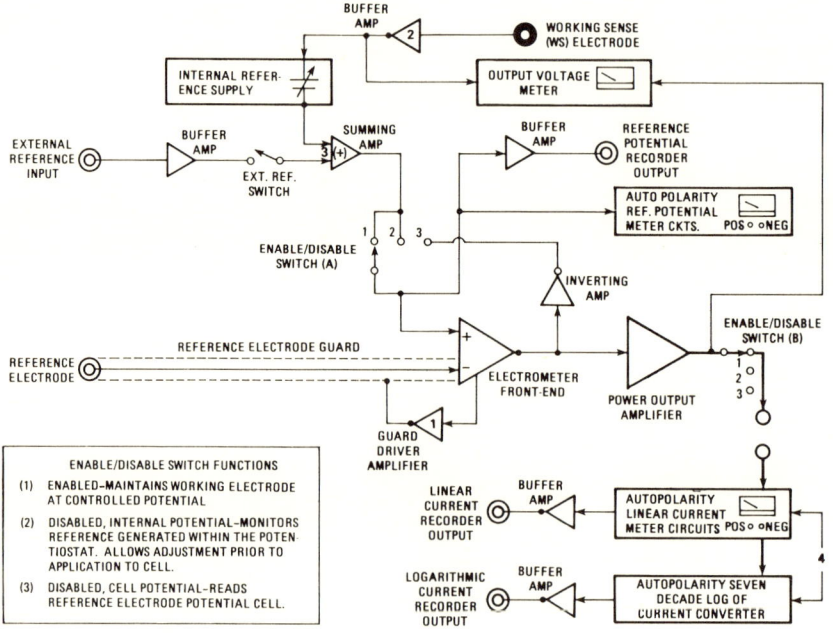

Figure 7-12. Function Diagram of Aardvark Model V-2LR Potentiostat.

(3) Reference-potential summing. Internal reference potentials are generated electronically (no batteries) and made available for summing with external potential sources such as scanners and function generators. This allows for dc offset calibrations which are especially useful in linear polarizations, Tafel plotting, and anodic polarizations.

(4) Simultaneous linear and logarithmic current outputs. Both sections may be operated independently and simultaneously for the calibration of recorders, establishment of limits, control functions, etc. The Log Converter boasts an extremely wide (seven decade) response with both polarities of current and no range adjustments.

Portable/Field Potentiostat

A potentiostat is a device which uses a current output to control the potential of a test specimen. In order to work properly, the instrument must be able to detect the potential of the test material; and if that potential is different from the desired potential, then the instrument must supply the right amount of current to that electrode to bring it to the desired potential. The Petrolite Potentiodyne Analyzer™ (Figure 7–13) is designed to control the potential of a test electrode automatically. The potential may be held at a fixed value (potentiostatic), or the potential may be changed continuously at a rate chosen by the operator.

The Potentiodyne Analyzer should be connected to a corrosion test cell as

Figure 7-13. Petrolite Model M-4100 Potentiodyne Analyzer.

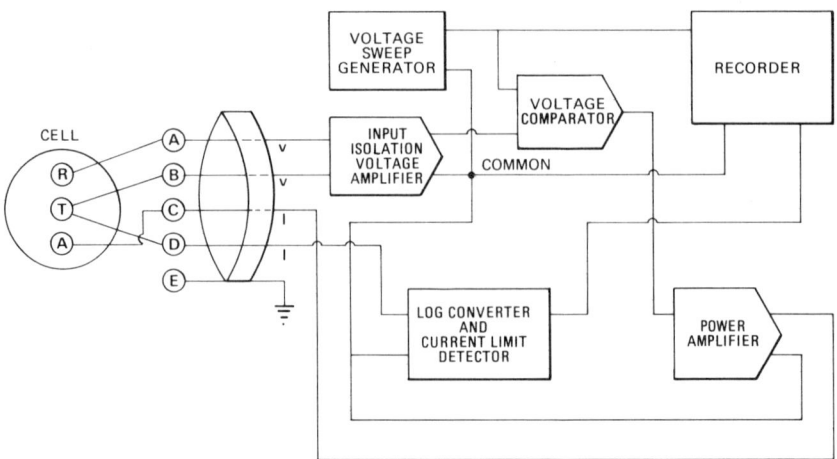

Figure 7-14. Schematic of Potentiodyne Model M-4100 components as related to the test cell.

shown in Figure 7–14. Terminals A and B are used to measure the potential between the reference and test (working) electrodes. Terminals C and D supply power to the test electrode (through the auxiliary electrode), which is used to control the potential of the test electrode.

A very-high-impedance isolation voltage amplifier is used to measure the potential difference between the reference and test electrodes. The output of this voltage-measuring circuit is continuously compared to the output of the voltage sweep generator in the voltage comparator. The signal from the voltage sweep generator may be constant or it may be changing at a programmed rate. In either case, a small imbalance in the voltage comparator drives the power amplifier to change the current in the test–auxiliary leg of the test cell in such a way that the measured cell voltage is equal to the output of the voltage sweep generator. As the voltage sweep generator sweeps between limits, the output of the voltage comparator is continuously driven to zero by the power amplifier. As this occurs, the current required to keep the cell voltage equal to the sweep-generator voltage is measured and recorded on the $x-y$ plotter as log current versus the sweep-generator voltage (cell potential).

Summary Comments on Potentiostatic Mode

A potentiostat is a device which controls the potential of the working electrode with respect to the reference electrode by varying the potential at a third counterelectrode, so as to correct for the voltage drop due to the solution resistance. To do this, it must compare the measured potential at the reference electrode with a control signal related to the desired applied potential, and must vary the counterelectrode potential to correct for any discrepancy. In addition, the potentiostat must be sufficiently fast to accomplish this correction quickly.

The rise-time specification of a potentiostat, as given in the manufacturers' literature, is supposed to be a parameter which describes the response speed of the device. This specification is of no value unless the nature of the load being driven is taken into account, both with regard to the current required and as concerns the characteristic time constant of this load. For example, any electrode has a characteristic double-layer capacitance associated with it, and if the potential of this electrode is changed, the capacitor must be charged. The rate at which the capacitor can be charged is governed by the rate at which current can be provided. Thus a potentiostat must have significant current-delivery capabilities, in addition to its speed, if it is to be at all useful.

Consider that a typical electrode may have a capacitance of the order of 1 μf. The time required to charge this electrode capacitance for a 1-V change in applied potential is equal to CV/I, so that for a potentiostat with a maximum current-delivery capability of 10 mA, the time required is 100 μs. So even if the potentiostat in question is capable of changing its output by 1 V in 100 ns when lightly loaded, it will be unable to drive a practical electrode system to the required voltage in less than 100 μs because of the limitations on its current-delivery capability. This means that a potentiostat must have significant current-delivery capabilities, even if it is to be used only for measurements of low currents.

Another limitation of many commercial potentiostats is low compliance voltage. This is especially important when the current demand is large relative to the solution resistance between the working and counterelectrodes and a large potential drop across the solution results. Under these circumstances, although the potentiostat can deliver the necessary current, it must also have sufficient compliance voltage available to maintain the working electrode at a constant potential, or an appreciable error will result. For example, if the solution resistance is 500 Ω (not unreasonable in an organic system) and the current required by the system is 100 mA, the voltage drop across the solution will be 50 V. For this system, the potentiostat must have a 100-mA capability and a compliance voltage of more than 50 V in order to provide the correct potential control.

Another factor which can affect potential control and rise time when driving a real cell is the uncompensated resistance. The use of three electrodes allows the potentiostat to compensate for the voltage drop across the resistance between the counterelectrode and the reference electrode. However, it is not possible to position the reference electrode exactly at the point of interest, and there can be significant resistance between the point at which the reference electrode actually resides and the point at which the potential should be measured. Significant voltage drops can develop, especially at high currents, when the wires between the electrodes and the instrument have notable resistances. This is because the design of the system assumes that the working electrode is being held at a potential equal to zero by the current-measuring amplifier, or else is actually grounded, so that all voltage drops appear across the cell. If the wires have resistance, voltage drops will appear across these wires and thus cause the working electrode to be slightly away from ground. The applied potential at the working-electrode/solution interface will thus again deviate from the desired value. Overcoming this problem requires a capability in addition to current and voltage availability and rise time; in a sense, the instrument must anticipate what this voltage drop will be and apply enough additional voltage to compensate for it.

This compensation can be accomplished by introducing a positive feedback between the output of the current-measuring amplifier and the input of the potentiostat control amplifier. If sufficient positive feedback were applied, the system would become regenerative. If this amount of feedback were adjusted to just below that which was necessary for regenerative action, application of a potential step at the counterelectrode would cause application of a compensating signal at the control amplifier input by the current-measuring amplifier.

It is essential to have the capabilities which were discussed above incorporated into any useful potentiostat to avoid the introduction of significant and occasionally indeterminate errors in measurement. It is important, however, to realize the limitations of even a system which includes these capabilities, because the laws of electricity sometimes prevent a system from performing as wanted, even if the world's best instrument is available. For example, a typical electrode has 0.1 μf of capacitance, and if the resistance in series with the electrode is 100 Ω, rather than nothing, it would, at first glance, seem impossible to charge this electrode in less than 100 μs. There would thus be no point in having a potentiostat with a rise time significantly better than 10 μs. Since, in fact, these parameters are quite common, potentiostats whose rise time is in excess of 10 μs seem to be providing far more than is usually required. Use of the positive-feedback IR-compensation technique can partially compensate for this, since it in effect reduces the 100-Ω resistance. Test-cell characteristics will often be the limiting factor, however.

The Potentiostat

There is little or no doubt that the potentiostat is the key to the successful application of anodic protection to most systems. This is true in spite of the facts reported elsewhere herein concerning applications in which it was not used. These applications took advantage of a providential faculty of a singular combination of corrosive and metal substrates to retain passivity over long intervals without incremental current additions. This persistence factor is used in some systems to an advantage, although not in the same manner as reported in the instances not requiring potentiostatic controls. In some systems, however, the persistence factor and the protection range is so limited that constant additions of current are required to maintain passivity.

Without the potentiostat or a similar instrument with the same ability to sense and react in response to electrochemical conditions at a corrosive-solution–metal interface, anodic protection would be a minor scientific curiosity, seldom or never used. It was this lack of a sufficiently responsive instrument that delayed so long the practical use of the reactions which were well understood early in the twentieth century, if not before. Thus the surface and interface reactions controlled by the potentiostat have been understood for a half a century or more.

The descriptions of the potentiostat in this chapter, and especially explanations of its application in anodic protection installations, are particularly important. Because the instrument is the key to successful applications, it behooves those responsible for system operations to know its attributes and how it functions. This does not mean that one needs to become an electronics expert who comprehends the circuitry and functions of components in detail. It does mean, however, that those in charge of anodic protection need to understand how the potentiostat works and what part it plays in the system.

The considerable detail given about this instrument will be useful to the persons who are responsible for its maintenance and repair, so this chapter should be accessible to them at all times.

It is also obvious that understanding the functions of the potentiostat improves appreciation of feedback signals which indicate the condition of the protected surface. These signals may indicate satisfactory controls or may indicate almost instantly when upset conditions have reduced the passivity of the protected surfaces. These warnings permit corrective action to be taken.

References

1. I. G. Murgulescu and O. Radsvici, *First International Congress on Metallic Corrosion*, London, 10–15 April 1961, Butterworths, London (1961).
2. A. Hickling, *Trans. Faraday Soc.* **28**, 27–33 (1942).
3. N. L. Conger and O. L. Riggs, Jr., *Corrosion* **23**, 181–184, June (1967).
4. J. J. Lingane and S. L. Jones, *Anal. Chem.* **22**, 1169, September (1950).
5. M. H. Roberts, *Br. J. Appl. Phys.* **5**, 351–352, October (1954).
6. A. Bewick, A. Bewick, M. Fleischman, and M. Liler, *Electrochim. Acta* **1**, 83–105 (1959).
7. D. N. Staicopoulos, *Rev. Sci. Instrum.* **32**, February (1961).
8. A. Hickling, *Electrochim. Acta* **5**, 161–168 (1961).
9. J. M. Matsen and H. B. Linford, *Anal. Chem.* **34**, January (1962).

10. A. Bewick and M. Fleischman, *Electrochim. Acta* **8** (1963).
11. G. Laver, H. Schlein, and R. H. Osteryoung, *Anal. Chem.* **35,** November (1963).
12. G. L. Bookman and W. B. Holbrook, *Anal. Chem.* **35,** November (1963).
13. D. Shaw and A. M. Edwards, *Corros. Sci.* **5,** 413–424 (1965).
14. B. J. K. Lengyl and J. D'Evay, *Acta Chim. Acad. Sci. Hung.* **68,** 61–64 (1971).
15. C. Lamy and P. Malaterre, *Electroanal. Chem. Interfac. Electrochem.* **32,** 137–151 (1971).
16. K. Lowe, *Natl. Inst. Metal. Repub. S. Afr. Rep.* **1509,** 5 June (1973).
17. P. T. Sawyer and J. L. Roberts, Jr., *Experimental Electrochemistry for Chemists,* Wiley–Interscience, New York, 1974.
18. F. Mansfeld and R. V. Inman, *Corrosion* **31**(1), 21 (1979).

Laboratory Tests and Procedures

So that respective plant processes and parameters influencing prospective application of an anodic protection system can be understood, specific laboratory tests must be conducted. These tests should be designed to include all pertinent conditions involved in the plant system under consideration. The data which result from these tests provides an overall corrosion-rate profile for the specific metal/electrolyte system over the preselected potential range. The tests can be conducted using either the potentiostatic or potentiodynamic mode. The potential range of interest can be scanned very slowly, or by rapid-scan techniques, the choice depending upon the system in question. Further, the tests should be designed so that the resulting data can be used to (1) provide a basis for quality control, (2) establish performance parameters, and (3) serve as a comparison standard for improved developments. In view of the diversity of natural conditions, very serious consideration should be given to laboratory testing to improve the reliability of, and confidence in, the resulting information.

An immense amount of research has been devoted to the study of metal/solution interface phenomena of which a considerable amount includes the study of the anodic-polarization phenomenon. Some of the work was carried out by electrochemists, but due to the complexities of processes in modern technology, others such as chemical engineers, metallurgists, physicists, and corrosion engineers have made major contributions to this discipline. Because of the increasing interest in anodic protection, many techniques, conditions, and test cells were introduced in order to accommodate a wide range of needs. It became important then to have some base for comparison of various potentiostatic data.

Wiring Sequence for Potentiostatic Experiments

A simplified block wiring diagram for potentiostatic anodic-polarization experiments is illustrated in Figure 8-1. This arrangement is more or less a

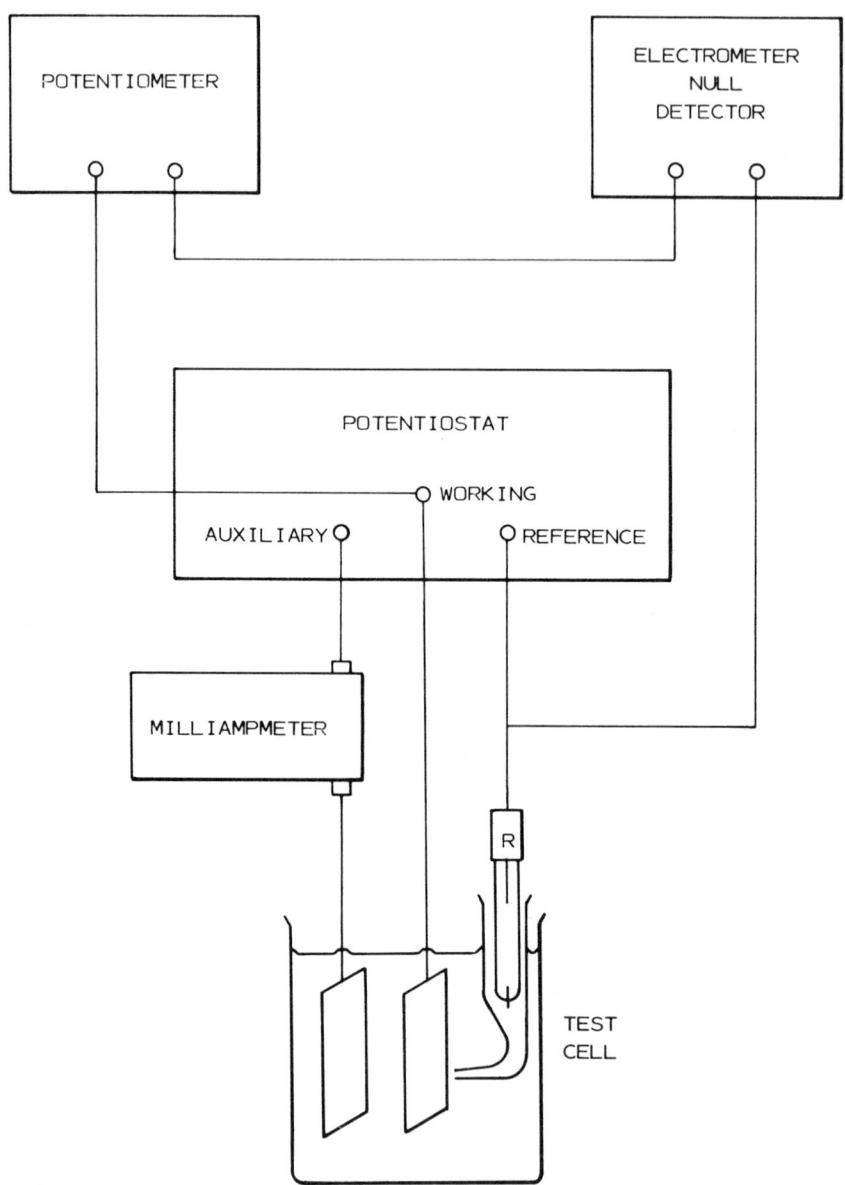

Figure 8-1. Schematic wiring diagram for potentiostatic anodic polarization experiments.

standard setup for routine potentiostatic data gathering. Although the potentiostat is automatic, the technique requires manual adjustment of the set-point potential at predetermined time intervals. This operation requires the full attention of an operator.

Many of the earlier investigators recognizing the man-hours consumed to obtain the anodic-polarization data, began to develop automatic recording systems. In 1969, Henry and Wilde[1] reported on their automatic polarization apparatus for electrochemical corrosion studies. The apparatus and its complete electrical circuit are shown in Figure 8-2. A variable voltage signal, generated by a Wenking Programmer (Model MP 165), was applied to the input of a Wenking electronic potentiostat (Model 68TS3). The potentiostat output was attached to the working electrode of the test cell. The reference-electrode potential and the applied-current density were simultaneously recorded on

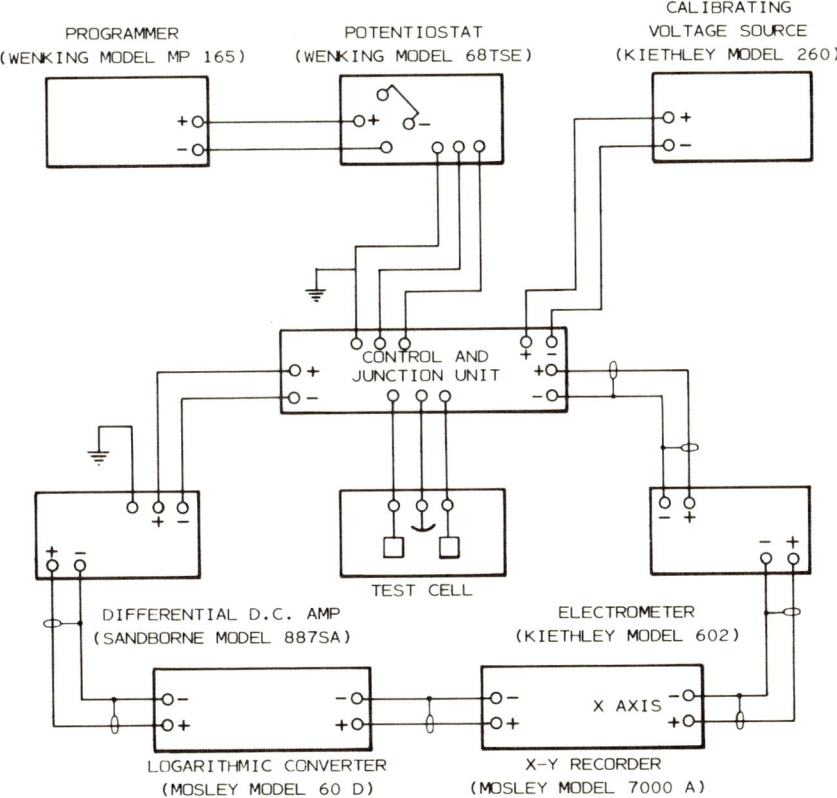

Figure 8-2. Simplified flow diagram of automatic polarization apparatus.[1]

semilogarithmic chart paper on a Mosely X4 recorder (Autograf Model 7000-A). The potential between the working electrode and a saturated calomel electrode (SCE) was applied to a high-input-impedance Keithly electrometer (Model 602), the output of which was attached to the recorder's y axis. The author's reasoning here was to ensure that no current was drawn from the working electrode/SCE cell during operation. Another precaution exercised here was the use of shielded coaxial cable throughout the system to minimize the electrical-noise level.

The applied current was measured by the potential developed across a precision resistor ($\pm 0.5\%$) in series with the auxiliary electrode, which was fed into a dc differential amplifier (Sandborne Model 8875-A). The output of the amplifier was used to drive the x axis of the $x–y$ recorder via a logarithmic convertor (Moseley Model 60D). The current-measuring resistor was placed in the auxiliary-electrode circuit because, with the resistor in this location, a deleterious effect on the potentiostat response time could be avoided. To obtain maximum stability of the logarithmic converter, a special impedance-matching plug was used on the x axis input terminals to effectively reduce the 1-MΩ input impedance of the $x–y$ recorder to the 20-kΩ output of the logarithmic converter. Since several improvements on the present apparatus were accomplished by the incorporation of the dc amplifier, the function of this unit will be described in more detail. For optimum operation, the logarithmic converter should have one side of the input connected to a ground. Since the working electrode is connected to a ground, the dc amplifier provides a convenient way of breaking the ground loop that would occur between the working and auxiliary electrodes. Further, since the amplifier is provided with a variable gain, the applied-current axis can be calibrated to read directly in terms of current per unit area, by means of a standard calibrating voltage and the amplifier-gain setting. Finally, the amplifier provides a convenient power source with which to drive the logarithmic converter instead of draining the potentiostat power circuit.

Early investigators used sophisticated test-cell systems, one of which[2] is shown in Figure 8-3. This circuit consisted basically of the following:

1. Anode used as test specimen.
2. Inert cathode—usually platinum.
3. Reference electrode—any standard half-cell such as calomel, silver chloride, or hydrogen cell.
4. Potential controller maintains any preset potential of the anode with respect to the reference electrode. The microcurrent flow between reference and anode acts as a signal to the controller to alter the current flow between anode and cathode.

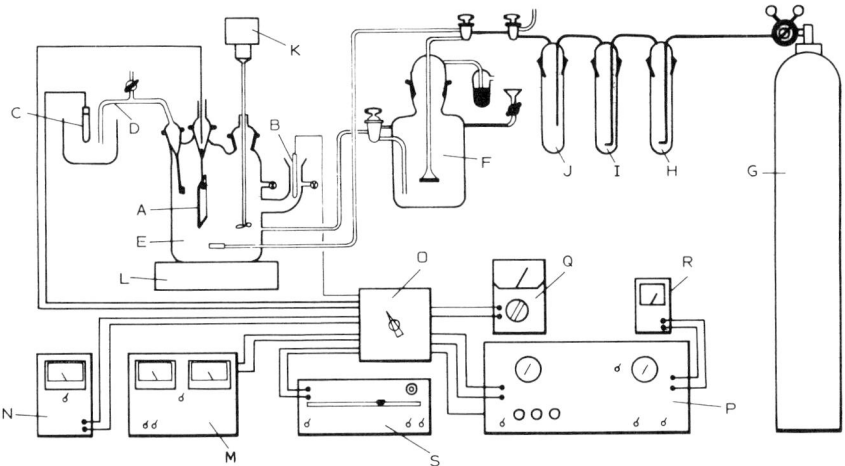

Figure 8-3. Schematic drawing of experimental anodic polarization equipment.[2]

The specimen which is to be anodically polarized is assembled at position (A). The platinum electrode (B) and the saturated calomel reference electrode (C) complete the electrochemical circuit. The calomel electrode and metal specimen are connected by an agar-KCl bridge (D). The corrosion test chamber (E) is filled with the selected corrosive fluid which has been properly treated (aerated, deaerated, chemical reagent added, etc.) in the solution-preparation cell (F). The gas (G) flows through a series of scrubbing bottles. The first bottle (H) contains 1 percent KOH. The second bottle (I) is filled with distilled water. The third bottle (J) serves only as a trap.

The test solution is stirred by use of an air-driven motor and glass-rod stirrer (K). Various elevated temperatures are reached by use of the heater (L) below the corrosion test chamber. The electronic portion of the experimental equipment is comprised of power supply (M), voltmeter (N), a junction-switch box (O) for all electrical leads (serves to switch in or out of the circuit either the power supply or the potentiostat), the potentiostat (P), a vacuum-tube voltmeter (Q), a microampere meter (R), and a recorder (S). The detailed experimental work was done by using a potentiostat design similar to one used by Hickling.[3]

Reproducibility of Potentiostatic and Potentiodynamic Anodic-Polarization Measurements

Investigators have found a rapidly increasing number of applications for electrochemical techniques in the study of corrosion. The electronic potentiostat, which is designed to maintain the potential of a metal in an electrolyte at a constant value, has been valuable for determining the variations in corrosion rate with potential. A review of the literature has indicated that experimental techniques vary widely among those conducting anodic-polarization measurements with this instrument. Since experimental procedures markedly influence polarization characteristics, it is difficult and often impossible to make valid comparisons of data from different sources. The continued growth and development of these techniques for evaluating corrosion resistance only magnifies the problem unless standard experimental practices are established and used.

For this reason, a task group was formed under the American Society for Testing and Materials (ASTM) Committee G-1 on Corrosion of Metals,[4] Subcommittee XI on Electrochemical Measurements in Corrosion Testing, to plan and to conduct an interlaboratory anodic-polarization test. A program was designed to determine the degree of experimental reproducibility possible when several laboratories followed a standard procedure for potentiostatic and potentiodynamic measurements.

A ferritic AISI Type 430 stainless steel of known chemical composition was selected for evaluation in hydrogen-saturated 1.0 N H_2SO_4 at 30°C. This alloy in this environment exhibits a wide current-density range with well-defined current maxima, minima, critical potentials, and inflection points. To eliminate, or at least minimize, variations in chemistry and metallurgical history, each distributed sample was cut from the same annealed rod, numbered sequentially, and then wet ground on both ends with 240-grit SiC paper.

Specific values were assigned to all experimental parameters that were known to influence potentiostatic and poteniodynamic anodic-polarization measurements, such as prepolarization immersion time and polarization rate. The values selected for the parameters and the techniques were considered representative of those in the literature.

Figure 8-4 is a graphical summary of potentiostatic anodic-polarization curves from the laboratories which followed the standard experimental procedure. These data are plotted in accordance with the ASTM Recommended Practice for Conventions Applicable to Electrochemical Measurements in Corrosion Testing [5] (G 3-68) (potential–ordinate, current density–abscissa). Curves have been drawn through most maximum and minimum extremes which enclose 93 percent of the data. These interlaboratory data show excellent reproducibility of critical potentials with only a factor of 1.5 to 2.0 variation in current density at these potentials.

The data in Figure 8-4 can be considered a standard potentiostatic anodic-polarization plot for the given Type 430 steel in N H_2SO_4 at 30°C when the experiment is conducted according to the standard procedure for this program. The reproducibility of this curve or curves for other systems, can be obtained by following standard ASTM procedure.

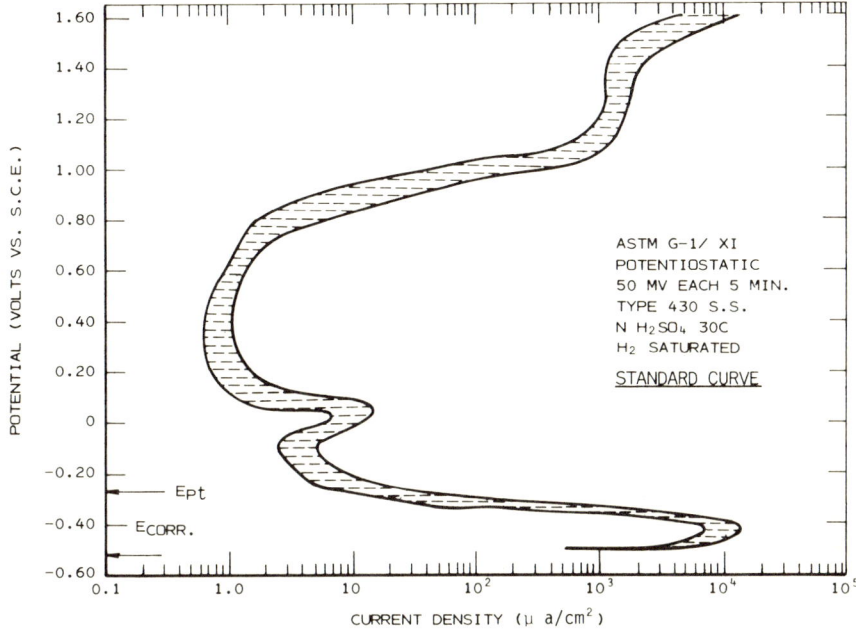

Figure 8-4. ASTM standard potentiostatic anodic polarization curve.

Polarization Cells

The essential features of the polarization cell which is common to all anodic-polarization systems is that it contains three electrodes: (1) the working (test) electrode, the (2) counter (auxiliary) electrode, and (3) the reference (half-cell) electrode. From there, the polarization cells vary considerably. Because of this variance and the role different cell designs possibly play in the data obtained, the ASTM G-1 committee adopted the Greene[6] cell as the design for standard testing. Figure 8-5 is a schematic representation of the polarization cell which describes the functional features. The specimen holder is further described in Figure 8-6, illustrating the electrical continuity from specimen to attachment post.

Further modifications of the Greene cell were reported by Morris and Scarberry[7] in order to obtain reliable high-temperature anodic-polarization data (Figure 8-7). The important feature was the water-cooled reference-electrode assembly, as illustrated in Figure 8-8.

Another variation[8] of a polarization cell which permits deviated temper-

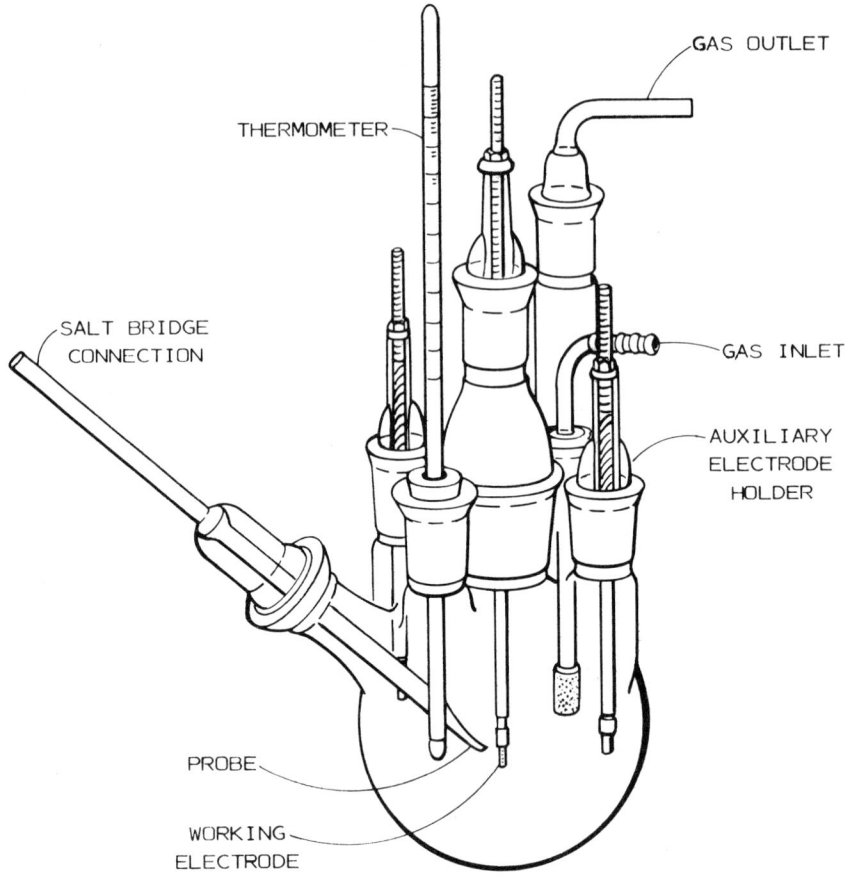

GAS OUTLET

THERMOMETER

SALT BRIDGE
CONNECTION

GAS INLET

AUXILIARY
ELECTRODE
HOLDER

PROBE

WORKING
ELECTRODE

Figure 8-5. Schematic diagram of polarization cell.[6]

ature and also contains a water-jacked cooling path between the working electrode and the reference electrode is shown in Figure 8-9. The distinguishing
difference is that the electrolyte of question is cooled prior to contacting the
salt-bridge cell for the reference electrode. The Morris and Scarberry cell sur-
rounds the reference electrode (exposed in the test electrolyte with cooled
walls). The Riggs cell accommodates a fluorocarbon specimen model (see Figure 8-10) which utilizes specimens shaped from sheet-form metals.[8]

The test specimen is forced between spring-tight brass jaws for its electrical test jack. The brass-test specimen junction is isolated from the electrolyte
due to the liquid-tight seal provided by the compressible chemical-resistant

"plug." The benefit of this holder is that it permits testing of sheet-form engineering materials actually scheduled for plant construction.

Elevated-Pressure Polarization Cells

Not all anodic-polarization studies are made under normal-pressure conditions. Anodic protection of systems which operate under high-vapor conditions has been studied. One such system was the NH_3—NH_4NO_3—H_2O solution. To accomodate the vapor pressure of the ammoniated solutions, an autoclave[9] (test cell capable of containing the test pressures) was designed with a pressure-indicating gauge. To forestall the corrosive effects of the test solutions, an inert thimble (i.e., glass, fluorocarbon) was housed inside the cell. The

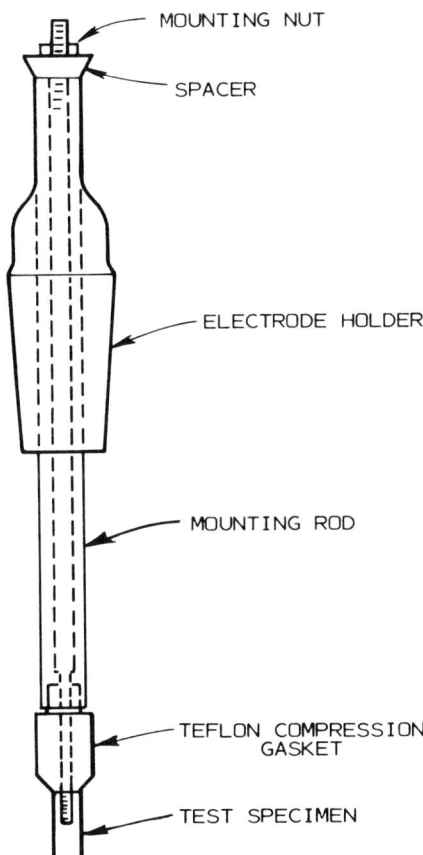

Figure 8-6. Specimen mounted on Greene electrode holder.

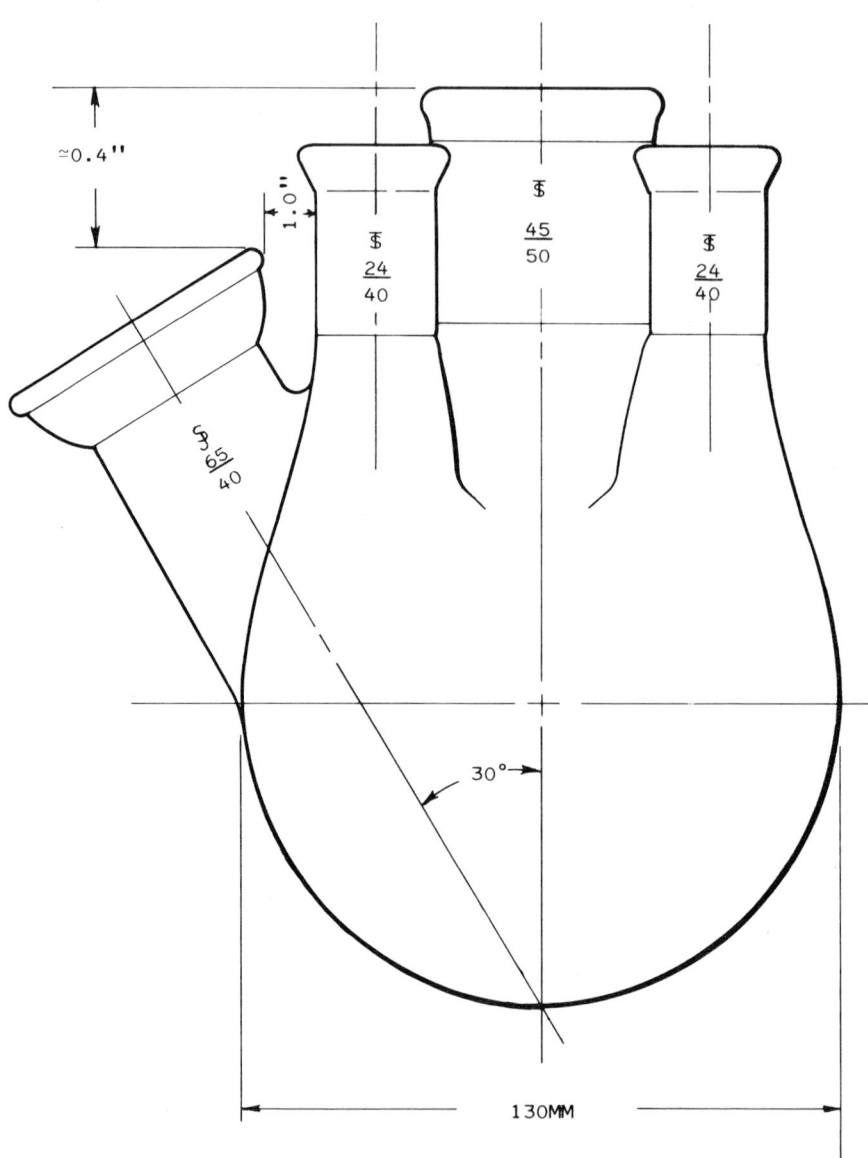

Figure 8-7. Side view of polarization cell.

reference electrode in the reported system was a Type 316 rod. Other such materials could be useful for equivalent systems as long as the pretest open-circuit corrosion potentials reach equilibrium in a reasonable time and remain at steady state during cell operation.

Figure 8-11 is a schematic representation of the anodic-polarization pressure cell. Note that all electrodes are electrically insulated from the AISI Type

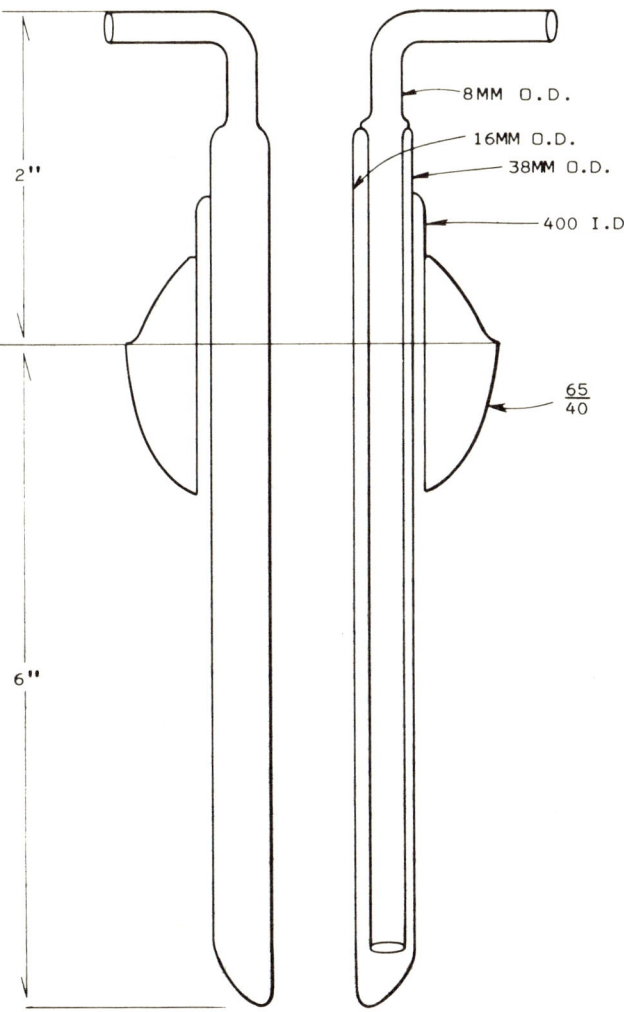

2"

8MM O.D.

16MM O.D.

38MM O.D.

400 I.D.

$\dfrac{65}{40}$

6"

Figure 8-8. Water-cooled reference-electrode assembly.

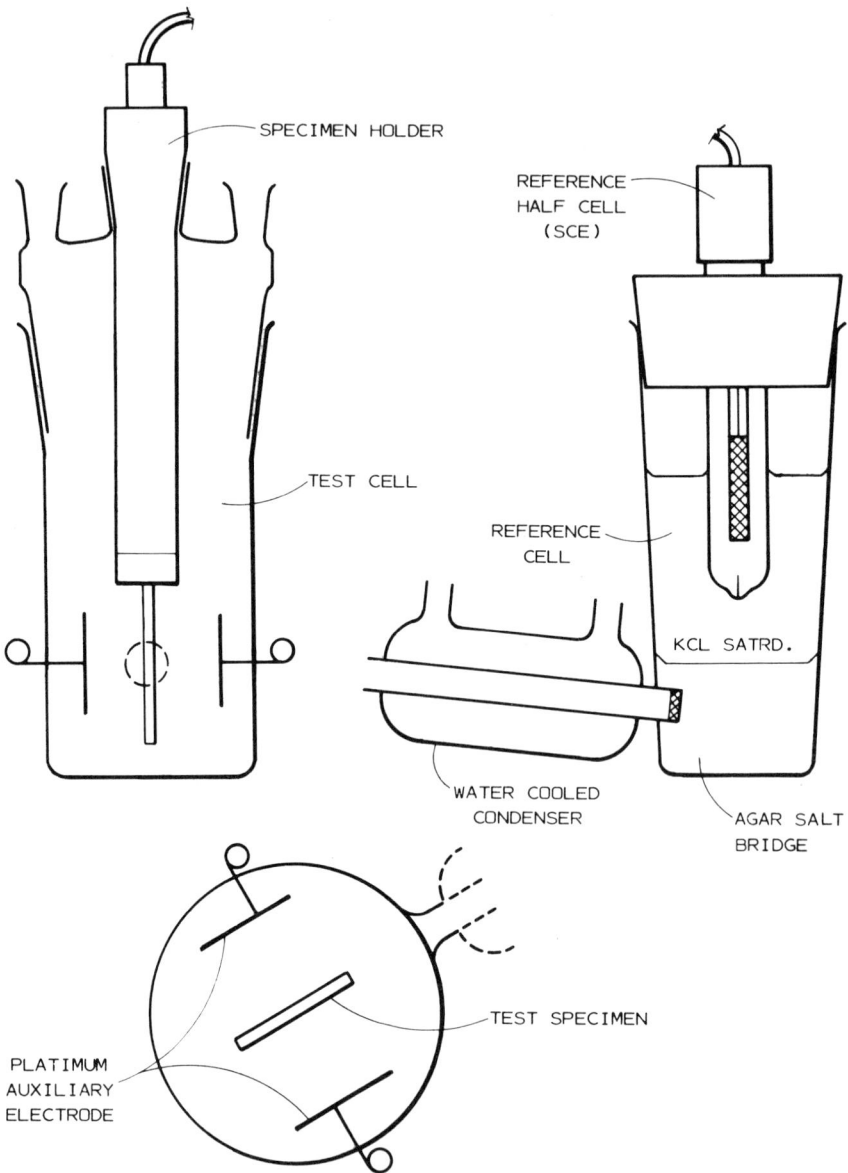

Figure 8-9. Electrochemical cell used for laboratory anodic protection experiments with temperature.

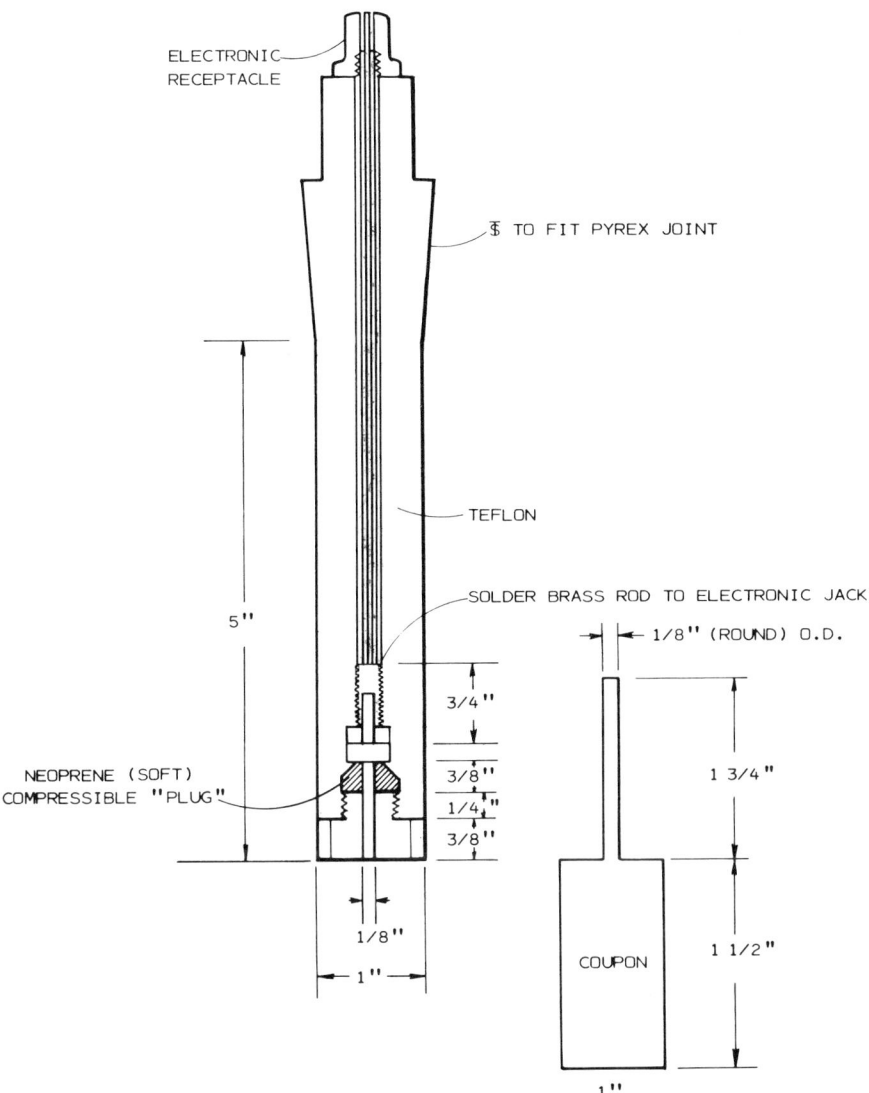

Figure 8-10. Riggs specimen holder for electrochemical cell.

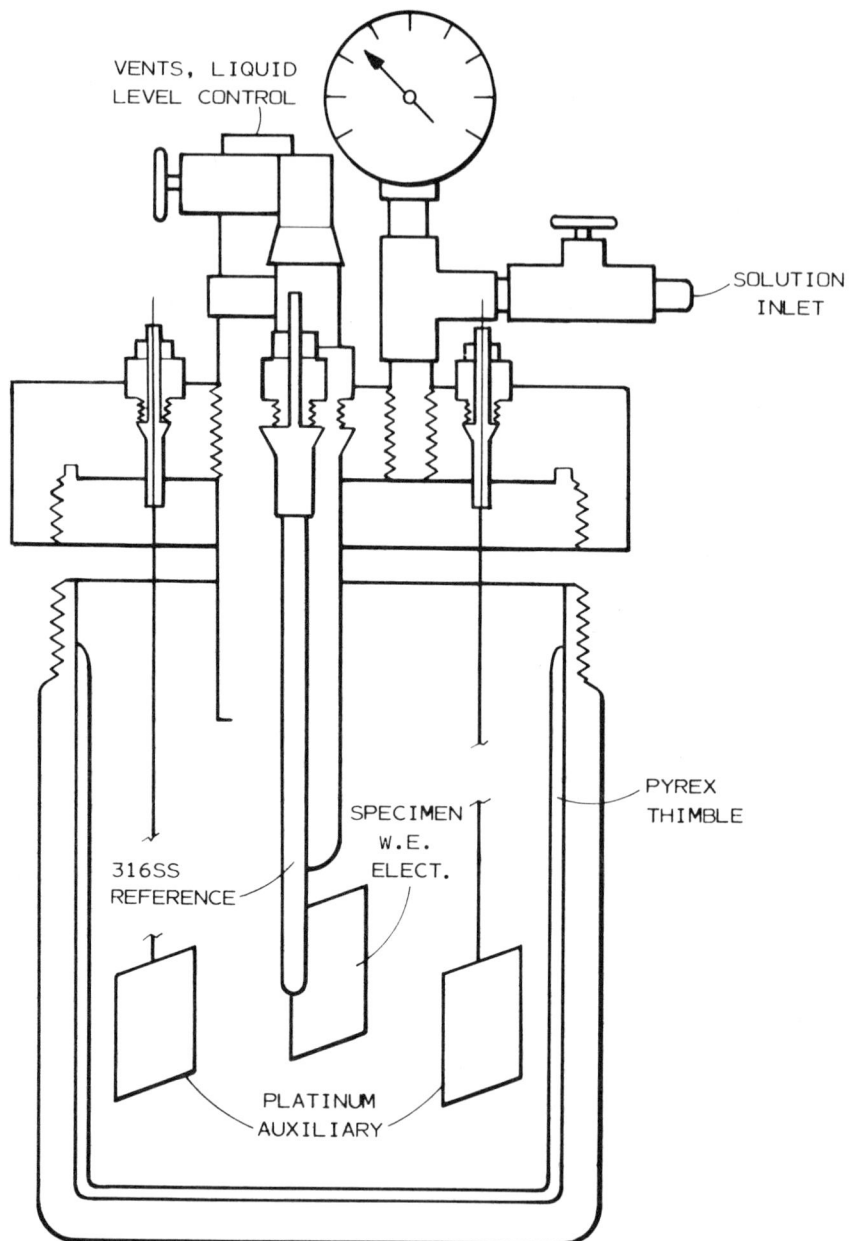

Figure 8-11. Schematic of pressure cell for anodic polarization experiments.

316 steel and are also fitted with a compressible plug to provide a pressure-tight, leak-proof seal.

Proposed Experimental Procedure

To accommodate the vapor pressure of the ammoniated solutions, an autoclave should be designed with a pressure indicator. Also, an inert thimble (i.e., glass) should fit inside the cell to contain the NH_3—NH_4NO_3—H_2O solution. Other details for the experiment are straightforward. A stainless steel rod could be satisfactorily used for obtaining electrochemical reference potentials. The instrumentation most fitted to this type of study would be those necessary for obtaining potentiostatic data.

Also, insight into the contributing causes of corrosion could be gained by using both cold-rolled and annealed steel test specimens. To improve reliability of results, several specimens should be corroded prior to the experiment by exposing them to the NH_3—NH_4NO_3—H_2O solutions and permitting them to remain in the open container exposed to air. Additional electrochemical data of real significance could also be gained by use of the galvanostatic technique.

Experimental Procedure

Figure 8-11 shows the test cell. It was constructed to retain vaporized NH_3 and to permit the use of accepted electrochemical techniques. Data were obtained potentiostatically (automatically traversing the potential range at a rate of 100 mV/min) and galvanostatically (manually adjusting the current every three minutes).

Three different surfaces of mill-machined carbon steel test specimens were studied. Briefly, they were (1) annealed in an inert gas furnace at $915 \pm 5°C$, (2) heavily rusted in the NH_3—NH_4NO_3—H_2O solutions, and (3) partially cleaned by exposure to dilute HCl (etched). These test specimens presented the effects of "clean" or "dirty" surfaces, and surfaces that were somewhere in between. The test solutions were as follows: NH_3, 25% wt; NH_4NO_3, 69% wt; and H_2O, 6% wt. Solution temperature was $20 \pm 1°C$.

High-Temperature Polarization Cells

Much of modern technology is concerned with high-temperature conditions, and because higher temperatures usually mean higher corrosion rates, it

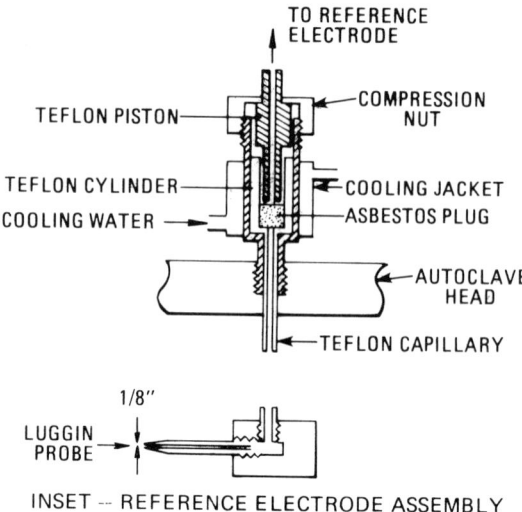

INSET -- REFERENCE ELECTRODE ASSEMBLY

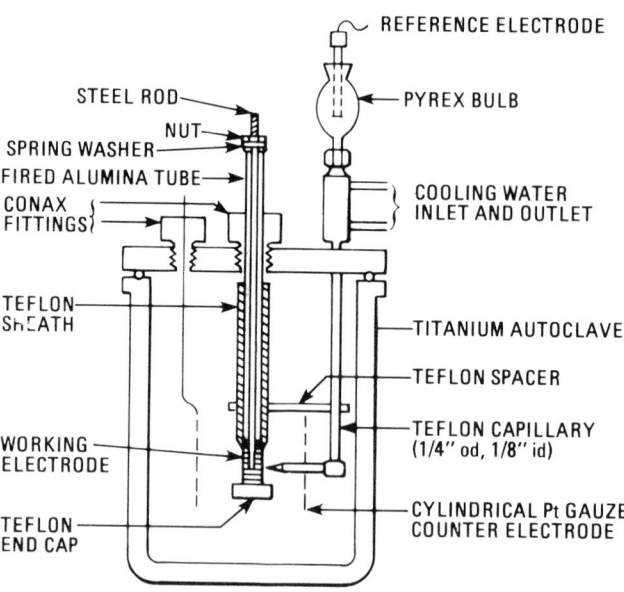

Figure 8-12. Diagram of the electrochemical cell. The 300°C temperature is maintained by a block oven (not shown) which is pulled down over the assembly.

is advantageous to use anodic protection on these systems when metals are involved that are amenable to the technique.

The titanium high-temperature electrochemical cell used for polarization studies[10] of nickel in aqueous solutions at temperatures to 250°C is shown in Figure 8-12. The cell was used without a liner since analysis of the autoclave contents by atomic-absorption spectroscopy after polarization experiments lasting up to 10 days failed to detect any titanium in solution (i.e., <0.5 ppm).

The working electrode was insulated from the autoclave body by a fired alumina tube. Direct contact between the alumina tube and the alkali solution was prevented by a Teflon sleeve to avoid contamination of the solution by aluminate ions. The working-electrode assembly was made pressure tight by compressing the spring washers at the top of the alumina tube. A coaxial relationship was maintained between the cylindrical platinum-gauze counterelectrode and the active working-electrode surface.

The reference-electrode assembly used is shown in Figure 8-12 inset. A porous asbestos plug was used to make electrolytic contact between the autoclave contents and the external saturated calomel electrode (SCE) which was maintained at room temperature (22 ± 2°C). By carefully adjusting the compression nut, the rate of flow of solution across the pressure drop (~ 3.8 ml/min at 250°C) could be reduced to less than 5 ml/day. The autoclave contents were periodically analyzed for Al and Mg by atomic-absorption spectroscopy and no evidence was found for contamination of the solution by decomposition of asbestos.

Test electrodes (12.7 mm long, 9.525 mm in diameter, and $3.80 \times 10^{m-4}$ m^2 apparent surface area) were machined from 99.99% spectroscopic-grade nickel metal. Initial surface treatment consisted of polishing with #400 and #600 grit silicon carbide paper. The electrode surface was then degreased in acetone and washed with distilled water. All solutions were purged at 22°C with 99.99% N_2 before polarization. Prior to heating, the autoclave was pressure tested with with 0.3 MN/m^2 of nitrogen. This pressurization also forced solution through the asbestos plug, thus forming electrolytic contact between the working electrode and the external reference electrode.

Crevice Corrosion Testing Polarization Cells

Before progressing very far into studies that lead the engineer into practical applications of anodic protection, instances of special corrosion forms such as pitting-, stress-, cracking-, and crevice-type attacks will be useful for background information. These processes require special cells for the study of electrochemical methods, including potentiostatic techniques for their investigation of crevice corrosion[11-19]. Apparently there are two major problems involved in designing an electrode suitable for crevice corrosion studies: (1) elimination of undesirable crevices but retaining an exposed, well-defined electrode area, and (2) a desired crevice which is accurately reproducible. The magnitude of this problem comes into focus when one remembers that the fissures involved in crevice corrosion of stainless steels are of the order of 10^{-3} inches (or less).

Lizlovs[20] designed a cell as shown in Figure 8-13 for crevice-corrosion testing using potentiostatic techniques. If an appropriate working electrode could be designed, he reasoned that the potentiostatic polarization current in the passive state could be utilized to measure the progress of crevice corrosion. With reproducible crevices, the relative resistance of alloys to crevice corrosion could be determined by comparing the polarization current to some preselected potential.

The polarization cell for the crevice-corrosion studies consisted of a 5-in. × 3.5-in. (12.7 × 9 cm) diameter glass tubing mounted between two rectangular polycarbonate blocks. The essential parts of the cell are shown in Figure 8-13. The cell was equipped with two 10/30 and two 24/40 standard-taper ground glass necks to accommodate the electrolytic bridge from the reference electrode, inlet and outlet tubes for purging gas, and the auxiliary platinum electrode, which was separated from the bulk of the solution by a fritted-glass disk. The glass tube was mounted to polycarbonate blocks by placing the tube

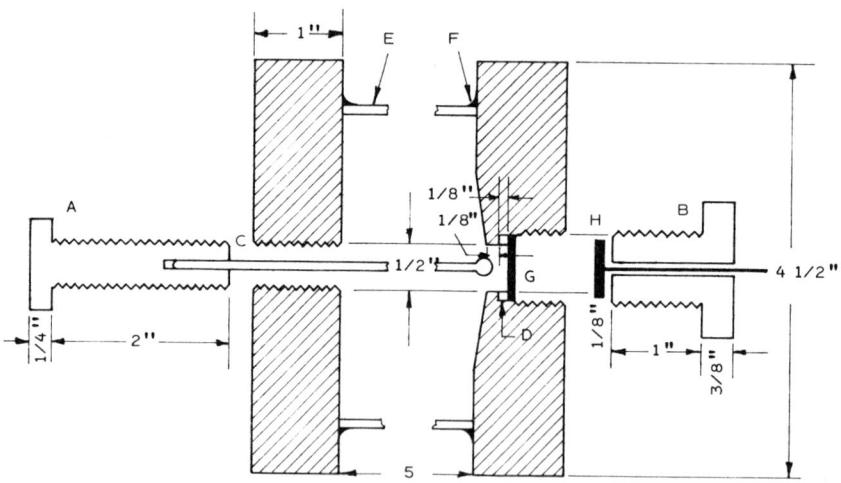

Figure 8-13. Cross Section through polycarbonate end-blocks and plastic parts of Crevice Corrosion Cell: A-Teflon bolt, ½-20 thread; B-PVC bolt, with hexagonic head, ⅞-14 thread; C-Glass rod one-eighth-inch diameter with 4-mm-diameter glass bead at its end; D-Teflon gasket, one-eighth inch thick × 0.810 inch OD; E-Glass tubing 3½ inch OD; F-Silicone rubber mount; G-Stainless Steel working electrode; H-Copper disc with copper lead wire.[20]

on the blocks in a proper position and applying silicone rubber (General Electric RTV-12 adhesive) around it and allowing the adhesive to solidify. One endblock that served as the holder for the working electrode was pressed against a Teflon washer by a polyvinyl chloride (PVC) bolt. A copper wire extending through the PVC bolt was attached to the back of a copper backup disc, completing the electrical circuit. A 0.5-in. (1.27 cm) hole, concentric to the working electrode opening, was contained in the opposite endblock. This opening held a Teflon bolt with a glass rod inserted into the threaded end. A 4-mm-diameter glass bead was attached to the opposite end of the glass rod. The crevice was produced by advancing the threaded bolt until the glass bead was pressed against the surface of the working electrode.

Summary

A sufficient presentation of circuit diagrams and equipment details has been made to make possible the organization of a laboratory to make polarization tests preliminary to installation of anodic protection. These make it possible for a company to make its own tests and also present the opportunity for commercial laboratories to offer services to those who do not wish to do their own testing.

Methods used successfully for polarization experiments are described and the various criteria of reproducibility, data accumulation, and analysis and other details are presented. Information is given also on pressurized polarization cells. Experimental procedures are outlined.

References

1. W. D. Henry and B. E. Wilde, *Corrosion* 25, 515–519, December (1969).
2. J. O. Sudbury, O. L. Riggs Jr., and D. A. Shock, *Corrosion* 16, 57t–62t, February (1960).
3. A. Hickling, *Trans. Faraday Soc.*, 38, 27 (1942).
4. ASTM Designation G-5, *Book of ASTM Standards*, Part 31, ASTM, Philadelphia (1970).
5. ASTM Designation G-3, *Book of ASTM Standards*, Part 31, ASTM, Philadelphia (1969).
6. N. D. Greene, *Experimental Electrode Kinetics*, Rensselaer Polytechnic Institute, Troy, (1965).
7. P. E. Morris and R. C. Scarberry, *Corrosion* 28, December (1972).
8. O. L. Riggs, Jr., *Corrosion* 20, 275t–281t, September (1964).
9. O. L. Riggs, Jr., *Corrosion* 26, June (1970).
10. D. D. MacDonald and D. Owen, *High Temperature, High Pressure Electrochemistry in Aqueous Solutions*, D. de G. Jones, J. Rater, and R. W. Staehle, eds., NACE, Houston (1976). pp. 513–523.
11. I. L. Rozenfeld and I. K. Marshakob, *Zavod. Lab.* 21, 1346 (1955).
12. R. R. Salem, *Zavod. Lab.* 26, 291 (1960).

13. I. B. Vlanovskii, *Zhur. Prikl. Khim.* **39,** 814 (1960).
14. G. Bombara, D. Sinigaglid, and G. Taccani, *Electrochim. Met.* **3,** 81 (1968).
15. W. D. France, Jr., and N. D. Greene, *Corrosion* **24,** 247 (1968).
16. R. J. Picard and N. D. Greene, *Corrosion* **30,** 393 (1974).
17. D. A. Jones and N. D. Greene, *Corrosion* **25,** 367 (1969).
18. E. A. Lizlovs and A. P. Bond, *J. Electrochem. Soc.* **116,** 574 (1969).
19. W. D. France, Jr., *J. Electrochem. Soc.* **114,** 818 (1967).
20. E. A. Lizlovs, *J. Electrochem. Soc.* **117,** 10, (1970).

Selected Examples of Anodic Protection

Among major decisions to be made in establishing the design and materials parameters of equipment that will be exposed to corrosive solutions is the choice of a material. For some materials sufficient data based on experience is available to permit unequivocal approval of one or more candidates. Thus the materials engineer must weigh the availability, design, and fabrication aspects of alternates concurrently with an economic analysis. Frequently these decisions are further complicated by consideration of the possibility that one or more of the candidates is amenable to electrochemical protection, or that their attributes with respect to inhibitors need also be taken into account. Many of these considerations are discussed elsewhere herein.

It is convenient and useful, however, to have data available with respect to performance under anodic protection conditions because these data may influence a decision for or against a material or class of materials. For example, if corrosion rates of a carbon steel can be kept low enough, it may become the material of choice for a system for which higher-cost alloys are also viable candidates. A parallel consideration also involves the possibility that a relatively low-cost alloy whose corrosion rate can be controlled by anodic protection may become the choice over a higher-cost alloy which may or may not be amenable to anodic protection. Both of these examples impinge on the economic analyses because extended life due to reduced corrosion rates becomes a significant factor. There are, of course, numerous other variations of this kind that influence materials and design decision.

If a decision to use anodic protection is in order, data in this chapter make it possible to discriminate to some extent among candidate materials on the basis of the probable current consumption to maintain passivity. Although current costs are not usually the largest economic factor involved, they become increasingly important as the cost of energy increases.

In consideration of these factors it is useful to know that reliable experimental data have been accumulated and much of it presented in this chapter, which makes materials selection easier. This does not mean that design decisions concerning current density and the like can be made from the data, but they narrow down the choices with respect to materials–corrosives combinations and reduce the amount of preliminary experimental work that must be done.

It probably will be the common procedure for engineers designing systems in which anodic protection is to be used to employ an outside laboratory to make preliminary tests using the design corrosive solution to identify the material or materials of construction suitable for the proposed system. Nevertheless, there is useful additional information with respect to some of the materials–corrosives combinations that may lead to solution modifications that either make possible or enhance the usefulness of anodic protection. This would be the case, for example, with respect to the reactions between methanol–sulfuric acid and nickel, in which addition of 1% water leads to passivation under anodic-polarization conditions.

Those who have adequate laboratory facilities to make the preliminary tests will find that the curves presented in this chapter are good starting places or checks against experimental error. Particularly useful are comments and information respecting the effects of temperature and the influence of experimental procedures on results.

Alloy Evaluation

Potentiostatic anodic-polarization curves are the most accurate method presently available for the development of new alloys and the determination of probable performance of existing metals. Studies of alloy systems that have increased stabilities of their passive states lead to the development of new materials for engineering which are more resistant to corrosion. Potentiostatic data identify alloy systems likely to have passivities stable in corrosive solutions.

Tomashov and Chernova[1] reported on factors which determine the anodic passivation of alloys. Factors which increase a metal system's corrosion resistance are:

1. A stronger passivating tendency in the action zone because of partially protective, relatively stable corrosion products.
2. A decreased limiting current density to obtain passivity.
3. A minimum current density to main passivity.
4. An initially more negative potential.
5. A more negative potential for the onset of passivity.
6. A more positive potential for passivity breakdown due to active (reductive) ions.

7. A more positive transpassive potential.
8. A "pitting" potential which is more positive than the transpassive potential.

The character of the anodic polarization curve is then governed by the composition of the metal and the type and characteristics (flow, temperature, concentration, time, etc.) of the corrosive environment. To illustrate this point, anodic polarization curves from several sources are given which represent the internal and external factors of alloy passivity.

Alloy Effects

Corrosion in the passive state is a property of a metal's passive film and the phase boundary between the passive film and the electrolyte, but it is not a direct property of the substrate material. However, the constituent elements of the substrate contribute to the formation of this passive film and indirectly render it more or less passive. Utilization of this concept has been made over the years by metallurgists in their attempts to provide more corrosion-resistant engineering materials. They have taken advantage of the contributions to passivity of such elements as nickel, copper, molybdenum, titanium, tantalum, chromium, zirconium, palladium, platinum, etc., to make the base (less costly, generally) metal more corrosion resistant.

Chromium is a very important alloying element for iron to increase its corrosion resistance. Oliver[2] demonstrated the beneficial effects of chromium on the passive nature of iron in sulfuric acid (see Figure 9-1). As chromium concentration increased from about 9 to 18%, it not only broadened the passive potential range, but also decreased the current density required to both obtain and maintain passivity. Oliver suggested that two clearly distinct regions could be discerned which had very marked differences in respective corrosion pressures. At potentials *less* positive than $+1.2$ V, E_h, the chromium dissolved as the Cr^{3+} ion, while in the region *more* positive than $+1.2$ V, E_h the chromium dissolved as the CrO_4^{2-} ion.[3] The corrosion-mechanism changes depend on the potentials which exist in the system. The corrosion current density therefore depends strongly upon the composition of the alloy. If a chromium alloy is to be anodically protected, the selection of the set-point potential (ESP) to maintain passivity becomes important.

Tomashov, Chernova, Ruscol, and Ayuyan[4] reported on the effect of alloying chromium and tantalum with titanium. Their potentiostatic anodic-polarization study (Figure 9-2) was made in 5 wt.% hydrochloric acid at 100°C. They showed that an increase in the chromium content in quenched (homogeneous) and annealed (two- or three-phase) alloys increased anodic currents

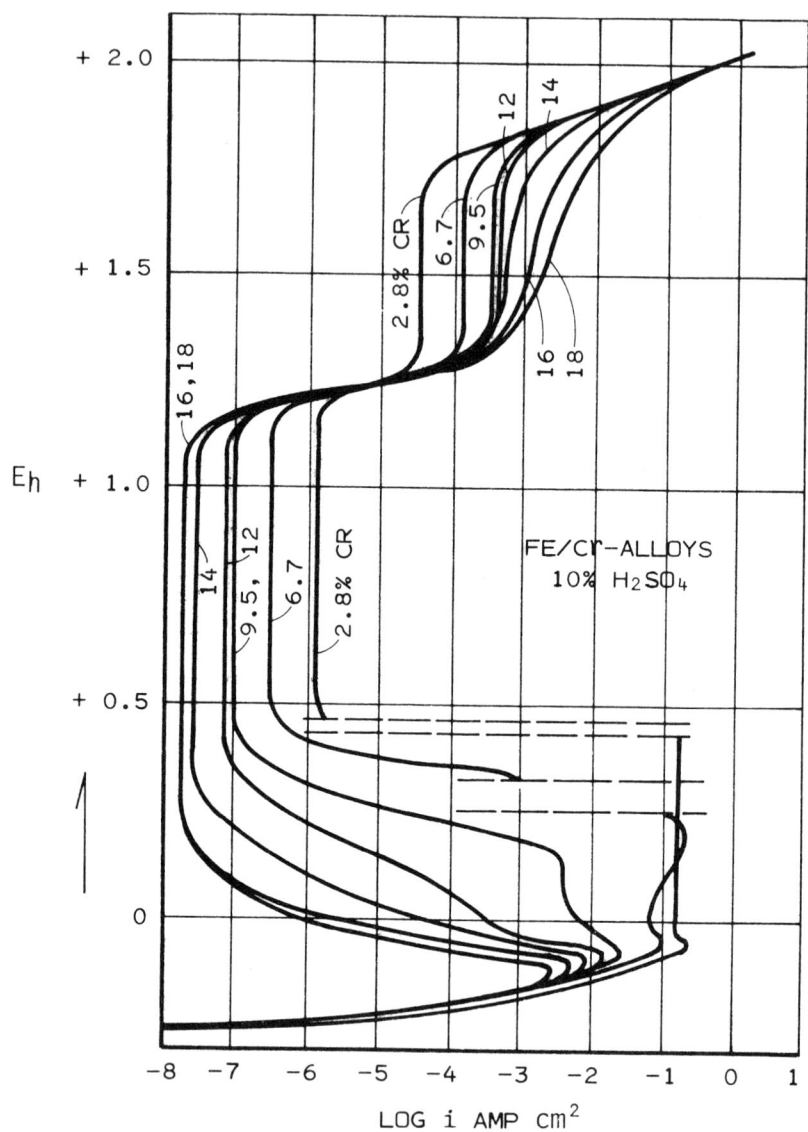

Figure 9-1. Static-potential current density versus potential curve for various iron–chromium alloys in 10% H₂SO₄.²

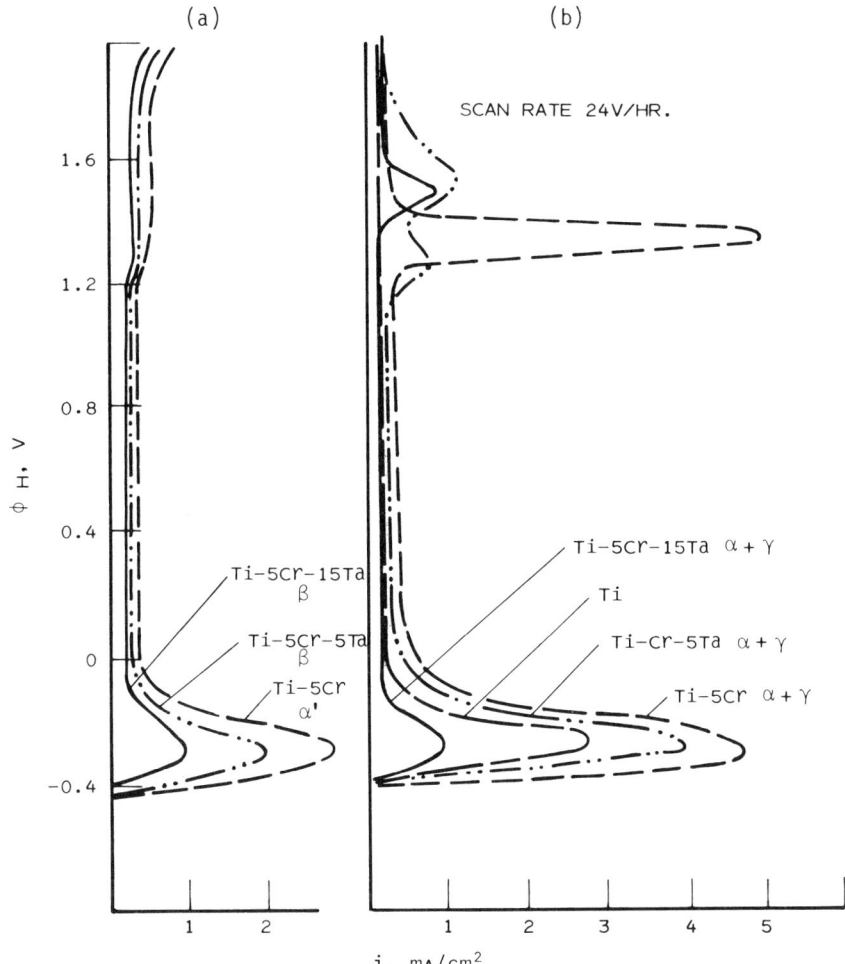

Figure 9-2. Anodic curves for (a) single-phase, and (b) two-phase TiCr/Ta alloys, 5% HCl, 100°C.[4]

in the active and passive regions, while an increase in tantalum decreased the anodic dissolution-current density.

Nickel is important both as an alloy addition and as a base material for corrosion-resistant alloys. Potentiostatic techniques were used during a study[5] in which aluminum was alloyed with nickel. The investigation was carried out in sulfuric acid at 22°C. Generally, precipitation-hardened aluminum–nickel-based alloys are investigated because of their excellent retention of strength at

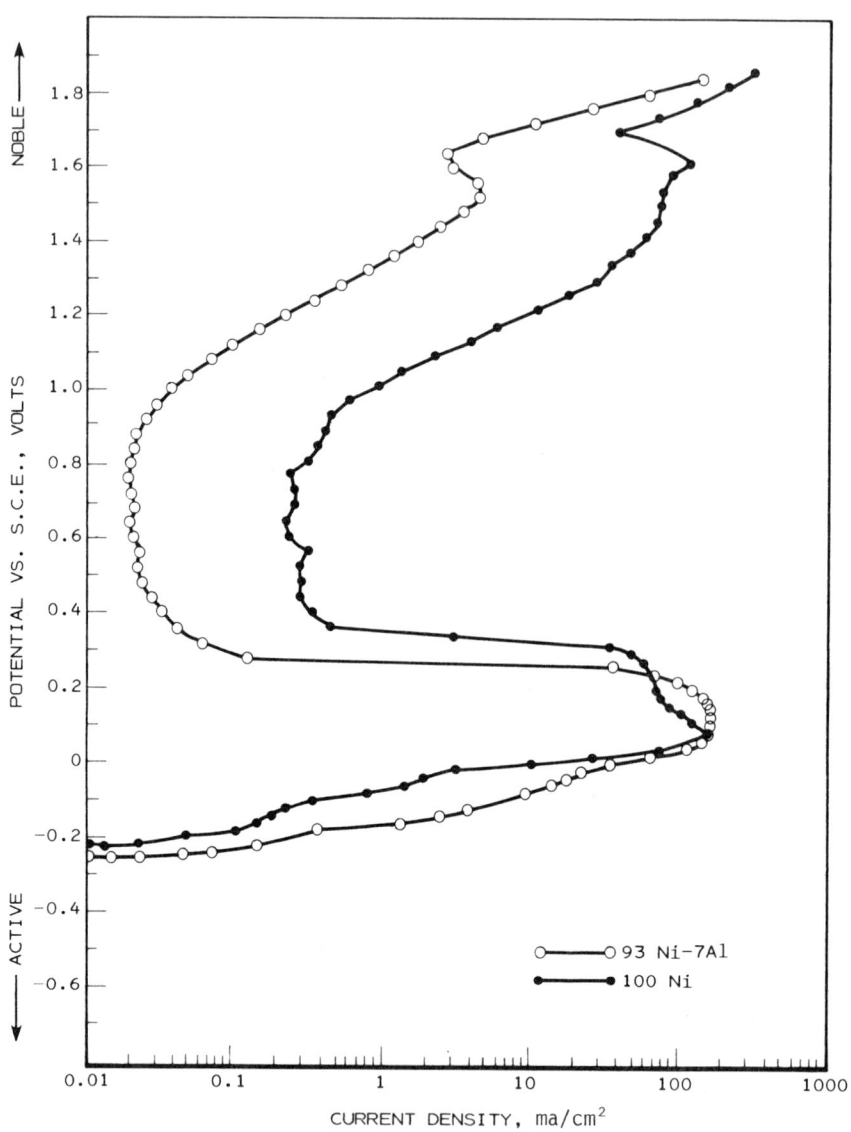

Figure 9-3. Potentiostatic anodic-polarization curves for pure nickel and 93Ni–7Al in hydrogen-saturated 5 *N* sulfuric acid solutions at 22°C.[5]

elevated temperatures. The effect on passive phenomena that aluminum exhibits on nickel is illustrated in Figure 9-3. The addition of 7 wt.% aluminum increased the critical-current density from about 110 mA/cm^2 for pure nickel to about 170 mA/cm^2. The aluminum also broadened the passive potential range by nearly 250 mm and decreased the current density required to maintain passivity from 0.22 mA/cm^2 to essentially 0.017 mA/cm^2 (over one order of magnitude). Finally, the addition of 7 wt.% aluminum decreased the oxygen overvoltage about 80 mV.

Hastelloy alloys are nickel-based commercial engineering materials. It is generally accepted that the B, C, and F series can be protected by applied anodic currents. Operational life of plant vessels constructed of these metals for service in sulfuric and hydrochloric acids can be lengthened by use of anodic protection.[6-8] Potentiostatic data obtained by Greene[9] with Hastelloy alloys B, C, and F in hydrochloric acid at room temperature are summarized in Figure 9-4. It is interesting to note that the acceptability of the alloys to potentiostatic induced passivity increased in the following order: F > C > B. The chemical compositions of these alloys are listed in Table 9-1.

Anodic-polarization data relative to passivity in hydrochloric acid predicts their relative performance in proportion to percent chromium. However, based on molybdenum concentration, their relative performance could be reversed (i.e., B > C > F). The effect of 1 N H$_2$SO$_4$ at room temperature on anodic

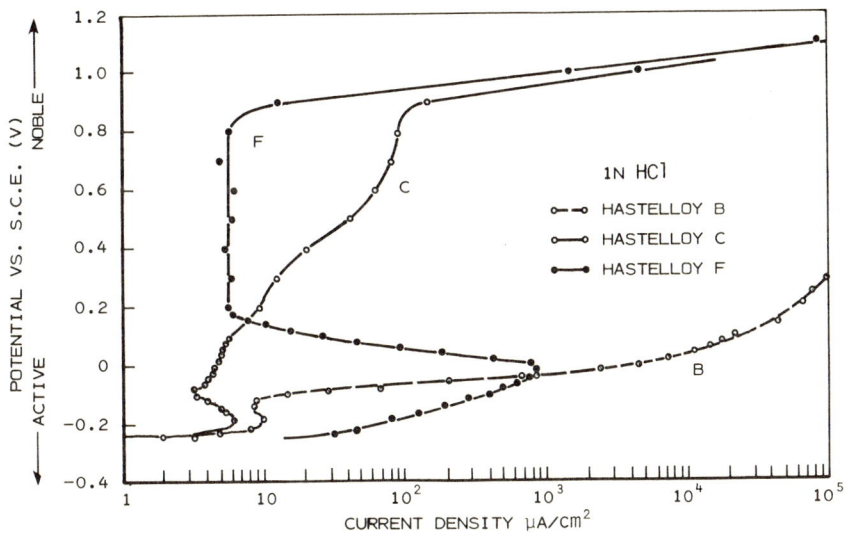

Figure 9-4. Anodic polarization of Hastelloy alloys in normal hydrochloric acid at room temperature.[9]

TABLE 9-1. Composition of Hastelloy Alloys

Hastelloy alloy	% Ni	% Cr	% Mo	% C	% Fe
B	62	—	28	0.10	5
C	56	15	17	0.08	5
F	47	22	7	0.05	17

TABLE 9-2. Anodic Polarization Parameters for Hastelloy Alloy F in Sulfuric Acid at Room Temperature[9]

H_2SO_4 normality	Reference potentials (mV)				Current density ($\mu A/cm^2$)	
	E_{corr}	E_{crit}	E_p	E_{tp}	i_{crit}	i_p
1	−265	−200	+90	+800	16	1.2
5	−210	−30	+435	+950	26	3.0
10	−150	+30	+600	+1030	30	5.2

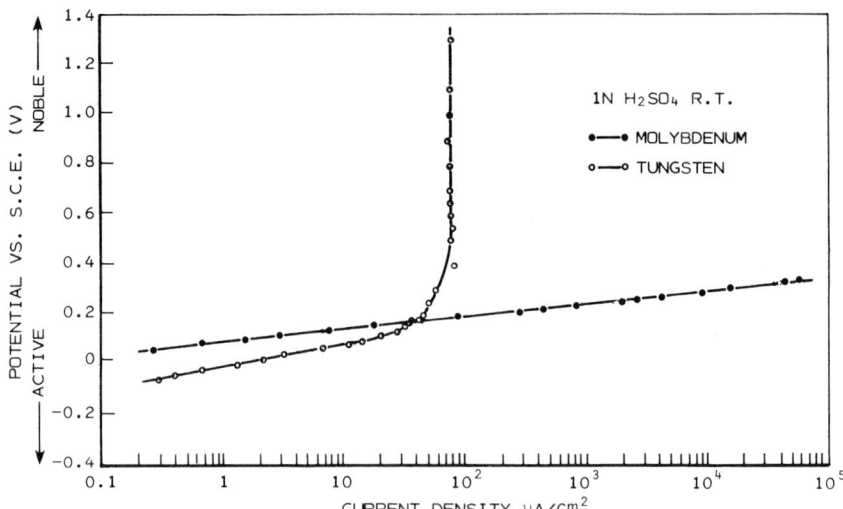

Figure 9-5. Anodic polarization of molybdenum and tungsten in normal sulfuric acid at room temperature.[9]

polarization of molybdenum is shown in Figure 9-5.[9] Check Table 9-2 for polarization parameters.

Concentration Effects

As discussed in Chapter 6, concentration of solution species plays a direct role in anodic polarization of active–passive metals. Greene[9] has demonstrated this in Figure 9-6 for Hastelloy Alloy F in various normalities of H_2SO_4 at room temperature. As the normality of H_2SO_4 increased, the critical-current density to passify Hastelloy Alloy F and the current required to maintain passivity increased, as shown in Table 9-3.

Also, as the normality (concentration) of H_2SO_4 increased, the corrosion potential (E_{corr}) of Hastelloy Alloy F became less negative. This was also true of the critical potential for passivation. The most critical parameter for application of anodic protection to Hastelloy Alloy F in H_2SO_4 is the current density to maintain passivity (i_p). As H_2SO_4 normality increased from 1 to 10 N, the i_p increased from 1.2 $\mu A/cm^2$ to 5.2 $\mu A/cm^2$. Although a very high level of protection can be obtained, the 10 N H_2SO_4 is almost five times as corrosive to Hastelloy Alloy F as is the 1-N solution.

Potentiostatic anodic-polarization curves for 97Fe–3Si alloy in hydrogen-saturated sulfuric acid at 22°C indicated concentration effects somewhat different from those found with the nickel alloy (Hastelloy F). Figure 9-7 summarizes the work of Crow, Meyers, and Jeffreys[10] for this alloy.

Ferrosilicon alloys have excellent corrosion resistance, as shown in Table 9-3. This is evidenced by the low i_p current density in 10 N H_2SO_4 (i.e., 0.06 mA/cm or 60 $\mu A/cm^2$ is equated to an approximate corrosion rate of about 24 mpy). In general, increasing sulfuric acid concentration shifted the corrosion potential to less negative values, increased the critical current, but decreased the current density to maintain the ferrosilicon alloy passive.

TABLE 9-3. Anodic Polarization Parameters for 97Fe–3Si Alloy in H_2-Saturated Sulfuric Acid at 22°C[a]

H_2SO_4 normality	Reference potentials (mV)				Current density (mA/cm^2)	
	E_{corr}	E_{crit}	E_p	E_{tp}	i_{crit}	i_p
1	−525	−355	+775	+1500	28	0.20
5	−465	−355	+650	+1550	60	0.09
10	−430	−350	+500	+1600	150	0.06

[a] Taken from Reference 10.

Sulfuric acid concentrations were used also in anodic-polarization studies for Kroll zirconium. A typical summary of polarization curves is given in Figure 9-8.[11]

Zirconium demonstrates the straightforward effects of increasing current density and more positive corrosion potentials (See Table 9-4) as the concentration of sulfuric acid increased from 5 to 20 wt.%. One additional and most significant feature with anodic protection: As H_2SO_4 concentration increased, the range of passive potentials decreased (narrowed).

Chromium and titanium carbide are important and constituents not to be neglected in stainless steels. The presence of chromium carbides ($Cr_{23}C_6$) in grain boundaries renders the alloy more susceptible to intergranular corrosion. This sensitized condition is unavoidable because it occurs during fabrication, where heat treatment is impractical. Titanium additions to stainless steels tend to decrease the susceptibility to sensitization-caused intergranular attack. Payer and Staehle[12] reported on the behavior of $Cr_{23}C_6$ and TiC during anodic

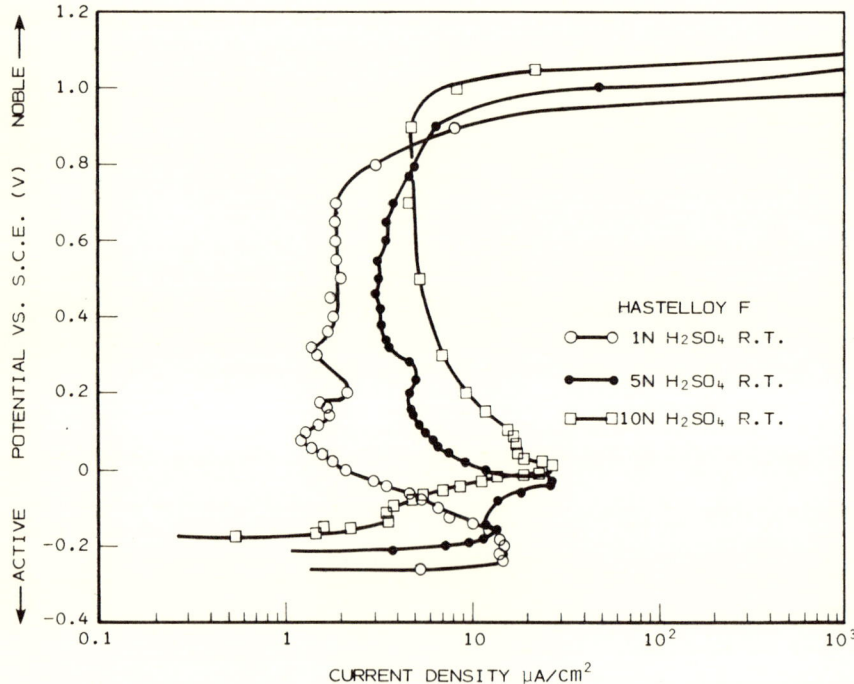

Figure 9-6. Effect of acid concentration on anodic polarization of Hastelloy Alloy F in sulfuric acid at room temperature.[9]

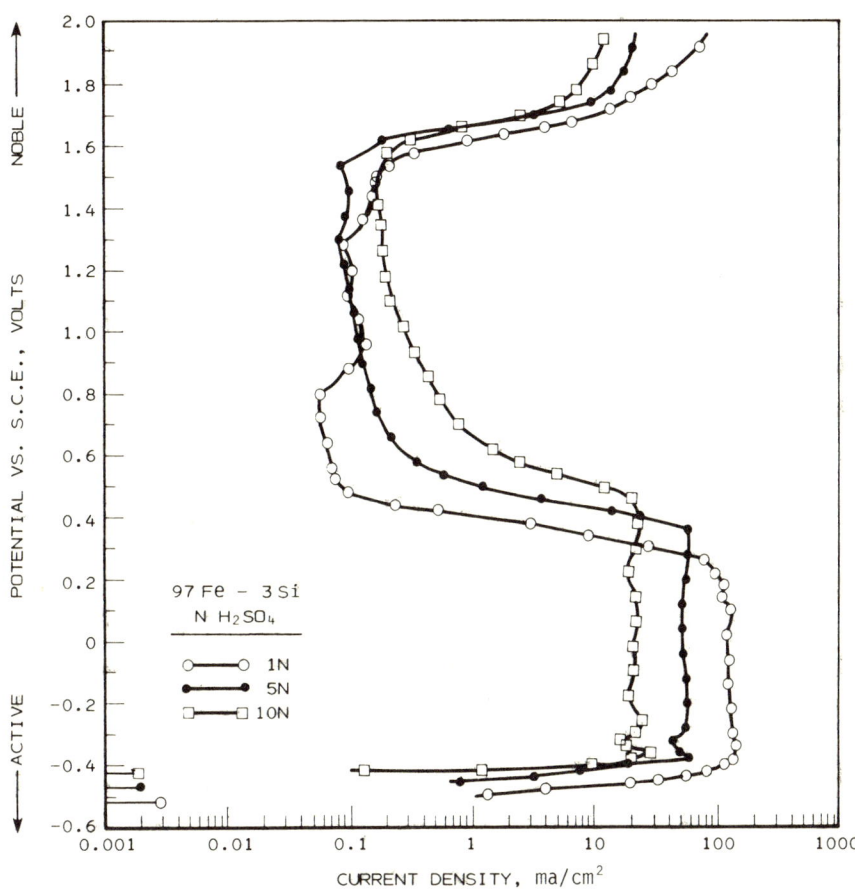

Figure 9-7. Potentiostatic anodic-polarization curves for 97Fe–3Si alloy in H_2-saturated sulfuric acid solutions at 22°C.[10]

TABLE 9-4. Anodic Polarization Parameters for Kroll Zirconium in H_2SO_4 at Room Temperature[a]

H_2SO_4 (% by wt)	Reference potentials (mV)				Current density (A/cm²)	
	E_{corr}	E_{crit}	E_p	E_{tp}	i_{crit}	i_p
5	-525	-305	$+200$	N.A.[b]	5×10^{-4}	1.15×10^{-5}
10	-500	-230	$+400$	$+1600$	1.8×10^{-3}	1.4×10^{-4}
20	-475	-190	$+900$	$+1600$	5×10^{-3}	2.2×10^{-4}

[a]K. Elayaperumal, S. S. Chouthai, and J. Balanchandra, Corros., **29**, February (1974).
[b]N.A. = not applicable.

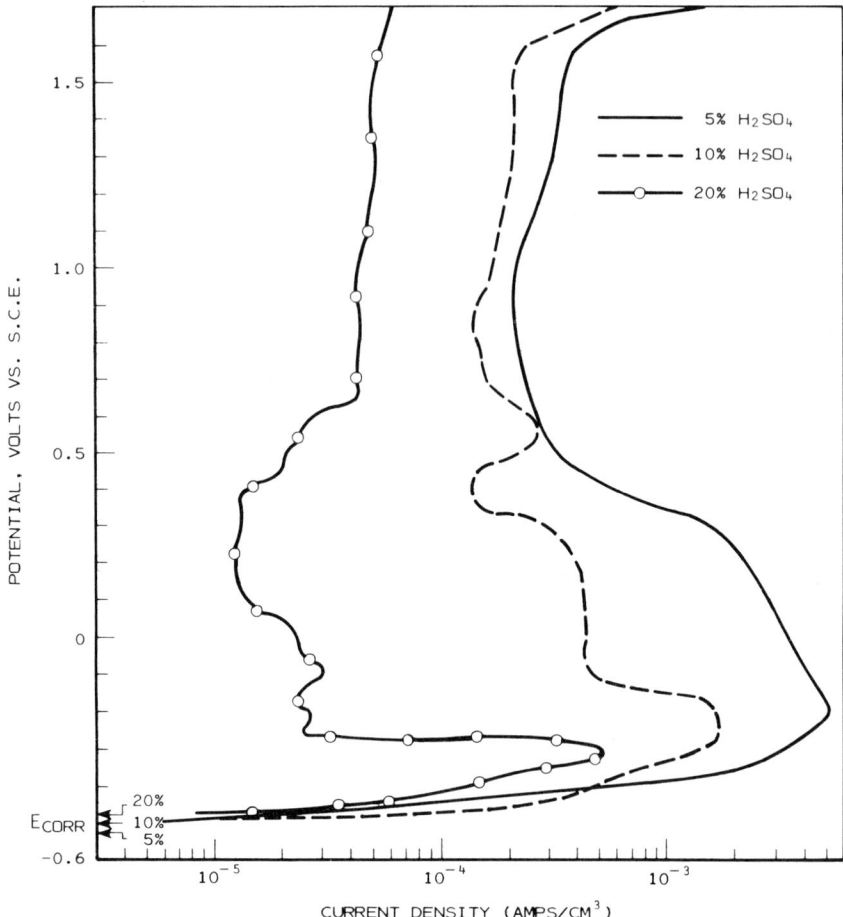

Figure 9-8. Potentiostatic anodic-polarization curves of Kroll zirconium in 5, 10, 20 Vol. % H_2SO_4 at room temperature.[11]

polarization in boiling (101°C) 2 *N* H_2SO_4. Figure 9-9 summarizes their anodic-polarization curves for both $Cr_{23}C_6$ and TiC.

Very dramatic differences are apparent between the reactions caused by the two carbides. Chromium carbide exhibits polarization characteristics as listed in Table 9-5, which indicates that it was much more resistant to the boiling sulfuric acid than is titanium carbide. For example, the corrosion rates for anodically passivated materials (i_p) were about 18 mpy for $Cr_{23}C_6$ and 1400 mpy for TiC.

TABLE 9-5. Anodic Polarization Parameters for $Cr_{23}C_6$ and TiC in Boiling ($101\,°C$) H_2SO_4

	Reference potential (mV)				Current density (mA/cm^2)	
Carbides of	E_{corr}	E_{crit}	E_p	E_{tp}	i_{crit}	i_p
Chromium	-190	-150	$+400$	$+1100$	1.1	0.02
Titanium	$+510$	$+1040$	$+1400$	$+1850$	45.0	2.0

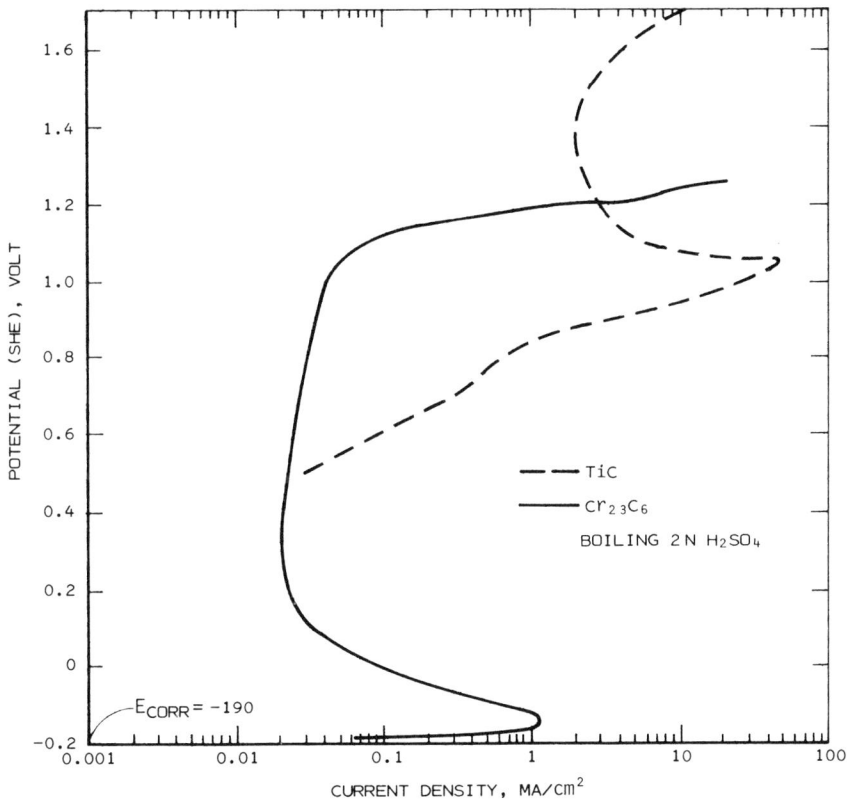

Figure 9-9. Anodic-polarization curve for $Cr_{23}C_6$ and TiC in boiling 2 N H_2SO_4 ($101\,°C$).[12]

Temperature Effects

Temperature is a major variable for the characterization of any anodic-polarization curve. These effects were discussed earlier (Chapter 6). Briefly, temperature increases the critical current to obtain or maintain passivity and generally reduces the passive potential range. Greene[9] reported on the effects of temperature on anodic polarization of Hastelloy Alloy C in H_2SO_4. These data are summarized for system temperatures of 24 and 90 °C in Figure 9-10. The current density at the primary passive potential peak increased, beginning at 28 $\mu A/cm^2$ at 90 °C. The passive potential range decreased from about 400 mV at 24 °C to 100 mV at 90 °C. The minimum current density (in the passive potential range) occurred at about −60 mV versus SCE for both curves. The current density (at −60 mV) increased from 1.9 $\mu A/cm^2$ at 24° to 10.2 $\mu A/cm^2$ at 90 °C.

Trout and Daniels[13] reported on the effect of temperature on the anodic polarization of solution-treated Inconel X-750 alloy. Figure 9-11 shows the effect in 1 N H_2SO_4 at 30, 40, and 50 °C. They reported that the passive-current density, the critical-current density, and the transpassive region were shifted in the direction of increasing current density at the higher tempera-

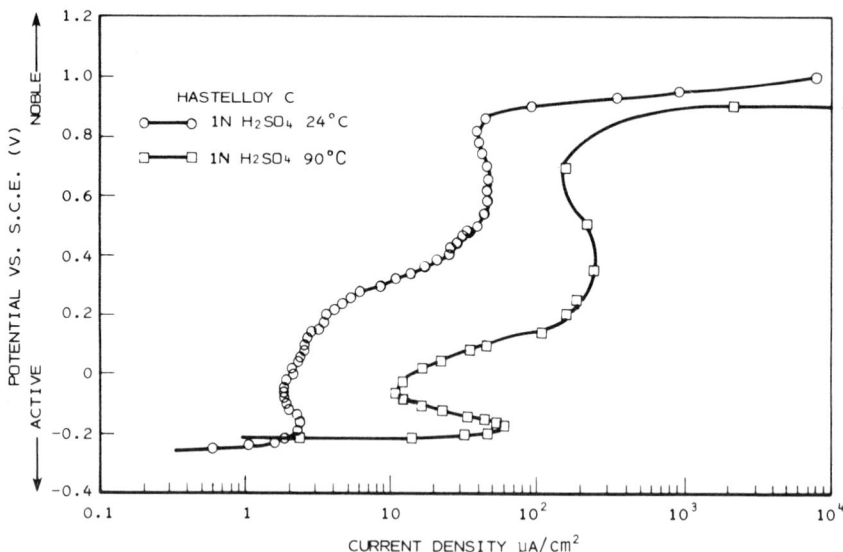

Figure 9-10. Effect of temperature on anodic polarization of Hastelloy Alloy C in 1 N sulfuric acid.[9]

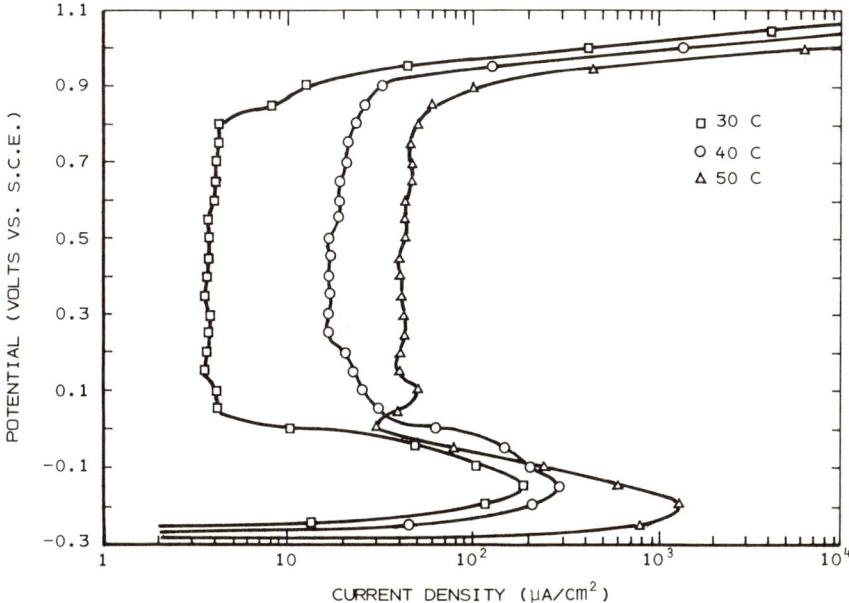

Figure 9-11. Potentiostatic anodic-polarization curves of solution treated Inconel X-750 in 1 *N* H₂SO₄ at 30, 40, and 50°C.[13]

tures. The corrosion potential, the primary passive potential, and the range of passive potentials were essentially unaffected by the increasing temperature.

These data (Table 9-6) suggest that the corrosion of Inconel X-750 (solution treated) in 50°C H₂SO₄ can be very effectively controlled anodically. Not only is the current density to maintain passivity very low (40 μA/cm² for less than 20 mpy), but the range of passive potentials is wide (almost 900 mV).

TABLE 9-6. Anodic Polarization Parameters for Inconel X-750 Alloy in 1 *N* H₂SO₄[a]

Temperature (°C)	Reference potentials (mV)				Current density (μA/cm²)	
	E_{corr}	E_{crit}	E_p	E_{tp}	i_{crit}	i_p
30	−250	−140	+400	+905	2×10^2	3.7
40	−265	−140	+400	+890	3×10^2	18.0
50	−290	−200	+400	+875	1.4×10^3	40.0

[a] Taken from Reference 13.

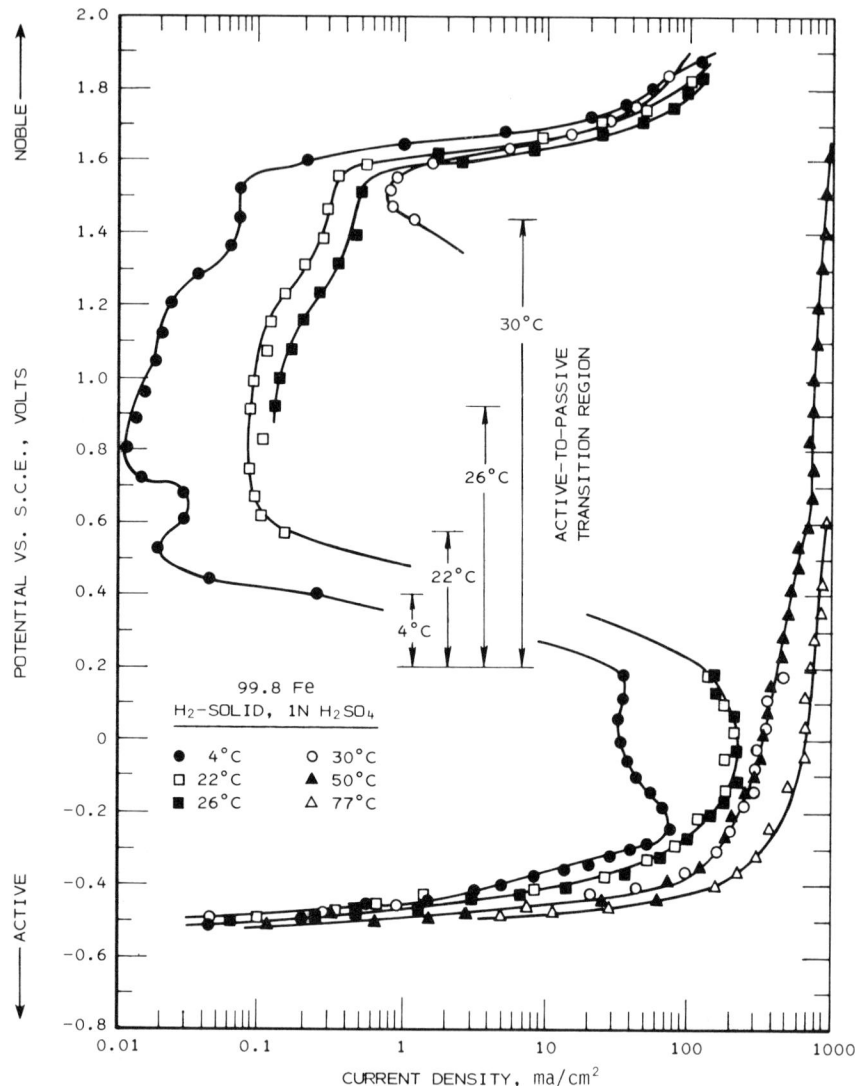

Figure 9-12. Effect of temperature on potentiostatic anodic polarization of iron in H_2-saturated 1 N H_2SO_4.[14]

Finley and Meyers[14] have demonstrated the effect of temperature on the anodic polarization of 99.8% Fe in a H_2-saturated 1 N H_2SO_4 solution maintained at 4, 22, 26, 30, 50, and 77°C (see Figure 9-12).

Iron anodically polarized in the 1 N H_2SO_4 (H_2 saturated) at 4 to 30°C exhibited active, active to passive, passive, and transpassive potential regions. At 50 and 77°C, the 99.8% Fe specimens remained active over the potential range studied. Temperature had no significant effect on either the corrosion potential or the overvoltage for visible oxygen evolution. Perhaps the most significant effect was that 99.8% Fe could be anodically protected in 1 N H_2SO_4 (H_2 saturated) at 4 to 30°C.

Care should be taken in the selection of a set-point potential. For example, if a E_{sp} + 800 E_{mv} versus SCE were selected for the 4–200°C range, passivity would be lost if the H_2SO_4 temperature were to increase 4°C (to 26°C) at any point within the system.

Environmental Effects

The form taken by an anodic polarization curve is a typically reproducible record of a particular system. The completed curve is essentially a "fingerprint," and with sufficient familiarity, the system could be identified no matter when the curve is run if all conditions are repeated. As previously discussed, the characteristics of the anodic-polarization curve are determined by internal and external effects. This section will present curves which have been obtained under the influence of various external effects.

Mansfeld[15] investigated the polarization behavior of nickel in methanolic solutions containing sulfuric acid. The addition of 1% H_2O leads to nickel passivation under anodic polarization techniques. Figure 9-13 shows the effect of H_2SO_4 concentration on the potentiostatic anodic-polarization of nickel in methanolic solutions. He reported that 0.1 N H_2SO_4 in CH_3OH contains 0.04% H_2O, while a solution of 1.0 N H_2SO_4 in CH_3OH contains 0.4% H_2O. The decisive role played by water in the passivation of nickel in methanolic solutions containing 2 N H_2SO_4 is apparent from these results.

MacDougall and Cohen[17] investigated the formation and growth of protective oxide films on nickel electrodes in a pH 8.4 Na_2SO_4 solution. A typical potentiostatic profile for the anodic polarization of nickel in an aqueous electrolyte shows that nickel dissolution occurs in the active potential region but that it is passivated at more anodic potentials. Figure 9-14 is an anodic-polarization curve obtained by potentiostatic stepwise procedures for an oxide-free nickel specimen in pH 8.4 Na_2SO_4 solution at 25°C. (The reference potential is versus the Hg_2SO_4 half-cell.)

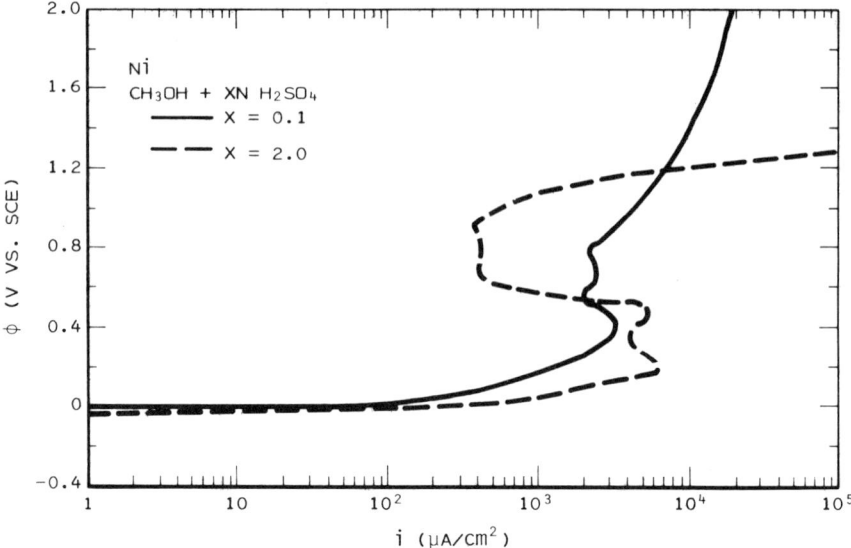

Figure 9-13. Potentiostatic polarization curves for nickel in 0.1 *N* H$_2$SO$_4$ in CH$_3$OH and 2 *N* H$_2$SO$_4$ in CH$_3$OH.[15]

A wide range of passive potentials for nickel, under the stated conditions, is apparent from the polarization curve (-655 to $+460$ E_{mV} versus Hg$_2$SO$_4$). The current density for maintaining the nickel specimen passive at -400 mV versus Hg$_2$SO$_4$ suggests a corrosion rate of about 0.8 mpy. Considering all of the information (Table 9-7), anodic protection could be effectively applied to a nickel system typical of this study.

Investigators[18] at the Naval Research Laboratory, in their studies of the potentiostatic passivation of mild steel in 300°C NaOH solutions, have suggested that (1) mild steel passivates with characteristic changes in the morphology of the surface oxide film, and (2) the corrosion mechanism changes

TABLE 9-7. Anodic Polarization Parameters for Oxide-Free Nickel in pH 8.4 Na$_2$SO$_4$ at 25°C

	Reference potentials (mV)			Current density (μA/cm^2)	
E_{corr}	E_{crit}	E_p	E_{tp}	i_{crit}	i_p
-1050	-850	-400	$+460$	18	2

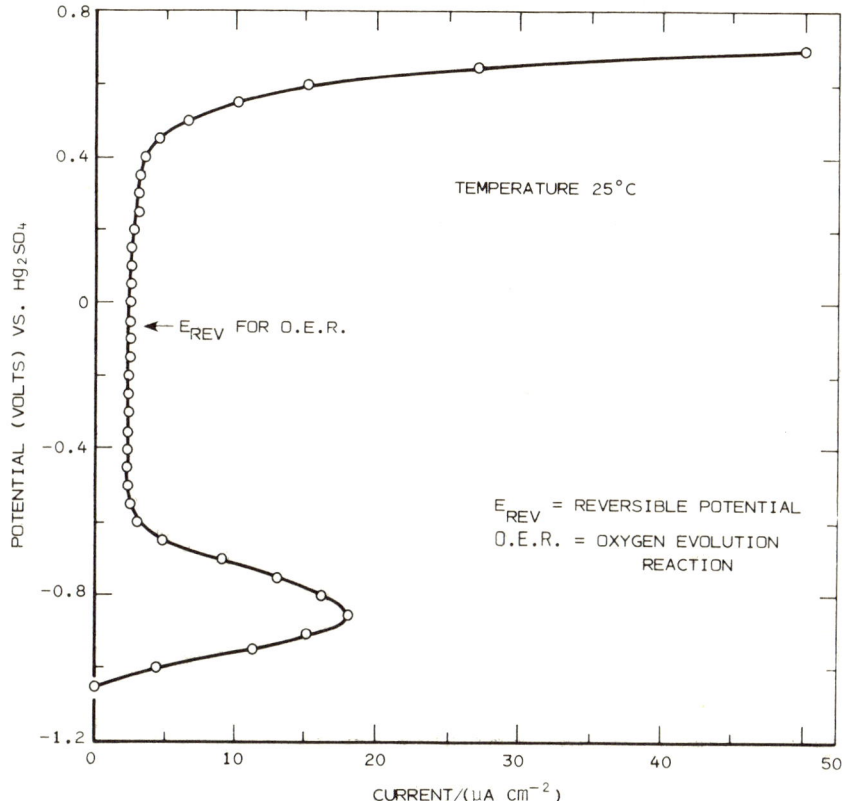

Figure 9-14. Anodic potentiostatic current—potential profile for a previously oxide-free nickel electrode in pH 8.4 Na_2SO_4 solution with potential increments of 0.05 V every 60 s. Measured current is that registered at the end of 60 s.[17]

from a passive to active mode as the region of passive film stability is exceeded with increasing NaOH. Figure 9-15 shows the classic passivation behavior of carbon steel in 5 N NaOH at 300°C. An interesting consequence was the secondary passivity which developed between about +1100 to 1250 mV.

The passive potentials range (about 500 mV) combined with the low current density to maintain passivity ($i_p = 400$ $\mu A/cm^2$) indicate that the corrosion rate of mild steel could be reduced to about 160 mpy. Also, the initial-current peak (just prior to the critical-current peak) suggests the dissolution of iron to Fe^{2+}. The anodic-polarization parameters are listed in Table 9-8.

The transportation and storage of fertilizer solutions, i.e., NH_3—NH_4NO_3—H_2O, usually calls for mild steel to store large quantities. For this solution, corrosion may become a problem of major economic proportions.

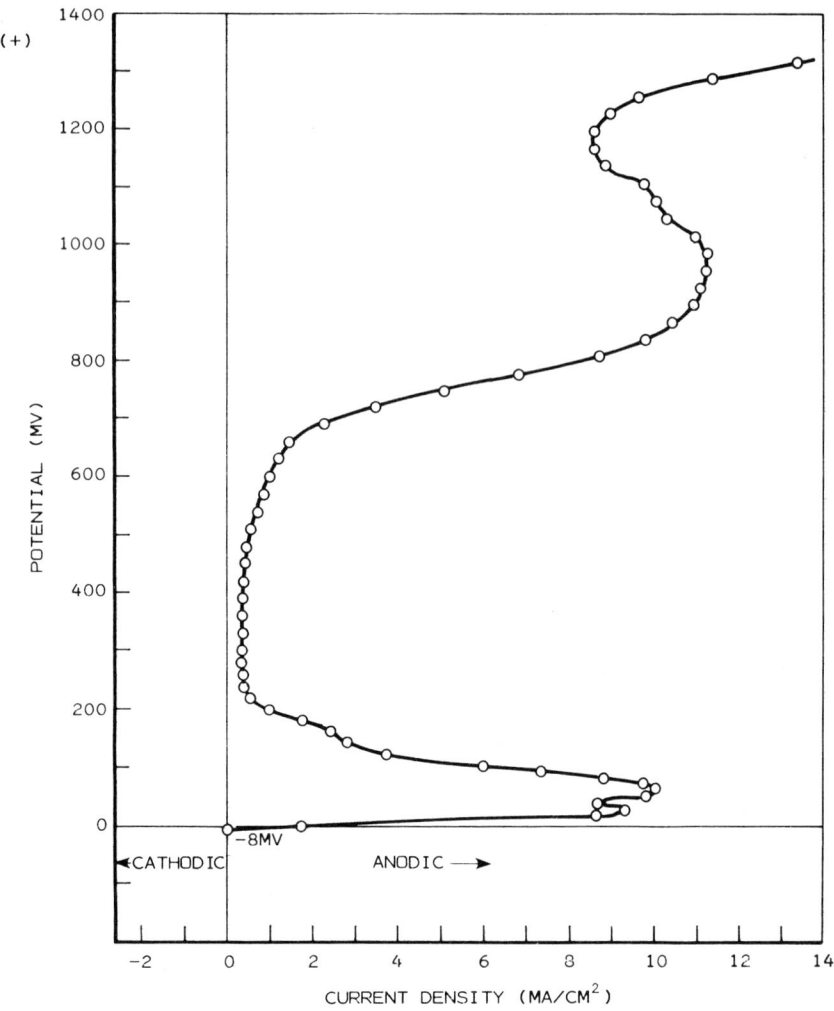

Figure 9-15. Current–potential curve for mild steel in 300°C 5 N NaOH. Potential measured against hydrogen electrode.[18]

TABLE 9-8. Anodic Polarization Parameters for Mild Steel in 5 N NaOH at 300°C

	Reference potentials (mV)			Current density (mA/cm²)	
E_{corr}	E_{crit}	E_p	E_{tp}	i_{crit}	i_p
−8	+70	+200	+700	10	0.4

Anodic protection is a means by which the severe economic losses from corrosion of steel tanks for fertilizer storage could be curtailed.

Preliminary Investigations Are Necessary

Arbitrary application of anodic protection, however, should not be made without first carefully investigating the proposed system. In one instance, Riggs[19] alerted potential users to a possible adverse result. The application of anodic protection to a rusty-surfaced carbon steel tank filled with NH_3—NH_4NO_3—H_2O was proposed with a set-point potential (E_{sp}) indicated as correct for the passivation of the rust. This potential is much too positive for the same system (if surfaces were not rusted) and could be well into the transpassive potential region and therefore hazardous.

Consequently, anodic-polarization curves were obtained for carbon steel in 69% NH_3—NH_4NO_3—H_2O showing the effects of rust on E—I characteristics.[20] Figure 9-16 illustrates the problem with potentiodynamic anodic-polarization curves. (The reference potentials were made versus a AISI Type 316

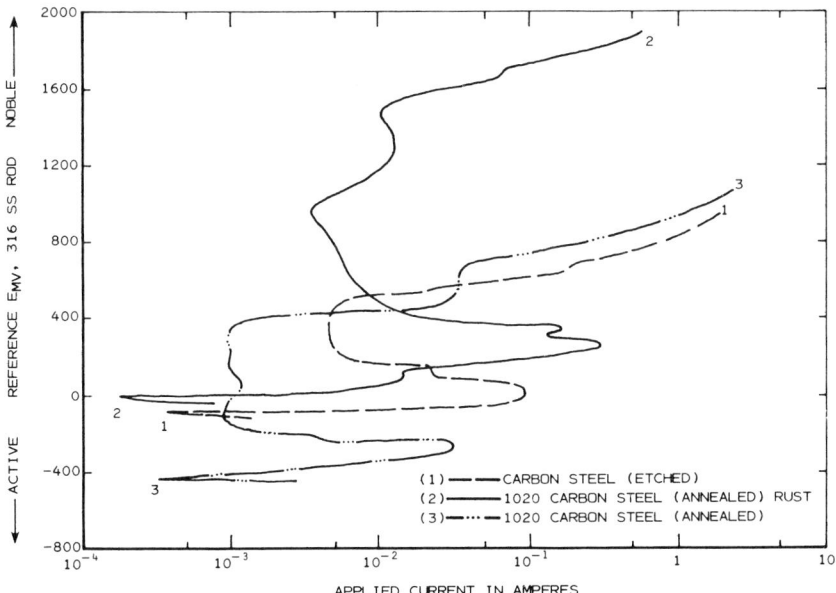

Figure 9-16. Potentiodynamic anodic polarization of specimens in 69 % wt NH_3-NH_4NO_3-H_2O at 20°C.[20]

steel electrode.) These results support the necessity for the investigation of prospective systems prior to application of anodic protection.

Time Effects

Potentiostatic polarization methods are generally, and conveniently, divided into two categories: potential-step and potential-sweep methods. The potential step is accomplished by a rapid change of electrode potential over a finite movement, and the corresponding current is measured after a predetermined time interval. The process is repeated as often as required to cover the potential range of interest. The potential-sweep method, referred to commonly as the potentiodynamic method, is accomplished by a continuous change of electrode potential over a predetermined potential range and at a preselected scanning rate, while simultaneously recording the current.

As can be seen, time is a variable of primary consequence. This can be true whether the information is gathered by the sweep or the step method. Early work[21,22] on the effects of time were approached by investigating the traverse (scanning) rate using step and sweep methods.

Leonard and Greene[21] reported on the effect of traverse rate for the potentiostatic polarization of Type 304 and titanium in 1 N H_2SO_4 at 25°C. Figures 9-17 and 9-18 are self-explanatory and effectively illustrate the role that time plays with regard to passivity. As slower traverse rates were used, a corresponding decrease in current densities was measured.

Additional information can be gathered from experimental results such as those reported by Leonard and Greene.[21] The decreased current density measured for the corresponding increasing traverse time to record each succeeding anodic-polarization curve suggests a time-dependence property for the passive layer. Further, the rate of current decrease can provide important information about passivity formation in a metal's relation to alternate engineering materials. When polarization curves for Type 304 and titanium are examined, it can be seen that the current density measured for each traverse rate at a potential of $+400$ mV versus SCE was plotted as a function of respective traverse rates.

Figure 9-19 illustrates the effect of the potential traverse rate on current density to maintain passivity for Type 304 and titanium in 1 N H_2SO_4 at 25°C. These data suggest that under the potentiostatic mode the passive layer formed on Type 304 was more stable in the 1 N H_2SO_4 than was the passive layer formed on titanium. Initially, the passive layer formed more quickly on the titanium, but at traverse rates which required more and more time the passive layer formed on Type 304 required less current to be maintained.

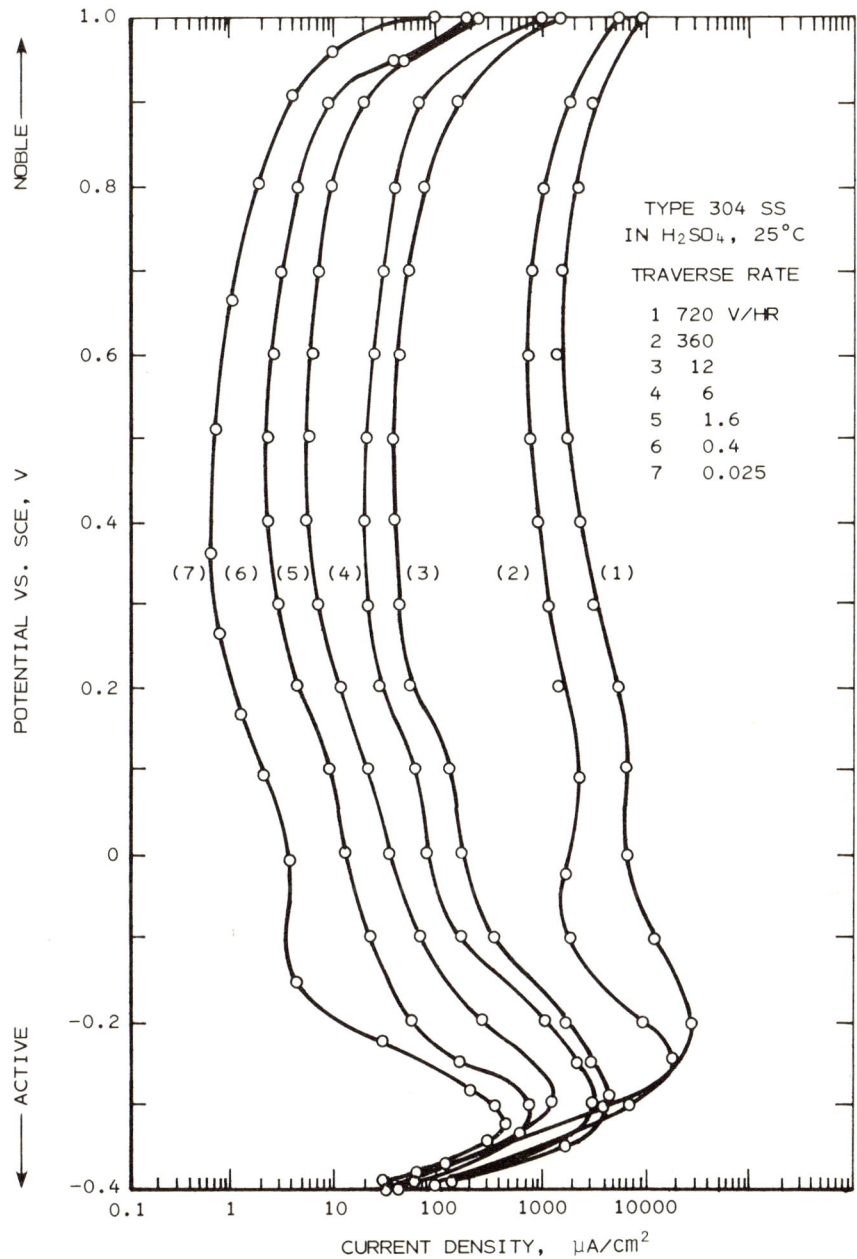

Figure 9-17. Potentiostatic anodic-polarization behavior of Type 304SS in 1 *N* H₂SO₄ at 25°C determined by potential-sweep method.[21]

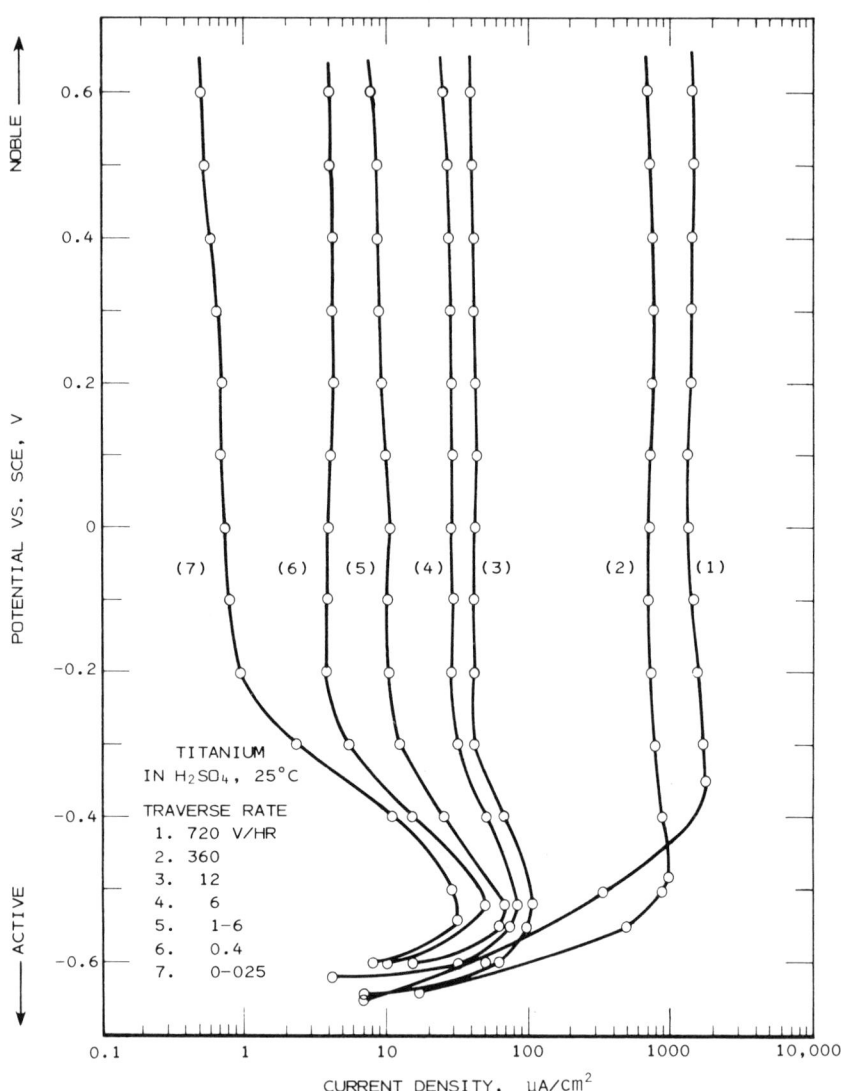

Figure 9-18. Potentiostatic anodic-polarization behavior of titanium in 1 *N* H_2SO_4 at 25°C determined by potential-sweep method.[21]

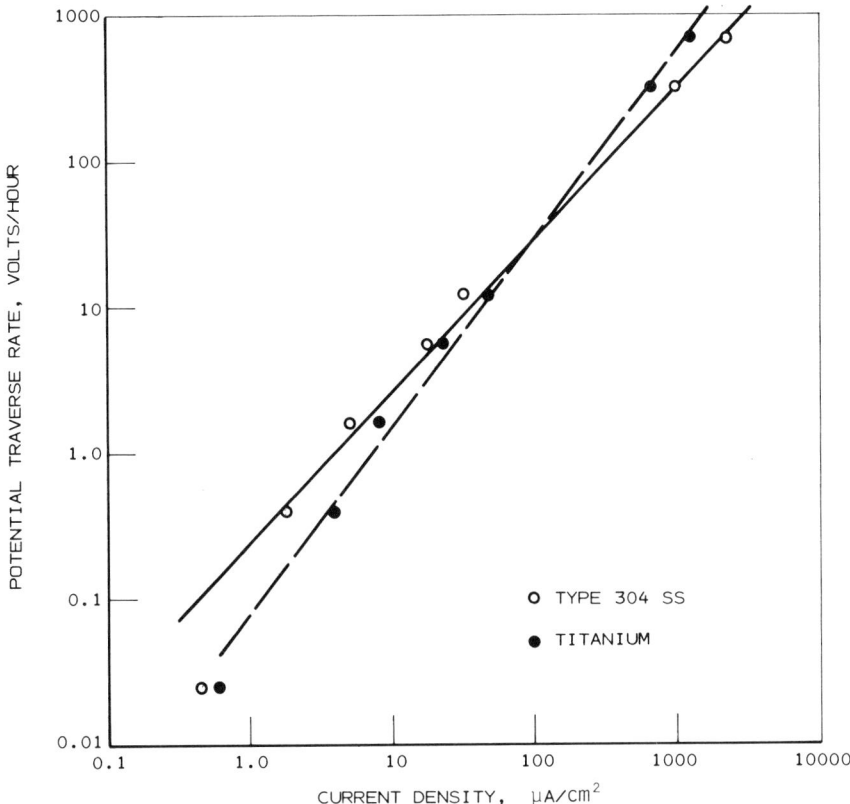

Figure 9-19. Effect of potential traverse rate on current density to maintain passivity for Type 304SS and titanium in 1 N H_2SO_4 at 25 °C.

References

1. N.D. Tomashov and G.P. Chernova, *Passivity and Protection of Metals Against Corrosion,* Plenum, New York (1967).
2. R. Oliver, Ph.D. Thesis, Leyden (1955).
3. M. Prazak, V. Prazak, and V. Cihal, *Z. Elektrochem.* **67**, 739 (1958).
4. N.D. Tomashov, G.P. Chernova, Y.S. Ruscol, and G.A. Ayuyan, *Electrochim. Acta* **19**, 159–172 (1974).
5. W.B. Crow, J.R. Meyers, and B.D. Marvin, *Corrosion* **27**, November (1971).
6. W.A. Mueller, Can. *J. Technol.* **34**, 162 (1956).
7. C. Edeleanu, *Metallurgia* **50**, 113 (1954).
8. J.D. Sudbury, O.L. Riggs, Jr., and D.A. Shock, *Corrosion* **16**, 47t (1960).
9. N.D. Greene, *First International Congress on Metallic Corrosion, London, 10–15 April 1961,* Butterworths, London (1961).

10. W.B. Crow, J.R. Meyers, and J.V. Jeffreys, *Corrosion* **28**, March (1972).
11. K. Elayaperumal, S.S. Chouthai, and J. Balachandra, *Corrosion* **29**, February (1974).
12. J.H. Payer and R.W. Staehle, *Corrosion* **31**, January (1975).
13. B.L. Trout and R.D. Daniels, *Corrosion* **28**, September (1972).
14. T.C. Finley and J.R. Meyers, *Corrosion* **26**, December (1970).
15. F. Mansfeld, *J. Electrochem. Soc.* **120**, February (1973).
16. F. Mansfeld, *J. Electrochem. Soc.* **121**, September (1974).
17. B. MacDougall and M. Cohen, *J. Electrochem. Soc.* **121**, September (1974).
18. R.L. Jones, L.W. Strattan, and E.O. Osgood, *Corrosion* **26**, October (1970).
19. O.L. Riggs, Jr., *Corrosion* **25**, April (1969).
20. O.L. Riggs, Jr., *Corrosion* **26**, June (1970).
21. N.D. Greene and R.B. Leonard, *Electrochim. Acta* **9**, 45–54 (1964).
22. R. Littlewood, *Corros. Sci.* **3**, 99–105 (1963).

Future Uses for Anodic Protection

When one attempts to project or predict anything in the future, it is much like sailing an uncharted voyage having informed the passengers of the high points which are anticipated prior to embarkation. This chapter is not too much different. However, if predictions are to be made, one trusts that some reasonable possibility exists for the events to occur. The possibility for occurrence becomes greater if the prediction is based on logic. Some of the conclusions on the following pages may seem impossible, or at best, ridiculous, but they do approach the realm of probability. Other information of practical merit will be sufficient or interesting that it will challenge the reader to examine "nontypical" systems for anodic protection applications. These objectives are consistent with the aim of this book to promote a wider use of anodic protection for the prevention of corrosion.

A Realm of Probability

Following is a list of probable applications whose merits are supported by experimental evidence.

Cooling-Tower Water

Because of increased emphasis on environmental pollution control, it is now common practice to find ways to discontinue the use of $Cr_2O_7^{2-}$ as an inhibitor of cooling-tower water corrosion. This suggests that an array of mesh-type carbon steel anodes could be installed in the blowdown circuit and that these anodes can be maintained at a predetermined set-point potential (E_{sp}) for the production of $Fe(OH)_2$. This iron hydroxide would be a reducing medium for $Cr^{2+} \rightarrow Cr^{3+}$ to remove it from the effluent.

Raney Catalyst

In time, activity of the Raney catalyst is impaired because of products on its surface which interfere with its activity. It is proposed that an anodic protection process could protect the catalyst from corrosion (being maintained at a protective set-point potential) while the deleterious contaminants would be retained in solution.

Prosthetic Implants

Research is underway to determine *in vitro* corrosion rates of various metal implants.[1-3] The most corrosion-resistant materials are generally used for this service. One of the difficulties with metallic prosthetic devices is the localized infectious disturbance of the tissue surrounding the implant, resulting from corrosion reactions. It is suggested that by combining proper alloy selection, with periodic reestablishment of passivity with anodic protection of the implant, a substantial reduction of this effect can be achieved.

Stainless Steel Pickling

During the processing of metals, a layer of millscale must be removed in most cases. This scale is usually removed by acid pickling. It is suggested that by combining the proper selection of pickling acid (suited to the respective metal) and anodically protecting the metal from the pickling solution, the millscale could be dissolved without significant attack on the base metal.

Molten Salt–Metal Technology

One of the very real problems in molten metal and salt technology is that the various heavy metals and salts in the molten state are severely corrosive to their respective containers. Recent discoveries[4] have indicated a reduction in Type 316 corrosion in molten-metal service by application of a zirconium nitride film. A possibility exists here for the application of anodic protection to prevent corrosion of the protective zirconium film. (See also Chapter 1, Reference 23.)

Metallography

Although little has been done in this area, it has been suggested[5,6] that anodic protection concepts are useful for controlled and selective etching of test specimens. This could be useful in promoting a better understanding of intercrystalline corrosion.

Coolant Systems

Periodic cleaning is required of heat exchangers in refineries, nuclear facilities, and elsewhere. When minimal base-metal removal is a prerequisite (i.e., nuclear-reactor coolant system), special cleaning solutions are required. It is here that anodic protection offers great possibilities. Successful applications would minimize radioactive waste disposal.

Solution-Potential Concept

During the process of acid digestion of ores, the resulting liquor often results in a negative solution potential (active corrosive). Often the active–passive steels in the plant experience severe dissolution.

It is suggested that these aggressively active liquors could be caused to pass through a properly designed electrolytic cell under anodic potential control, and the solution potential would be shifted to more positive (less active) values. The positive-solution-potential liquor would be rendered less corrosive, promoting longer plant life of the metallic units.

Some of the acidic solutions required for the digestion of uranium ores are extremely corrosive to even the most corrosion-resistant alloys. Even moderate rates of corrosion cause an undesirable contamination of the end product. Corrosion of metallic units used in the solution-extraction processing of uranium could be reduced to a minimum with the application of anodic protection systems.[7]

Sulfide Minerals

Controlling the oxidation rate of sulfide minerals in aqueous suspension is a prerequisite for the efficiency of many ore-treatment operations. Control of the oxidation rates is important to leaching of sulfide ores, cyanidation of gold ores, or leaching of uranium ores.[8] The application of anodic protection to specially designed processing units could not only protect the processing units, but provide the controlled potential necessary to promote predetermined oxidation rates for the processed ore.

Enzyme Production

Enxyme production is related to either oxidizing or reduction media. Because the most efficient production of a specific enzyme occurs at unique potentials, it is reasonable to assume that the most effective enzyme-production system would be that which was carried out under potentiostatic conditions. If so, then it is reasonable to assume that a metallic unit could be designed to affect the desired oxidation while the system is under anodic protection.

Surface-Area-Ratio Concept

Galvanic corrosion is the result of dissimilar metal contact in an electrolyte. Generally, this is a condition to be avoided. However, without this property, sacrificial anode types of cathodic protection could not be experienced.

Another galvanic condition which offers a type of anodic protection is the surface-area-ratio concept for large anodes and small cathodes. It is possible to rid the engineering design of galvanic corrosion by causing the dissimilar metals to work together for preservation. This is accomplished by designing the anode surface to be much greater than the cathode surface (a rule of thumb could be 100 to 1).

There are undoubtedly many more processes that could benefit from the use of anodic protection. It is now useful to consider possibilities which are close to the realm of possibility.

The Sufficiently Real Possibilities

During the early years of anodic protection application, essentially every unit protected was concerned with sulfuric acid in some manner. A side result of great economic importance was the improvement of product quality due to decreased concentrations of metal ions.[9] These metal contaminants went into solution (end product). Other contaminants were the coatings, scales, or deposits which adhere to the surface of the metal (end product). In the same way that anodic protection reduced solution contamination to provide a higher-quality product, it can safely and efficiently remove deposited contaminants.

Potentiostatic Conditioning of Electrodes

A primary problem with the electrodeposition of metals is the subsequent removal of electrodeposited metal from the cathodes. Residual amounts of deposited metal remain on cathodes after "harvesting," an amount which usually increases with successive cycles until such time that it becomes necessary to electropolish them because of surface roughening. The cycle is then repeated.

An accepted method for extending the interval before electropolishing is necessary to also help ease the removal of the deposited metal involves the chemical treatment of the cathode prior to cell assembly. The chemicals create a film-type barrier between the cathode and deposited metal to facilitate subsequent removal of the metal. This procedure is time consuming, requires additional labor, and is only partially effective.

A new concept has been reported[10,11] for conditioning the electrolytic cell following the harvest of the electrodeposited metal. The concept suggests that while the cathode is maintained at a predetermined reference potential (in a noncorroding state), the procedure forces a more rapid removal of residual electrodeposited metal. The method is both time and labor saving, as well as one which more effectively prepares the cathode for cell use without the need of additional chemicals. In addition to being applicable to copper electroplating, it can be used also in systems for plating cobalt, chromium, and nickel.[12]

The Cell Cathode

Several apparent problems involving the electrolytic cathode following the removal of the electrodeposited metal were considered. Generally, the cathode is moved to an acid dip bath where it is "soaked" until such time as all residual deposited metal has been removed. In some cases the cathode may be exposed to an increasingly corrosive environment, because the acid bath becomes more strongly reducing due to the dissolution of the electrodeposited residual metal. Most systems of this type are chemically treated to control this type of corrosion of the cathode.

Also, the residual metal deposit may have a geometry such that a concentration cell is formed on the cathode. The fissures or cracks in the deposit through which the electrolyte reaches the cathode promote a potential difference between the bulk cathode surface (higher O_2 access), essentially a crevice concentration cell. This promotes accelerated, localized corrosion and usually causes pitting of the cathode surface.

The Electrochemical Conditioning Concept

The potentiostatic technique can maintain the cathode at a passive potential while forcing the removal of the more electroactive deposited metal from its surface. Figure 10-1 presents a case where the active–passive metal undergoes anodic polarization (heavily lined curve), while the more electroactive metal deposit is under anodic activation polarization (fine-lined curve). The active–passive metal (the electrolytic cathode material) experiences anodic passivity. However, the electrodeposited metal (being active) cannot be anodically passivated. Remembering the previous discussion regarding the anodic-polarization curve, A and B are representative of potential regions (active and transpassive) which are undesirable in this case. The passive potential range is

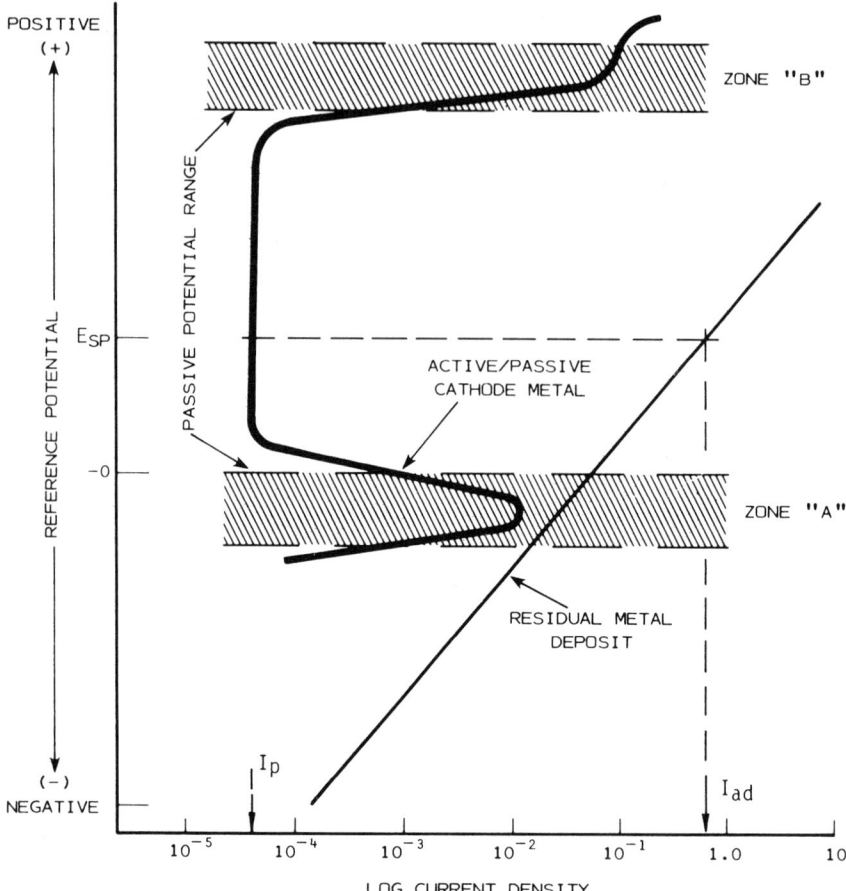

Figure 10-1. Anodic polarization of the active–passive metal or alloy and the corresponding electroactive metal deposit.

broad and provides a wide potential limit from which the desired set-point (E_{sp}) keeps the electrolytic cathode in a stable passive state and causes the potential of the electroactive metal deposit to be maintained at the same potential value, but forces the current to increase considerably (I_{ad}).

The path for this active dissolution current (I_{ad}) is through the deposits and this accelerates their removal from the passive electrolytic cathode surface. As the electrodeposited residual metal dissolves (is removed), the current decreases. The real benefits of this potentiostatic concept are as follows: (1) It provides automatic control of cathode cleaning and conditioning, eliminating manual operations and chemical additions, (2) it increases the dissolution rate of the metal deposit while maintaining the cathode in the passive state (non-corroding), and (3) the passive film formed improves the metal removal during the harvesting cycle.

So the last chapter has been concluded. It is the sincere hope of the authors that it does not represent the end. Rather, it is our hope that the reader might assume the philosophy underscored in a statement credited to Sir Winston Churchill during World War II. "This is not the end. It is not even the beginning of the end. But it is, perhaps, the end of the beginning."

References

1. B. C. Syrett, in *Electrochemical Techniques for Corrosion,* R. Baboian, ed., NACE, Houston, p. 93 (1977).
2. B. C. Syrett, *Corrosion* **33,** 221 (1977).
3. R. Bandy and J. R. Cahoon, *Corrosion* **33,** (6) (1977).
4. C. A. Rigby, *Anticorrosion* **XX,** 16, August (1977).
5. C. Edelenau, J. *Iron Steel Inst. London* **185,** 582, April (1957).
6. V. Cihal and M. Prazak, *Corrosion* **16** 530t-532t (1960).
7. F. P. A. Robinson and L. Golante, *Corrosion* **20,** 239, August (1964).
8. K. Lowe, National Institute of Metallurgy, Report No. **1509,** 5 June (1973).
9. O. L. Riggs, Jr., J. D. Sudbury, and D. A. Shock, *Corrosion* **16,** 475 (1960); O. L. Riggs, Jr., J. D. Sudbury, and D. A. Shock, *Corrosion* **16,** 55t (1960); O. L. Riggs, Jr., M. Hutchison, and N. L. Conger, *Corrosion* **16,** 58t (1960).
10. O. L. Riggs, Jr., and L. S. Surtees, *Met. Trans.* **7B,** 245, June (1976).
11. O. L. Riggs, Jr., and L. S. Surtees, U. S. Patent No. 3,826,724, July 30, 1974.
12. A. T. Kuhn, *Industrial Electrochemical Processes,* Elsevier, New York, 1971.

Appendixes

Considerable data have been presented within the foregoing chapters of this book and, where pertinent, the works of other scientists and engineers has been carefully referenced. Since it is essentially impossible to include the works of all scientists the world over, we have included as a unique aspect of anodic protection the following section for additional reading. It includes an abbreviated list of historical events which lead to U.S. patents relating to anodic protection, and a temporal collection of publications on anodic protection which have been accumulated over the past 25 years.

We are not unaware of the possibility of having overlooked pertinent literature. We welcome any and all articles and U.S. patents related to anodic protection which have not been included.

Electrochemical Principles of Corrosion

Corrosion

Metallic corrosion is the transformation of a metal into compounds formed by interaction with anions of the environment in which the metal is exposed. Since metals (alloys) are electronic conductors whose atoms incorporate cations and electrons that are readily dissociated, or produce ionically conductive species, almost all aqueous corrosion reactions are considered to be electrochemical in nature. This destructive process of the crystalline lattice of a metal in an aqueous environment is controlled by the kinetics of the electrochemical reaction which occurs at the metal–solution interface. This corrosion rate is the faradaic equivalent to the current flowing between the anodes and cathodes.

Aqueous corrosion of a metal occurs by two partial processes: An example is an overall anodic-oxidation reaction in which matrix iron atoms are ionized and go into solution leaving behind an electron(s),

$$Fe_{(metal)} \rightarrow 2e^-_{(metal)} + Fe^{2+}_{(aqueous)}$$

and the overall cathodic reduction reaction

$$\frac{1}{2} O_2 + H_2O + 2e^-_{(metal)} \rightarrow 2OH^-_{(aqueous)}$$

However, for iron corroding in a deaerated acid solution, the anode reaction is the same, but the cathode reaction is the reduction of hydrogen ion to hydrogen gas,

$$2H^+_{(aqueous)} + 2e^-_{(metal)} \rightarrow H_{2(gas)}$$

As soon as the electrons are liberated by the anodic reaction, they are consumed by the cathodic reaction which maintains the charge neutrality in the metal. The potential, then, at which the sum of the anodic and cathodic reaction rates are equal to zero,

$$i_a + i_c = 0$$

is defined as the corrosion potential, E_{corr}.

Electrode Terminology

Before going on further into corrosion processes, the confusion surrounding electrolytic cells and the related terminology should be discussed. Generally, the problem arises with the use of the terms anode and cathode, and positive and negative when referring to the electrodes of any electrochemical cell. The terms anode and cathode will be used in accordance with these definitions:

Anode—the electrode from which electrons leave the cell (oxidation).
Cathode—The electrode from which electrons enter the cell (reduction).

These definitions apply to both electrolytic cells, where current is passing through the cell from an external source, and to voltaic cells, which supply current to an external circuit.

The electrolytic cell is shown schematically in Figure A-1(A). The anode (electrons leave the cell) is connected to the positive terminal of the external power supply. The cathode (electrons enter the cell) is connected to the negative terminal of the external power supply. With the power on (forcing the flow of current), the positively charged cations migrate to the cathode, and the negatively charged anions migrate to the anode.

The schematic of a voltaic cell is shown in Figure A-1(B). Used as an emf source, the cathode in this type of cell is the positive electrode and the negative electrode is the anode. In any practical cell used as a source of power, the terms cathode and positive electrode, anode and negative electrode, are synonymous.

A secondary cell (battery) is a voltaic cell when supplying current, and an electrolytic cell when it is being charged. The electrode which was the cathode when the cell was discharging becomes the anode when the cell is on charge.

Figure A-2 further extends these concepts as they apply to the electrochemistry of metallic corrosion. As shown, the anode is attached to the positive terminal of the emf source. Also, the anode can be attached to the terminal of a high-impedance voltmeter whose other terminal is attached to a reference electrode (i.e., saturated calomel). Steady-state, open-circuit corrosion potentials can then be obtained. The cathode becomes the negative pole and is attached to the negative terminal of the emf source. In the scheme of electro-

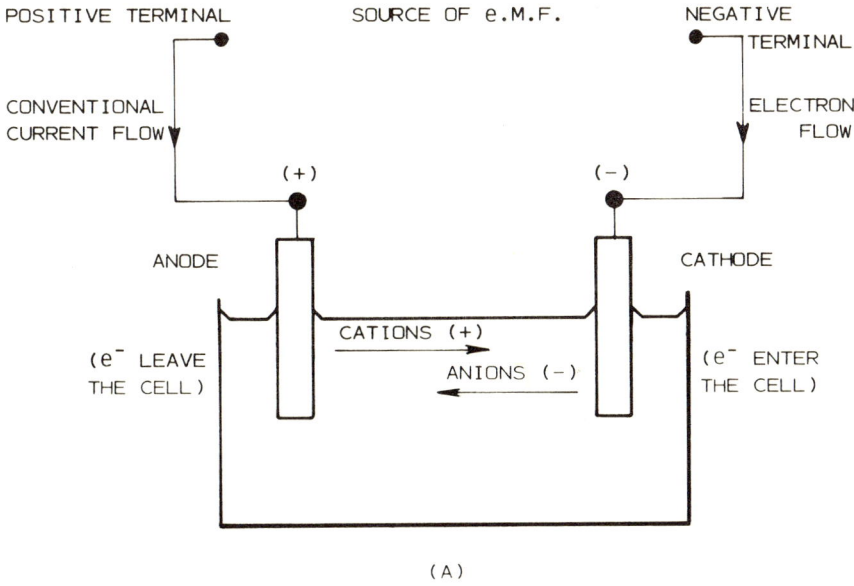

(A)

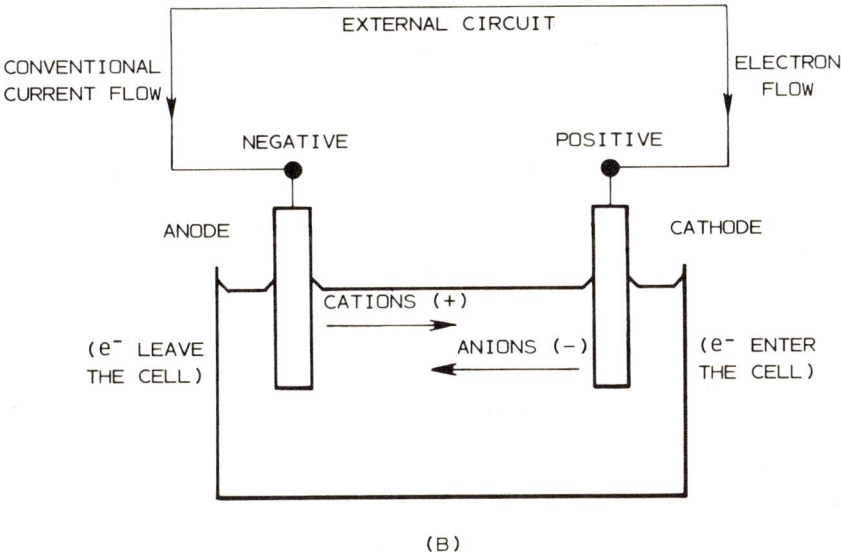

(B)

Figure A-1. Electrode terminology for the electrolytic and voltaic cells: (A) electrolytic cell; (B) voltaic cell.

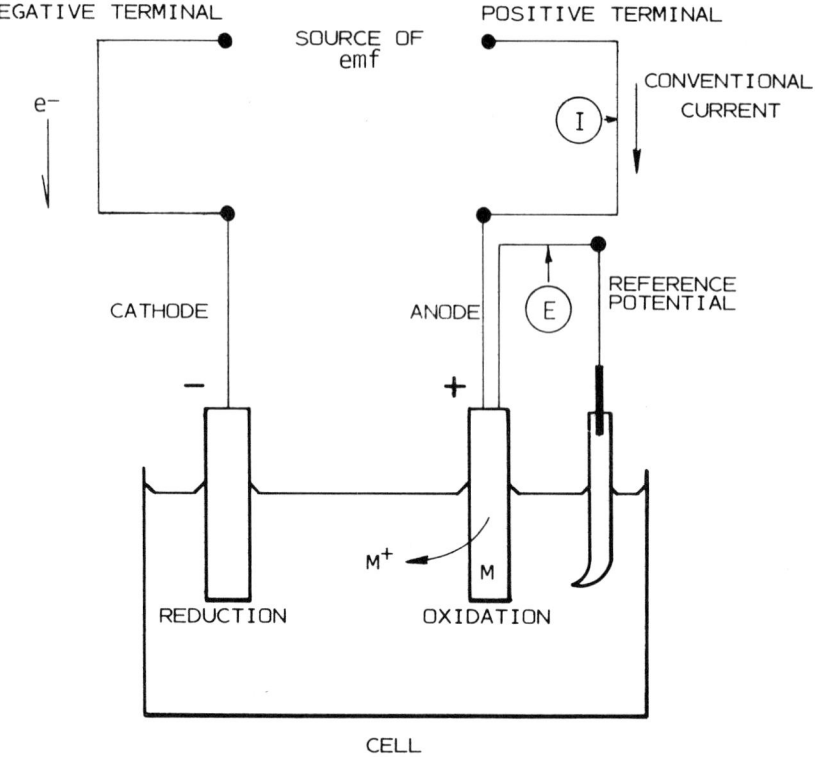

Figure A-2. Schematic of terminology for electrolytic corrosion of metals.

chemical corrosion processes, it is useful to remember that "oxidation always occurs at the anode" ($M \rightarrow M^+ + e^-$).

Potential Series

By immersing two electrodes, such as a strip of copper and a strip of iron, in a solution of the salts of both metals, an electrical potential difference can be created and current will flow between them when they are electrically connected.

It is convenient to consider, then, that the total emf of the cell is the difference of two "single-electrode potentials." That is, the two half-cell reactions

$$Fe^{2+} + 2e^- = Fe$$
$$2H^+ + 2e^- = H_2$$

or simply,

$$E_{cell} = E_{ox} - E_{red}$$

where E_{ox} is the single-electrode potential of the electrode forming the positive pole of the cell, and E_{red} is the single-electrode potential of the electrode forming the negative pole of the cell.

If a substance is more easily oxidized than hydrogen, its E_{ox} is assigned a positive value and its E_{red} a negative value. If a substance is not oxidized as easily as hydrogen, its E_{ox} is negative and E_{red} is positive in sign. The two electrodes will probably reach a potential equilibrium at different electrical potentials and this difference can be measured by the usual potentiometric techniques. A specific potential value will be measured, the value of which is related to the electrode composition. From these values, Table A-1 can be prepared,

TABLE A-1. Standard Potential Series[1,2]

Electrode	Reaction	E_{red}° (V)
Li^+, Li	$Li^+ + e^- \rightarrow Li$	-3.024
K^+, K	$K^+ + e^- \rightarrow K$	-2.924
Ca^{2+}, Ca	$Ca^{2+} + 2e^- \rightarrow Ca$	-2.87
Na^+, Na	$Na^+ + e^- \rightarrow Na$	-2.714
Mg^{2+}, Mg	$Mg^{2+} + 2e^- \rightarrow Mg$	-2.34
Ti^{2+}, Ti	$Ti^{2+} + 2e^- \rightarrow Ti$	-1.75
Al^{3+}, Al	$Al^{3+} + 3e^- \rightarrow Al$	-1.67
Mn^{2+}, Mn	$Mn^{2+} + 2e^- \rightarrow Mn$	-1.05
Zn^{2+}, Zn	$Zn^{2+} + 2e^- \rightarrow Zn$	-0.761
Cr^{3+}, Cr	$Cr^{3+} + 3e^- \rightarrow Cr$	-0.71
Fe^{2+}, Fe	$Fe^{2+} + 2e^- \rightarrow Fe$	-0.441
Co^{2+}, Co	$Co^{2+} + 2e^- \rightarrow Co$	-0.277
Ni^{2+}, Ni	$Ni^{2+} + 2e^- \rightarrow Ni$	-0.250
Sn^{2+}, Sn	$Sn^{2+} + 2e^- \rightarrow Sn$	-0.140
Pb^{2+}, Pb	$Pb^{2+} + 2e^- \rightarrow Pb$	-0.126
Fe^{3+}, Fe	$Fe^{3+} + 3e^- \rightarrow Fe$	-0.036
H^+, H_2	$2H^+ + 2e^- \rightarrow H_2$	0.000
Saturated calomel	$Hg_2Cl_2 + 2e^- \rightarrow 2Hg + 2Cl^-$ (saturated KCl)	0.244
Cu^{2+}, Cu	$Cu^{2+} + 2e^- \rightarrow Cu$	0.344
Cu^+, Cu	$Cu^+ + e^- \rightarrow Cu$	0.522
Hg_2^{2+}, Hg	$Hg_2^{2+} + 2e^- \rightarrow 2Hg$	0.798
Ag^+, Ag	$Ag^+ + e^- \rightarrow Ag$	0.799
Pd^+, Pd	$Pd^+ + e^- \rightarrow Pd$	0.83
Hg^+, Hg	$Hg^+ + e^- \rightarrow Hg$	0.854
Pt^{2+}, Pt	$Pt^{2+} + 2e^- \rightarrow Pt$	1.2 (ca.)
Au^{3+}, Au	$Au^{3+} + 3e^- \rightarrow Au$	1.42
Au^+, Au	$Au^+ + e^- \rightarrow Au$	1.68

where electrodes are arranged in a sequence such that each is negative in potential to the ones below.[1,2]

In this table, all electrode-potential values are measured with respect to the hydrogen electrode. The hydrogen electrode was arbitrarily assigned a value of zero volts. Potential measuring frequently uses other half-cells which are more easily accommodated to specific test systems. A common half-cell of this type is the saturated calomel electrode (SCE). Any half-cell, however, has certain limitations, i.e., temperature, common-ion effect, contamination, and dilution.

Nernst Equation

The standard hydrogen electrode (SHE) was arbitrarily assigned a potential value of zero. The potentials of the half-cells listed in Table A-1 are given compared to this electrode. The potentials of the electrodes listed in Table A-1 are for unit ion activities and a temperature of $25\,°C$.

The equilibrium potential of an electrode at other ionic activities and temperatures may be calculated using the Nernst equation. As an example, for the electrode reaction

$$M^{b+} + be^- \rightarrow M$$

$$E = E° - \frac{RT}{bF} \ln\left(\frac{a_M}{a_M{}^{b+}}\right)$$

where $E°$ = equilibrium potential from Table A-1, R = gas law constant, T = temperature, b = number of electrons transferred, F = faraday (96,500 coulombs/equivalent), $a_M a_M{}^{b+}$ = metal and ionic activities, a_m = 1 by definition. Therefore, for an iron electrode,

$$Fe^{2+} + 2e^- \rightarrow Fe$$

$$E = -0.441 - \frac{RT}{2F} \ln\left(\frac{1}{a_{Fe^{2+}}}\right)$$

We can use the Nernst equations written for each half-cell reaction in an electrochemical cell to calculate the overall cell potential. As an example, consider corroding iron on which the dissolution of iron occurs at the localized anodes and hydrogen is evolved at the cathode.

By convention, the cell is written

$$Fe, Fe^{2+} \,//\, H^+, H_2$$

From the IUPAC convention the cell potential is calculated by subtracting the potential of the left-hand side from the right-hand side:

$$\phi = E_{RHS} - E_{LHS}$$

The Nernst equation for each half-cell written in the reduction form is

$$\text{Iron: } Fe^{2+} + 2e^- \rightarrow Fe$$

$$E_{Fe} = -0.441 - \frac{RT}{2F} \ln \left(\frac{1}{a_{Fe^{2+}}} \right)$$

$$\text{Hydrogen: } 2H^+ + 2e^- \rightarrow H_2$$

$$E_{H^2} = 0 - \frac{RT}{2F} \ln \left(\frac{a_{H_2}}{a_{H^+}^2} \right)$$

[Note $a_{H_2} = P_{H_2}$ (partial pressure).]

$$\phi = E_{H^2} - E_{Fe}$$
$$= +0.441 - \frac{RT}{ZF} \ln \left(\frac{P_{H_2} a_{Fe^{2+}}}{a_{H^+}^2} \right)$$

If the overall cell potential is calculated in this is positive, the reactions will be spontaneous, as written. If it is negative, the reactions will spontaneously occur in the direction opposite to the way they are written, or an external voltage in excess of the calculated cell potential must be applied to force the reactions in the desired direction.

The standard hydrogen electrode is not a convenient experimental tool. Many other half-cells have been developed and are used widely by experimenters. The calomel cell is one of the most widely used and is based on the reaction

$$HgCl_2 + 2e^- \rightarrow Hg + 2Cl^-$$

The Nernst equation would be

$$E = E^\circ - \frac{RT}{2F} \ln \left(\frac{a_{Hg} a_{Cl^-}^2}{a_{HgCl_2}} \right)$$

By definition, $a_{HgCl_2} = 1$ because it is a solid and $a_{Hg} = 1$ because it is a pure metal. In many cases the activity is approximated by concentration, even though this is not exactly correct. Using this approximation and factoring the squared term,

$$E = E° - \frac{RT}{F} \ln M \qquad M = Cl^- \text{ normality}$$

Figure A-3 is a plot of the calomel-electrode potential as a function of the logarithm of the potassium normality, which illustrates the Nernst-equation relationship.

The ability of a metal to resist corrosion can be estimated to some extent by its relative position within the standard potential table. Another guide to the corrosion resistance of metals and alloys can be obtained from the so-called galvanic series in Table A-2. When two metals are coupled together and immersed in an electrolyte, an electrode potential difference occurs that results in an exchange of ions and electrons. A series can be prepared listing metals and alloys with positions relative to the potential measured in a given electrolyte. These potentials include both steady-state and reversible values, which permits both alloys and passive metals to be included.

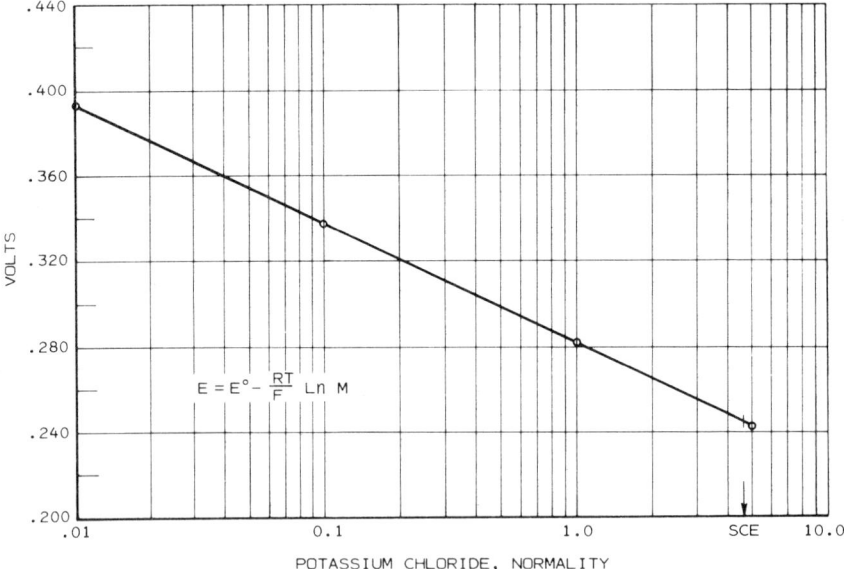

Figure A-3. Oxidation/reduction standard potential for calomel as a function of potassium chloride normality.

TABLE A-2. An Electromotive Series of Metals and Alloys in Seawater (Galvanic Series)[a][b]

Anodic end (least noble)	
Magnesium	Manganese bronze
Magnesium alloys	Naval brass
	Nickel (active)
Zinc	Inconel (active)
Galvanized steel	
Galvanized wrought iron	Yellow brass
	Admiralty brass
Aluminum 52SH, 45, 35, 25, or 535-T	Aluminum bronze
Alclad	Red brass
	Copper
Cadmium	Silicon bronze
	Ambrac
Aluminum Al7S-T, 17S-T, 24S-T	70–30 Copper nickel
Mild steel	Nickel (passive)
Wrought iron	Inconel (passive)
Cast iron	
	Monel
Ni Resist	
	18–8 stainless steel, Type 304 (passive)
13% chromium stainless steel Ind-410 (active)	18–8–3 stainless steel, Type 316 (passive)
50–50 lead tin solder	Silver
18–8 stainless steel, Type 304 (active)	Titanium
18–18–3 stainless steel, Type 316 (active)	Graphite
Lead	Gold
Tin	Platinum
Muntz metal	*Cathodic end* (most noble)

[a]For details see LaQue and Cox, Proceedings for the American Society for Testing Materials, Vol. 40, 17 (1940).
[b] Several galvanic series can be prepared according to changes in specific environment or differences in the tendency to form types of surface layers (or films).

The Electrical Double Layer

When a metal is exposed in an electrolyte, a double layer is formed at the metal–solution interface. There exists an excess charge on the metal side of the interface and an excessive concentration of one kind of anion on the solution side. The charge on the solution side is opposite the charge on the metal. Zero-

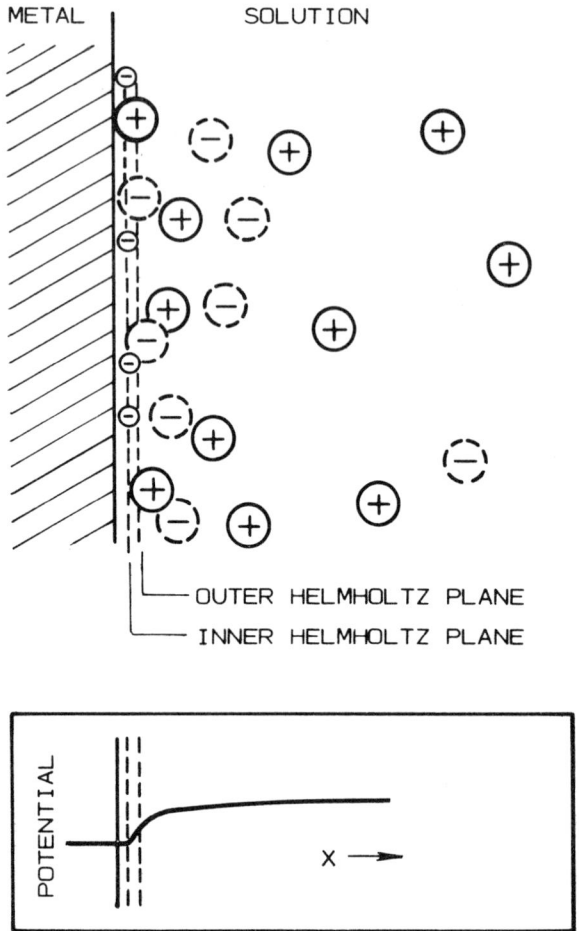

Figure A-4. Schematic representation of the electrical double layer at the potential of the electrocapillary maximum. Small circles represent adsorbed ions. Dotted circles represent "ghosts," ions which would be present if the double layer were not there (Ref. 5).

charge potential is then defined as the potential at which the metallic side of the double layer has no excess charge. Gileadi[3] has shown that the zero-charge potential is not only a property of the metal, but depends on the detailed composition of the whole system. Antropov[4] says that a distinction should be made between the null point of metals and the potential of zero charge. Grahame[5] and others[6-8] have used schematic diagrams of the electrical double layer very effectively to illustrate the boundaries, charges, and ionic concentrations. It is generally accepted that the electrical double layer consists of a diffuse (outer) layer called "outer Helmholtz plane" and an inner layer called the "inner Helmholtz plane."

In Figures A-4 through A-6, the large circles represent an excess of solvated ions. The dotted circles represent an ion type with concentration deficiency. The small circles (inner Helmholtz plane) represent nonsolvated excess ions. The positive or negative signs on the metal represent electron deficiencies,

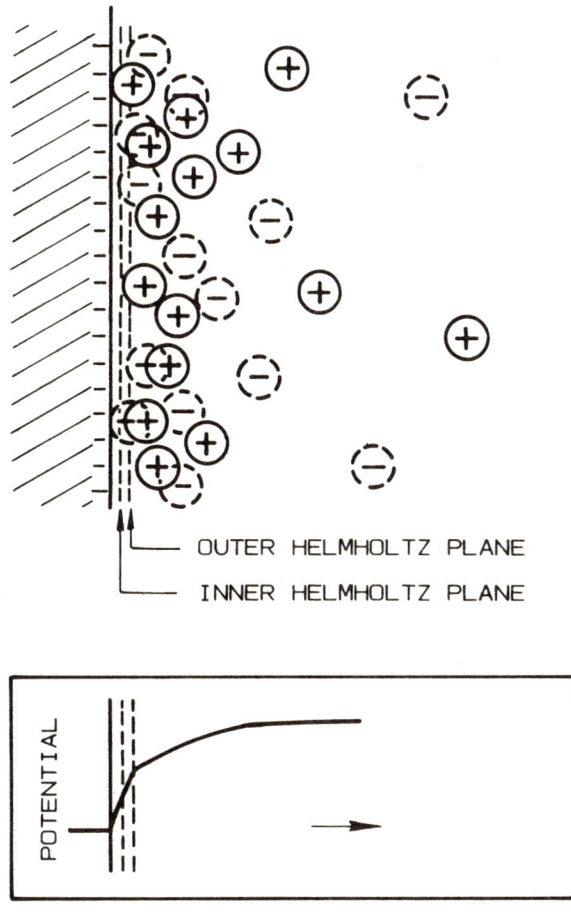

Figure A-5. Schematic representation of the electrical double layer with negative polarization. Note absence of adsorbed ions and increased concentration of positive ions as compared with Figure A-4. The concentration of "ghosts" is also increased (Ref. 5).

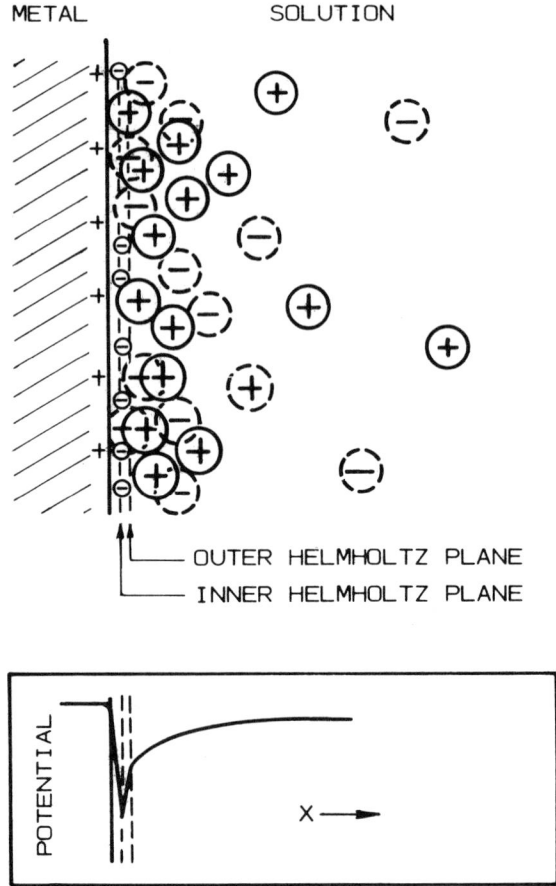

Figure A-6. Schematic representation of the electrical double layer with positive polarization. Note the presence of adsorbed anions. Diffuse double layer is identical with that depicted in Figure A-5 (Ref. 5).

or electrons. The change in potential as a function of distance is drawn schematically in the box below each figure. Figure A-4 diagrams the electrical double layer at zero-charge potential of the metal surface. Figures A-5 and A-6 show schematically the double layer with negative and positive polarization, respectively. These comparisons were schematically presented in Figure A-7, illustrating the relationship of both potential position and structure for the negative and positive surfaces with respect to the zero-charge potential surface. The processes (electron discharge and ionization) are related to the ion transition from a "hydrate" (aqua-ion) to a surface-absorbed atom and the reverse.

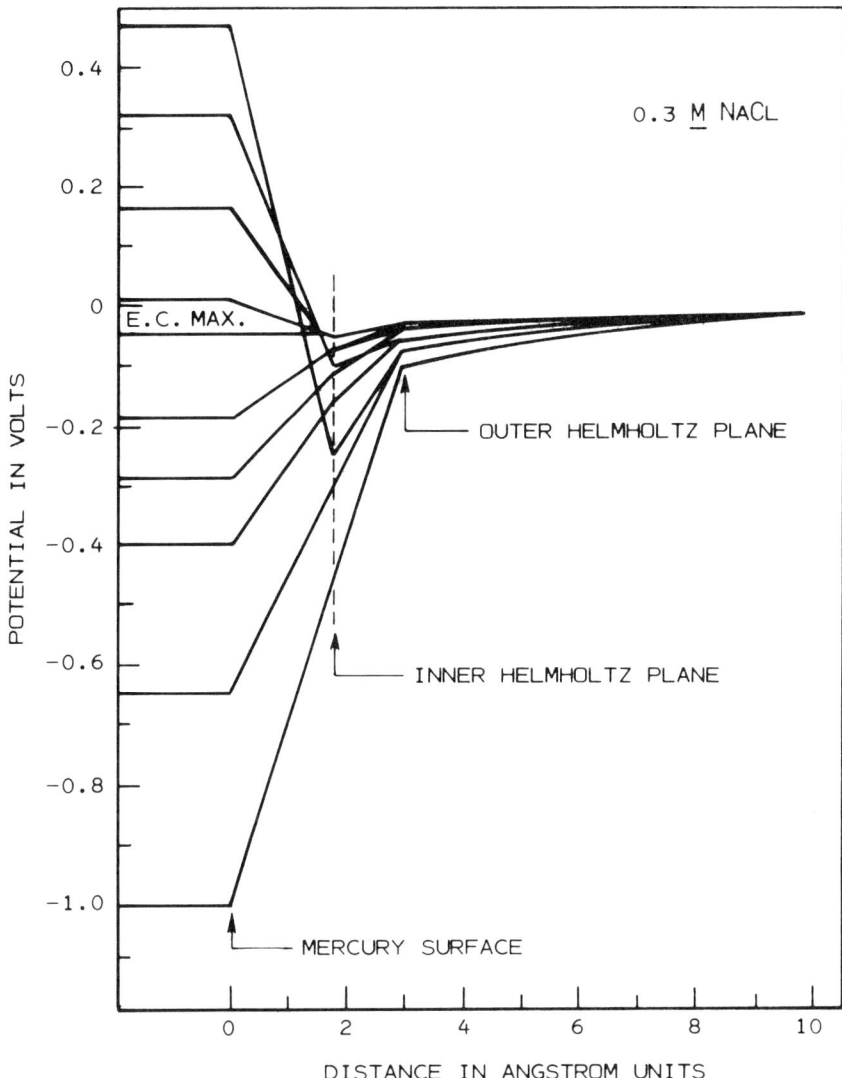

Figure A-7. Potentials in the electrical double layer between mercury and aqueous 0.3 M sodium chloride at 25°C at various polarizing potentials. Note that the potential of the inner Helmholtz plane reaches a maximum as the polarizing potential is varied from one extreme to the other (Ref. 5).

The electrical field within the double layer controls these directional processes. Levine, Mingins, and Bell[9] gave a detailed review of the electrical double layer with attention focused on the discreteness of charge, or discrete ion effect.

Free Energy

All changes in the nature of materials are caused by their tendency to reach a state of maximum stability comparable to the conditions of the system. Once this state has been reached (equilibrium), the tendency for further change is removed and the system is said to be stable. This tendency towards change increases in proportion to the distance of the system's state from equilibrium.

When considering a chemical system from the point of view that its constituents are electrically charged (ions or electrons), any change in charge distribution will require work. The system can perform maximum work only when the change is carried out reversibly. The force which causes the change is the maximum work difference between the final and initial states, which is the greatest energy possible from the process.

Every chemical entity possesses chemical free energy, $\overline{G}$. The electrically charged entity will also contain electrical energy $q\phi$, so the total energy can be expressed as

$$\overline{G} = G + q\phi$$

The electrical potential ϕ is the work expended in moving a unit of positive charge from infinity. The electrical charge is q. The quantity G is the electrochemical free energy, and for an electrically uncharged chemical substance it will be equal to the chemical free energy. Complete derivation is available from several sources.[10-14]

Morse curves effectively profile the chemical free energy of ions pulled out of the metal surface and then solvated. Figure A-8 illustrates the metal dissolution process when an ion (M) is "pulled out" of the surface into a polar solvent (water). The deep energy well for the metal ion bound to the metal surface corresponds to a second energy well for the metal ion surrounded by the primary solvation sheath. The M^{z+} aq (aqua-ion) can have up to six solvating water molecules. The ions or molecules that make up the primary solvation sheath depend on the kind of solvent, i.e., hydroxyl, ammonia, sulfate hydrate, hypophosphite, cyano, and others. The compressed solvation sheath is rigidly oriented about the metal ion and, in this manner, tends to shield it from further complexing ions. This, then, is the type of metal surface boundary that occurs during metallic corrosion and represents the electrical charges which must be accommodated for polarization (inhibition).

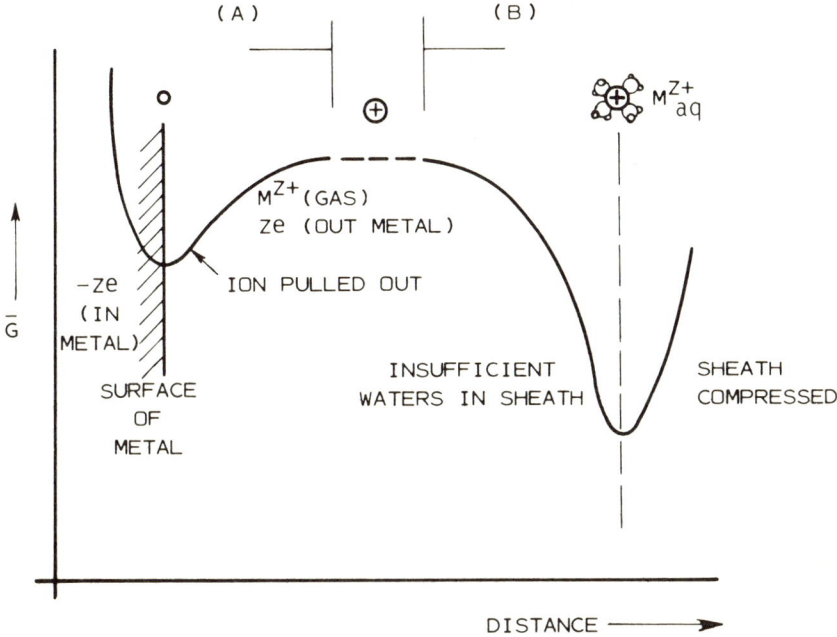

Figure A-8. A schematic morse curve showing the chemical free energy of ions pulled out of the metal (A) and subsequently solvated (B) (Ref. 11).

Polarization Diagrams

In 1905, Tafel[15] reported an empirical relationship between current density and overvoltage. He discovered that

$$\eta = a + b \log ip$$

where a is an empirical constant and b is the slope. Detailed derivations[16-19] of the electrochemical kinetics of activation polarization are reduced for simplicity and understanding of the mathematical determination of Tafel slopes from experimental data to

$$\frac{d\eta}{d \log ip} = \beta$$

The Tafel slope β is obtained by plotting the log of applied current ip versus the reference potential η. These slopes can be measured for either cathodic polarization (β_c) or anodic polarization (β_a).

When a metal electrode is in equilibrium, the anodic (i_a) currents and cathodic (i_c) currents are equal and no net reaction occurs. In other words, during the corrosion of an electrically isolated metal sample, the total rate of *oxidation* must equal the total rate of *reduction*. If this equilibrium is altered by imposing an external emf, the metal surface may become polarized. The metal can be either anodically polarized (electrons are withdrawn from the metal and a net anodic current will flow), or cathodically polarized (electrons are pushed into the metal and a net cathodic current will flow).

Polarization is associated with two processes: a net flow of current and a net shift of the electrode potential from equilibrium. These phenomena are basic in the electrochemical kinetics of corrosion. Application of these basics permits electrochemical polarization to be divided into (1) activation polarization and (2) concentration polarization.

Another term pertinent to the electrochemistry of the corrosion process is polarization resistance R. This resistance is related to the ease with which an electron may be transferred across the solution–electrode interface.

Stern[19] published an equation which expresses the β_a, β_c, and R as pertinent functions in determining corrosion current (i_{corr}):

$$I_{corr} = \frac{\beta_a \beta_c}{2.303\,R(\beta_a + \beta_c)}$$

For a given metal–solution interface, the Tafel slopes β_a and β_c can be assumed as constants and the corrosion current is the reciprocal of the polarization resistance $1/R$.

The various functions are schematically shown in the polarization curves of Figure A-9.

The rates of the cathodic and anodic corrosion reactions have been well defined by Hackerman and Hurd[20] as

$$i_{c,H} = k_{c,H} a_{H_3O^+}^y \exp\left(-\frac{F}{RT}\left[\alpha_{c,H}\Psi_c + (1 - \alpha_{c,H})\Psi\right] \right)$$

for the cathodic reaction of the hydrogen-ion discharge, and, as indicated in Eq. 8, for the anodic oxidation of iron,

$$i_{a,Fe} = k_{a,Fe} a_{OH^-}^x \exp\left(\frac{F}{RT}\alpha_{a,Fe}n(\Psi_a - \Psi) \right)$$

$i_{c,H}$ and $i_{a,Fe}$, expressed in electrical units, are the rates of the cathodic and anodic corrosion reactions, $\alpha_{c,H}$ and $\alpha_{a,Fe}$ are the transfer coefficients, Ψ is the potential drop in the outer boundary of the electrochemical double layer, Ψ_c or

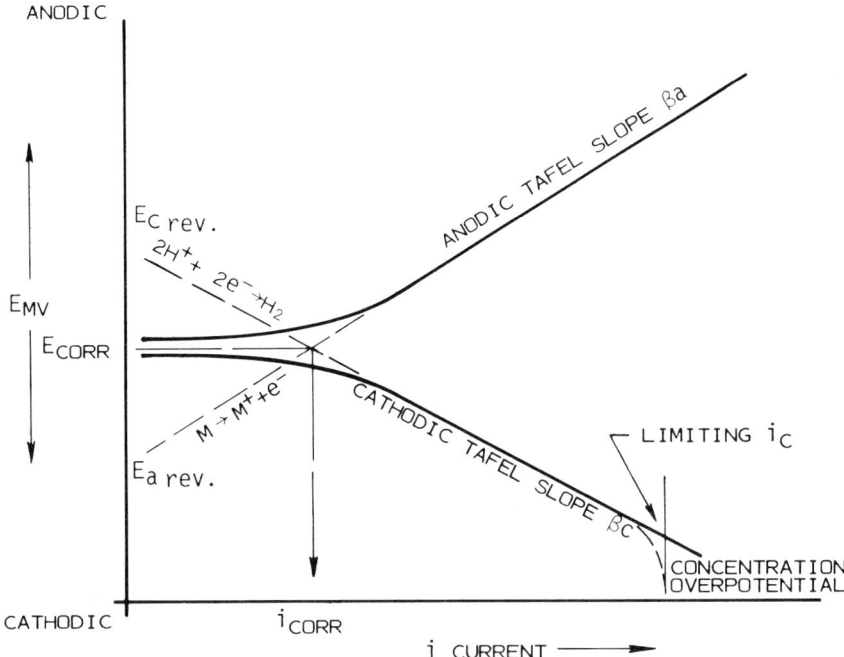

Figure A-9. Schematic activation polarization curves showing the pertinent processes for the metal–solution under impressed emf conditions.

Ψ_a is the electrode potential with respect to the solution potential, $a_{H_3O^+}$ and a_{OH^-} are the activities of the H_3O^+ and OH^- species, and x and y are the electrochemical reaction orders with respect to H_3O^+ and OH^-. A very useful review and derivation were published by Makrides.[21] The chemical portion of the standard free energy of activation is included in terms k_c and k_a of Eq. 7 and 8.

Figure A-10(A) illustrates the significant terms for a freely corroding metal. The line E_aD represents anodic reactions, line E_cD represents cathodic reactions. The point of intersection of anodic and cathodic reactions (D) establishes the open-circuit corrosion potential E_{corr} of the metal and indicates the magnitude of corrosion at i_{corr} (corrosion current).

Figure A-10(B) is a schematic diagram showing the relation of metallic corrosion (D), and cathodic (G) and anodic protection (F). Generally, an immersed metal may be corroding by reactions under anodic control (E_aF) or under cathodic control (E_cG). It is readily apparent that both reactions can be controlled by either anodic or cathodic protection.

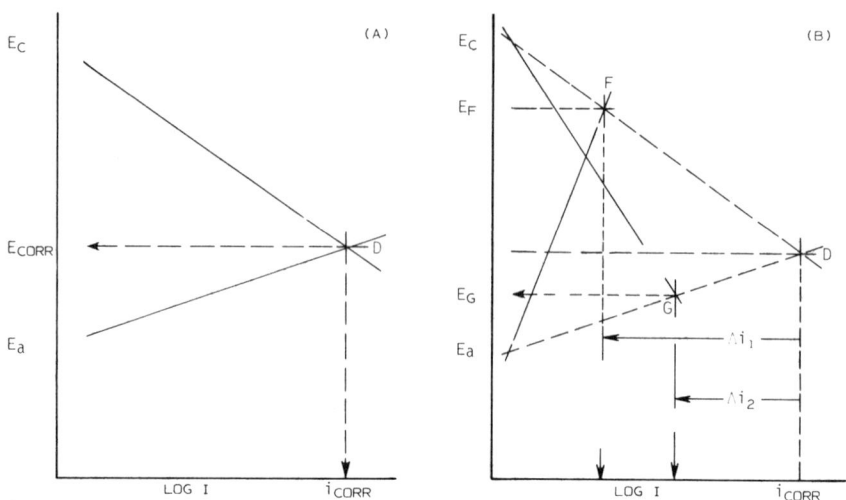

Figure A-10. Anodic- and cathodic-polarization curves. (A) Illustrates significant terms for freely corroding metal. (B) Schematic diagram showing relation of metallic corrosion with cathodic and anodic protection.

Electrochemical Equivalents of Metals

The electrochemical equivalent of an element is based on its atomic weight, the change of valence in the reaction being considered, and the faraday value. The value of the faraday, however, is based on the atomic weight of silver, because the definition of the international ampere is that amount of current which will deposit 1.1180 milligrams of silver in one second. The atomic weight of silver is 107.870 ± 0.003 (by C^{12} analysis) and this converts to 107.870/0.0011180 = 96,484.8 coulombs as the value of the faraday.

The accuracy of this value is claimed by the Bureau of Standards to be ± 10 coulombs, but, for convenience, is generally rounded off to 96,500 coulombs. This is the value upon which these calculations are based.

Conversion Factors

The combination of the faraday law with the equation for metal loss due to corrosion, equated to the corrosion rate, is in mils per year (mpy) and the following is developed for iron:

$$C.\ R.\ (mpy)\ =\ \frac{1\ A}{cm^2}\ \times\ \frac{gew}{96,500\ C}$$

$$\times\ \frac{\frac{1}{2}\ 55.85\ g}{gew(Fe^{2+})}\ \times\ density\ (Fe)\ \frac{cm^3}{g}\ \times\ \frac{1000\ mil}{2.54\ cm}$$

$$\times\ \frac{3600\ s}{1\ h}\ \times\ \frac{24\ h\ \times\ 365\ days}{yr}$$

Cancelling like units and collecting terms, this equation reduces to the electro-chemical corrosion factor (Table A-3) for iron of

$$1\ \mu A/cm^2\ =\ 0.46\ mpy$$

This factor combined with the Stern–Geary equation (using corrosion current) permits the direct calculation of the corrosion rate:

$$C.\ R.\ (mpy)\ =\ \frac{0.46\ \times\ 10^6 \beta_a \beta_c}{2.303\,R(\beta_a\ +\ \beta_c)}$$

Since the surface area A (in cm^2) for the respective electrode is not included in the equation, the resulting values must be divided by A. This procedure permits the use of electrodes of various surface areas.

The primary function in the calculations is

$$\frac{at.\ wt.}{valence\ \times\ 96,500}\ =\ mg/C$$

It is of no consequence whether the valence is a positive or negative value. The results are identical in either case. However, since generally the reactions of metals (electrochemical corrosion) are the chief concern, the valence value will be positive.

Partial Electrochemical Equivalent

A conceptual portion of the electrochemistry of corrosion is the "partial electrochemical equivalent," which teaches that the corrosion rate of an alloy metal is the sum total of the contributing rates of metallic constituents.

The faradaic weight-loss equation for corrosion-rate calculations (after collecting terms and cancelling units) reduces to

$$1 \ \mu A/cm^2 = 0.128 \left(\frac{\text{at. wt. of element}}{\text{valence} \times \text{density}} \right) \text{mpy}$$

The valence value is the initial oxidation state of the element in question. Generally speaking, this should be its lowest oxidation state. However, the more oxidizing a solution potential is, the greater will be the tendency for the next higher oxidation state to control the corrosion process. There also is a possibility that higher valences will be produced.

Electrochemical corrosion-rate conversion factors have been prepared for several metals and alloys based on the partial-electrochemical-equivalent concept (Table A-3). For the alloys, the conversion factor was obtained by (for each element)

$$1 \ \mu A/cm^2 = 0.128 \left(\frac{\text{at. wt. of element}}{\text{valence} \times \text{density}} \right)$$
$$\times \text{(percent of element in alloy)} \times \text{mpy}$$

For example, AISI Type 316 conversion factor was calculated as shown in Table A-4.

TABLE A-3. Conversion Factors for Selected Metals

Metal	Density	Outer-shell valence	Atomic weight	Conversion factor
Titanium	4.50	2	47.90	0.68
Vanadium	5.96	2	50.95	0.53
Chromium	7.10	1	52.01	0.93
Manganese	7.20	2	54.93	0.48
Iron	7.86	2	55.85	0.45
Cobalt	8.90	2	58.94	0.42
Niobium	8.57	1	92.91	1.39
Molybdenum	10.20	1	95.95	1.20
Nickel	8.90	2	58.69	0.42
Copper	8.92	1	63.54	0.91
Aluminum	2.702	3	26.98	0.43
Magnesium	1.74	2	24.32	0.89
Zinc	7.14	2	65.38	0.58
Zirconium	6.40	2	91.22	0.91
Tantalum	16.60	2	180.88	0.69
Type 304 SS	8.02	—	—	0.52
Type 316 SS	8.02	—	—	0.54

TABLE A-4. Electrochemical Conversion-Rate Factors for
AISI Type 316 Metal

Elements	Percent	Valence	Density	At. wt.
Cr	18	1	7.1	52.01
Ni	8	2	8.9	58.69
Mo	3	1	10.2	95.95
Fe	70	2	7.86	55.85

Substituting values for terms in the partial-electrochemical-equivalent expression,

$$1\ \mu A/cm^2 = \left(\frac{0.128(52.01)}{1(7.1)}\right) 0.18 + \left(\frac{0.128(58.69)}{2(8.9)}\right) 0.08$$
$$+ \left(\frac{0.128(95.95)}{2(10.2)}\right) 0.03 + \left(\frac{0.128(55.85)}{2(7.86)}\right) 0.70\ mpy$$
$$= 0.169 + 0.034 + 0.018 + 0.318\ mpy$$

$1\ \mu A/cm^2 = 0.54$ mpy for Type 316SS.

Another approach utilizes the electrochemical equivalent (mg/C),

$$(\text{electrochemical equivalent, mg/C}) = \left(\frac{12.42}{\text{density } d \text{ of metal}}\right) \% \text{ of element}$$

The conversion factor then for AISI Type 316 is calculated as follows:

$$1\ \mu A/cm^2 = \left(\frac{0.53896(12.42)}{d_{Cr}}\right) 0.18 + \left(\frac{0.30409(12.42)}{d_{Ni}}\right) 0.08$$
$$+ \left(\frac{0.28933(12.42)}{d_{Fe}}\right) 0.70 + \left(\frac{0.99430(12.42)}{d_{Mo}}\right) 0.03\ mpy$$

$1\ \mu A/cm^2 = 0.160 + 0.034 + 0.318 + 0.018$ mpy

$1\ \mu A/cm^2 = 0.54$ mpy for Type 316SS.

References

1. N.A. Lange, *Handbook of Chemistry*, 11th ed., McGraw-Hill, New York, 1967.
2. *The Oxidation States of the Elements*, W.M. Latimer, Prentice-Hall, New York, 1938.
3. E. Gileadi, *Electrosorption*, Plenum, New York, 1967.

4. L.I. Antropov, *First International Congress on Metallic Corrosion,* Butterworths, London, 1961.

5. D.C. Grahame, *Chem. Rev.* **41**, 441 (1947).

6. R.J. Watts-Tobin, *Philos. Mag.* **6**, 133, January (1961).

7. R. Smolouchowski, *Phys. Rev.* **60**, 661, November 1 (1941).

8. M.A.V. Devanathan, *Trans. Faraday Soc.* **50**, 373 (1954).

9. S. Levine, J. Mingins, and G.M. Bell, *J. Electroanal. Chem.* **13**, 280–329 (1967).

10. J.A.V. Butler, *Electrical Phenomena at Interfaces,* Methuen, London, 1951.

11. J.M. West, *Electrodeposition and Corrosion Processes,* D. Van Nostrand, London (1965).

12. C.F. Pruton and S.H. Maron, *Fundamental Principles of Physical Chemistry,* p. 309, Macmillan, New York (1947).

13. F.H. Getman and F. Daniels, *Outlines of Physical Chemistry,* 7th ed., p. 132, John Wiley and Sons, New York (1947).

14. S. Glasstone, *Textbook of Physical Chemistry,* 2nd ed., p. 229, D. Van Nostrand, New York (1948).

15. J. Tafel, *Z. Phys. Chem.,* p. 641–712 (1905).

16. J.O'M. Bockris, *Modern Aspects of Electrochemistry,* p. 180, Butterworth Scientific, London (1954).

17. J.A.V. Butler, *Trans. Faraday Soc.* **19**, 729–734 (1924).

18. K.J. Vetter, *Electrochemical Kinetics,* Academic, New York (1967).

19. M. Stern and A.L. Geary, *J. Electrochem. Soc.* **104**, 1 (1957).

20. N. Hackerman and R.M. Hurd, First International Congress on Metallic Corrosion, p. 313, Butterworths, London (1962).

21. A.C. Makrides, *Corrosion* **18**, 338t–349t, September (1962).

Suggested Reading

J.C. Scully, *The Fundamentals of Corrosion,* Pergamon Press, Oxford (1966).

L.L. Shreir *et al., Corrosion, Vol. I and Vol. II,* Butterworth–Newnes, London (1976).

J.M. West, *Electrodeposition and Corrosion Processes,* D. Van Nostrand, New York (1965).

H.H. Uhlig, *Corrosion and Corrosion Control,* Wiley, New York (1963).

N. Tomashov, *Theory of Corrosion and Protection of Metals* (Translation), MacMillan, New York (1966).

M.G. Fontana and N.D. Greene, *Corrosion Engineering,* 2nd ed., McGraw-Hill, New York (1978).

John O'M. Bockris and Amulya K.N. Reddy, *Modern Electrochemistry,* Plenum, New York (1970).

K.J. Vetter, *Electrochemical Kinetics—Theoretical and Experimental Aspects* (Translation), Academic, New York (1967).

U.R. Evans, *The Corrosion and Oxidation of Metals,* Edward Arnold, London (1960).

C.A. Coulson, *Valence,* 2nd ed., Oxford University Press, Amen House (1961).

A.K. Vijh, *Electrochemistry of Metals and Semiconductors,* Marcel Dekker, New York (1973).

Glossary

Activation overpotential: That part of polarization that exists across the electrical double layer at an electrode–solution interface and thus directly influences the rate of the electrode process by altering its activation energy.

Active: Freely corroding.

Anaerobic: Free of air or uncombined oxygen.

Anion: A negatively charged ion; it migrates to the anode in a galvanic or voltaic cell.

Anode: The electrode of a galvanic or voltaic cell where the positive current flows from the electrode to the solution.

Anode polarization: A reduction of the corrosion rate by making the potential of the metal sufficiently more electropositive by an external source of emf. (and maintaining the potential) so that the metal becomes passive.

Anodic protection: A technique to reduce corrosion of a metal surface under some conditions by passing sufficient anode current to it to cause its electrode potential to enter and remain in the passive region.

Anolyte: The electrolyte of an electrolytic cell adjacent to the anode.

Auxiliary electrode: An electrode commonly used in polarization studies to pass current to or from a test electrode. It is usually made out of a noncorroding material.

Cathode: The electrode of a galvanic or voltaic cell where positive current flows from the solution to the electrode (by transfer of cations from solution to electrode).

Cathodic protection: A technique to reduce corrosion of a metal surface by passing sufficient cathodic current to it to cause its anodic dissolution rate to become negligible.

Cation: A positively charged ion; it migrates to the cathode in a galvanic or voltaic cell.

Cell: Electrochemical system consisting of an anode and a cathode immersed in an electrolyte. The anode and a cathode may be separate metals or dissimilar areas on the same metal. When the electrodes are in electrical contact with each other they develop a difference in potential which causes current to flow and produce corrosion at the anode. A cell involving an electrolyte in the corrosion process is referred to as an electrolytic cell.

Corrosion: The transformation of a metal from the elementary to the combined state by reaction with a non-metal.

Corrosion fatigue limit: The maximum stress endured by a metal without failure in a stated number of stress applications under defined conditions of corrosion and stress.

Corrosion potential: The potential of a corroding surface in an electrolyte, relative to a reference electrode. Synonyms: rest potential, open-circuit potential, freely corroding potential.

Corrosion rate: The rate at which corrosion proceeds; generally expressed in terms of inches per year (ipy) or mils per year (mpy).

Current: The rate of transfer of a charge; unit current is the ampere (A) which is at the transfer of 1 C/s.

Current density: The current per unit area (geometric) of the surface of an electrode expressed in units of mA/cm^2, mA/in^2, etc.

Depolarization: Elimination of the electrode polarization needed to produce a specified current.

Deposit: A foreign substance, which comes from the environment, adhering to the surface of a material.

Double layer: The interface between the electrode and the electrolyte where charge separation takes place. The simplest model is represented by a parallel-plate condenser of 2×10^{-8} cm in thickness. In general the electrode will be positively charged with respect to the solution.

Electrochemical potential, electrochemical tension: The partial derivative of the total electrochemical free energy of a constituent with respect to the number of moles of this constituent where all factors are kept constant. It is analogous to the chemical potential of a constituent except that it includes electric, as well as chemical, contributions to the free energy.

Electrode: An electron conductor by means of which electrons are provided for, or removed from, an electrode reaction.

Electrode potential: The difference in electrical potential between an electrode and the electrolyte with which it is in contact. It is best given with reference to the standard hydrogen electrode.

Electrokinetic potential: This potential, often called the zeta potential, is a potential difference in the solution caused by residual, unbalanced charge on an adjacent surface. This charge causes a countercharge distribution in the adjoining solution, producing a double layer. The electrokinetic potential causes the effects of electrophoresis, electro-osmosis, streaming potential, and sedimentation potential. The electrokinetic potential is different from the electrode potential in that it occurs exclusively in the solution phase; i.e., it represents the reversible work necessary to bring a unit charge from infinity in the solution up to the interface in question, but not through the interface.

Electrolysis: The chemical change in an electrolyte resulting from the passage of electricity.

Electrolyte: A substance which in solution gives rise to ions.

Electrolytic cleaning: The process of degreasing or descaling a metal by making it an electrode in a suitable bath.

emf. series: A table of the standard equilibrium electrode potentials of systems of the type $M^{z+} + ze^- = M$ relative to the standard hydrogen electrode.

Environment: The surroundings or conditions (physical, chemical, mechanical) in which a material exists.

Equilibrium potential: The electrode potential of an unpolarized reversible electrode.

Erosion corrosion: A corrosion reaction accelerated by velocity and air/particle abrasion.

Exchange-current density (i_0): The rate of exchange of electrons (expressed as current per cm^2) between the two species concerned in a reversible electrode process at the equilibrium potential.

Flade potential: The potential at which a metal which is passive becomes active.

Galvanic corrosion: Corrosion associated with the current resulting from the coupling of dissimilar electrodes in an electrolyte. Also known as couple action.

Galvanic series: A list of metals and alloys arranged according to their relative corrosion potentials in a given environment.

Galvanostatic: Refers to the constant-current technique of applying current to a specimen in an electrolyte. Synonym: intentiostatic.

General corrosion: A form of deterioration that is distributed more or less uniformly over a surface.

Half-cell: One of the electrodes, and its immediate environment, in a electrolytic cell.

Hydrogen overvoltage: Overvoltage associated with the liberation of hydrogen gas.

Immunity: A state of resistance to corrosion or anodic dissolution caused by the fact that the electrode potential of the surface in question is below the equilibrium potential for anodic dissolution.

Intergranular corrosion: Preferential corrosion at grain boundaries of a metal or alloy. Also called intercrystalline corrosion.

Ion: An electrically charged atom or couplet of atoms.

Limiting current density: The current density at which change of polarization produces little or no change of current density.

Luggin probe or Luggin–Haber capillary: A scheme for measuring the potential of a specimen with a significant current density imposed on its surface. The purpose of the probe is to minimize the *IR* drop that is included in the measurement without significantly disturbing the current distribution on the specimen.

Mil: One thousandth of an inch (1 mil = 0.001 in.).

Mill scale: The heavy oxide layer formed during hot fabrication or heat treatment of metals and alloys.

Mixed potential: A potential resulting from two or more electrochemical reactions occurring simultaneously on one metal surface.

Nernst layer and Nernst thickness: The diffusion layer or the thickness of this layer as given by the theory or Nernst. It is defined by

$$i_d = n \, F \, D \, \frac{C^0 - C}{\delta}$$

where i_d = the diffusion-limited current density, D = the diffusion coefficient, C^0, = the concentration at the electrode surface, and δ = the Nernst thickness. It is a hypothetical thickness which has been found to be 0.05 cm in many cases of unstirred aqueous electrolytes.

Noble: Referring to the positive direction of electrode potential, thus resembling noble metals such as gold and platinum. Antonym: active.

Noble potential: A potential more cathodic (positive) than the standard hydrogen potential.

Overvoltage: The displacement of the equilibrium electrode potential required to cause a reaction to proceed at a given rate.

Oxidation: Loss of electrons by a constituent of a chemical reaction.

Oxygen concentration cell: A galvanic cell resulting from the difference in oxygen concentration between two locations.

Passive–active cell: A cell, the emf of which is due to the potential difference between a metal in an active state and the same metal in a passive state.

Passivity: A metal or alloy which is thermodynamically unstable in a given electrolyte solution is said to be passive when it remains visibly unchanged for a prolonged period.

Pits: Corrosion of a metal surface, confined to a point or small area, which takes the form of cavities.

Polarization: The difference of the potential of an electrode from its equilibrium or steady-state potential.

Potential: Base—A potential towards the negative end of a scale of electrode potentials. *Negative*—A potential more negative than the potential of the standard hydrogen electrode. *Noble*—A potential towards the positive end of a scale of electrode potentials. *Positive*—A potential more positive than the potential of the standard hydrogen electrode.

Reference electrode: A half-cell of reproducible potential by means of which an unknown electrode potential can be determined on some arbitrary scale.

Reversible electrode: An electrode in which a small increase or decrease in potential can reverse the direction of the electrode reaction.

Reversible potential: See under Equilibrium Potential.

Rust: Corrosion product consisting primarily of hydrated iron oxide—a term properly applied only to iron and ferrous alloys.

Scaling: The formation at high temperatures of thick corrosion-product layers on a metal surface. The deposition of water-insoluble constituents on a metal surface.

Standard electrode potential: The reversible potential for an electrode process when all products and reactions are at unit activity on a scale in which the potential for the standard hydrogen half-cell is zero.

Steady-state potential: The potential of an electrode that is operating under steady-state conditions of zero- or constant-current density.

Tafel line, Tafel slope, Tafel diagram: When an electrode is polarized, it will frequently yield a current–potential relationship over a region which can be approximated by

$$\eta = \pm B \log (i/i_0)$$

where η = change in open-circuit potential, i = the current density, B and i_0 = constants. The constant B is also known as the Tafel slope. If this behavior is observed on a plot of semilogarithmic components, the curve is known as the Tafel line and the overall diagram is termed a Tafel diagram.

Throwing power: The relationship between the current density at a point on the specimen and its distance from the counterelectrode. The greater the ratio of the surface resistivity shown by the electrode reaction to the volume of resistivity of the electrolyte, the better is the throwing power of the process.

Transpassive: The noble region of potential where an electrode exhibits a higher than passive current density.

Transpassivity: The active behavior of a metal at potentials more positive than those leading to passivity.

Voltaic cell: A term sometimes used for an electrochemical cell; it is sometimes used to refer to a cell in which chemical changes are caused by the application of an external emf.

Working electrode: The test or specimen electrode in an electrochemical cell.

Historical Development

An abbreviated list follows which records the electrochemical events which were significant in the development of anodic protection.

Year	Person(s)	Contribution
1790	Keir	Discovered that when iron was first placed in concentrated nitric acid, it was so "altered" that the violent reaction did not occur when the altered iron was subsequently placed in dilute nitric acid.
1792	Fabbroni	Hypothesis of the galvanic nature of corrosion processes.
1834	Faraday	Presented the basic law of electrochemical equivalence.
1836	Schönbein	Also experimented with the iron/nitric acid reaction and in corresponding with Faraday suggested the new name "passive iron."
1868	Kohlrausch	The measurement of electrolytic conductance.
1887	Arrhenius	The theory of electrolytic conductance caused by ions.
1889	Nernst	Derived the equation which laid the foundation for all potentiometric disciplines.
1898	Haber	Demonstrated the importance for the control of electrode potential in electrochemical preparations.
1907	Sand and Fischer	Showed that selective deposition could be achieved by proper choice of the controlled electrode potential.
1917	Grower	Technique for determining the thickness of corrosion products (layers) by practical application of Faraday's principle of electrochemical equivalence.

Year	Person(s)	Contribution
1834	Beckman	The first commercial pH meter.
1938	Wagner and Traud	Mixed potential effect on the electrochemistry of corrosion.
1942	Hickling	An electrical device to control potential—potentiostat.
1945	Lingane	Developed controlled-potential coulometry.
1949	Shockley	Invention of the transistor.
1954	Roberts	An electronic potentiostat with the output current reversible through zero.
1954	Edeleanu	Demostrated in the laboratory that the Hickling potentiostat could control corrosion.
1957	Stern and Geary	Theoretical analysis of the shape of polarization curves.
1957	Riggs and Conger	The first precision rapid-rise analog potentiostat.
1958	Riggs, Conger, and Hutchinson	The first commercial application of anodic protection to industrial processes.
1958	Sudbury, Riggs, and Shock	Laboratory evaluation of anodic-polarization-induced passivity.
1958	Shock, Riggs, and Sudbury	Chemical-industry applications of anodic protection.

United States Patents Relating to Anodic Protection

U. S. patent no.	Title and inventor(s)
1,335,209	Process for the prevention of selective corrosion of tubes and machinery parts of copper or copper containing alloys—Franz von Wurstemberger
1,335,210	Method and device for the prevention of selective corrosion of tubes and machinery parts of copper containing alloys—Franz von Wurstemberger
1,436,686	Method of preventing the formation of scale in system boilers, evaporators, economizers and the like—K. Schnetzer
1,558,647	Protection of metallic surfaces against incrustation and Corrosion—Walter Thalhofer
1,568,714	Prevention of scaling and encrustation—Charles E. Bonine
1,568,728	Process for protection of metallic surfaces—George C. Freeman
1,736,986	Protection of metallic surfaces against incrustation and corrosion—Walter Thalhofer
1,816,487	Electrolytic system for the protection of condensers—Alexander Kirkaldy
1,825,477	Scale prevention in boilers or the like—William M. Reichart
1,891,004	Means for preventing scaling and corrosion of metal units—George S. Neeley
1,944,778	Method of protecting lead against corrosion—Henpi Benit
2,021,519	Corrosion preventative—Herbert Spencer Polin
2,149,617	Method and apparatus for handling acidic solutions—Paul L. Menaul
2,366,796	Preventing corrosion of ferrous metals by ammoniacal solutions of ammonium nitrate—Charles K. Lawrence and Robert F. Engle

U. S. patent no.	Title and inventor(s)
2,377,792	Preventing corrosion of ferrous metals by solutions of electrolytes—Charle K. Lawrence and Robert F. Engle
2,508,523	Device for the protection of the cathodes of electrolytic cells—Eduard Krebs
2,576,680	Method for increasing the resistance to corrosion of stainless steel—Louis Guitton
2,584,623	System and method for protecting pipes and other current conducting structures against electrolytic corrosion—William R. Schneider
2,762,767	Method and means for the prevention of electrolytic corrosion—William C. Noddings and Merrill A. Mosher
2,916,429	Device for the electrolytic protection of a ship's metal skin against corrosion—Ernst Vossnack and Jan Hendrik Visscher
2,963,413	Electrolytic system—Rolland C. Sabins
3,009,865	Anodic protection of kraft digesters—Walter A. Mueller and Thomas R. B. Watson
3,102,086	Method of improving the corrosion resistance of titanium metals—Joseph Bernard Cotton and Sutton Coldfield
3,126,328	Electrolytic bridge assembly for the anodic passivation of metals—Merle Hutchinson, Olen L. Riggs, Jr., and John D. Sudbury
3,127,337	Anodic passivation system—Norman L. Conger and Olen L. Riggs, Jr.
3,135,677	Durable anode protective system—H. C. Fischer
3,147,204 Re. 26,261	Anodic prevention of hydrogen embrittlement of metals—Spencer W. Shepard and Charles K. Aldrich
3,152,058	Electrolytic bridge assembly for the anodic passivation of metals—Merle Hutchison, Olen L. Riggs, Jr., and John D. Sudbury
3,182,007	Electrode assembly for the anodic passivation of metals—Merle Hutchison, Olen L. Riggs, Jr., and John D. Sudbury
3,201,335	Corrosion protection—Adrian J. Mac Nab and Richard S. Treseder
3,208,925	Anodic protection against corrosion—Merle Hutchison, Olen L. Riggs, Jr., and John D. Sudbury
3,216,916	Anodic passivation of wetted wall vessels—Carl E. Locke
3,265,601	Process for protecting metals against corrosion at elevated temperatures—Francis Schein, Rueil-Malmaison, and Bernard le Boucher

U. S. patent no.	Title and inventor(s)
3,280,020	Anodic polarization system—Norman L. Conger
3,288,694	Methods and apparatus for anodic protection of vessels—William P. Banks
3,317,415	Cathode for anodic protection system—John F. Delahunt
3,345,278	Anodic passivation of metals—Matthew Mekjean
3,346,471	Use of composite D.C. power in anodic protection—Zisis Andrew Foroulis
3,347,768	Anodic protection for plating system—Wesley L. Clark, Bruce Griggs, Darrell D. Hays, and Germaine F. Jacky
3,354,061	Method and apparatus for anodic protection—Zisis Andrew Foroulis
3,354,062	Prevention of corrosion in alkali metal halide solutions by ammonia addition—Louis M. Dvoracek
3,371,023	Method and apparatus for automatically regulating the passivation potential of metals—William P. Banks, Richard L. Every, and Norman L. Conger
3,375,183	Apparatus for minimizing corrosion of metals—William P. Banks and Merle Hutchison
3,378,472	Anodic passivation using stainless steel reference electrode—William P. Banks and Merle Hutchison
3,379,629	Method and apparatus for automatically controlling corrosion of process vessels—William P. Banks, Norman L. Conger, and Carl E. Locke
3,409,526	Method and apparatus for corrosion protection—William P. Banks, Norman L. Conger, and Eddie C. French
3,409,530	Electrode for anodic protection—Carl E. Locke and Gerral D. Harral
3,414,496	Controlled potential of metallic vessel—latex solution systems—William P. Banks and John D. Sudbury
3,425,921	Methods and systems for protecting metal structures—Leon P. Sudrabin
3,442,779	Anodic protection of metals—Gordon Randolph Hoey
3,458,413	Method of inhibiting fouling of sea water conduits and the like by marine organisms—Kenji Ueda Etal and Minoru Hirata
3,461,051	Method and apparatus for protecting walls of a metal vessel against corrosion—John B. Vrable
3,462,353	Reference electrodes of particular utility in anodic corrosion protection systems—Richard L. Every and William P. Banks

U. S. patent no.	Title and inventor(s)
3,477,930	Method and system for preventing electrolytic corrosion of pipes—Virgil C. Crites
3,483,101	Control system and method for anodic protection—J. F. Delahunt and P. A. Haish
3,496,079	Corrosion prevention—James F. Norton
3,525,702	Method of increasing the activity and stability of Raney type catalysts—F. Von Sturm and A. Thieleking
3,623,967	Electrolytic apparatus for the production of alkali metal chlorate with grounding means—Richard M. O. Maunsell
3,694,334	Acid pickling of stainless steel—G. A. Bombara
3,766,032	Method for control of marine fouling—Andrew S. Yeiser
3,769,926	Marine galvanic control circuit—Richard T. Race
3,826,724	Method of removing a metal contaminant—O. L. Riggs and L. S. Surtees
3,841,250	Electrical chlorinator—Paul Trevor Davies and John Harold Morgan
3,951,207	Heat exchange arrangement—Hans Baumann, Manfred Halfmann, and Rudolf Janke
3,197,755	Apparatus for detecting and correcting malfunction of a standard reference electrode—N. L. Conger
3,272,731	Erosion resistant reference electrode assembly—M. Hutchison and W. P. Banks

V

Bibliography

1790

J. Keir, *Philos. Trans. R. Soc. London* **80,** p. 359 (1790).

1836

M. Faraday, *Philos. Mag.* **9,** 53, 57 (1836).

C. Schonbein, Weitere Beobachtungen über das Verhalten des Eisens gegen Salpetersäure (Further observations on the behavior of iron in nitric acid), *Pogg. Ann.* **37,** 590–594 (1836).

C. F. Schonbein and M. Faraday, *Philos. Mag.* **9,** 53,57,122,153 (1836).

1844

M. Faraday, *Experimental Researches in Electricity, Vol. II,* University of London (1844).

1903

W. R. Whitney, The corrosion of iron, *J. Am. Chem. Soc.* **25,** 394 (1903).

1911

F. Flade, Beiträge zur Kenntnis der Passivität (The passivity of electrodes), *Z. Phys. Chem.* **6,** 513–559 (1911).

1924

V. Rothmund, Uber den Einfluss der Anionen auf die Passivierbarkeit der Metalle (Influence of anions on the passivatability of metals), *Z. Phys. Chem.* **110,** 384–393 (1924).

1926

G. Grube, R. Heidinger, and L. Schlecht, Uber das elektrochemische Verhalten des Chroms (The electrochemical behavior of chromium, communication I. The anodic behavior of electrolytic chromium), *Z. Elektrochem.* **32,** 70–79 (1926).

1927

G. Grube, Die Passivität der Metalle bei Anodischer Polarisation (The passivity of metals in anodic polarization,) *Z. Elektrochem.* **33,** 389–399 (1927).

1932

U. R. Evans and T. P. Hoar, The velocity of corrosion from the electrochemical standpoint, *Proc. R. Soc.* **137,** 343–365 (1932).

K. Georgi, Über das Anodische Verhalten des Nickels. I. (Anodic behavior of nickel), *Z. Elektrochem.* **38,** 681–688 (1932).

1933

K. Georgi, Das Anodische Verhalten des Nickels. III. (Anodic behavior of nickel), *Z. Elektrochem* **39,** 736–742 (1933).

1934

W. J. Shutt and A. Walton, The anodic passivation of gold, *Trans. Faraday Soc.* **30,** 914–926 (1934).

1937

E. Muller and V. Cupr, Zur Passivität des Chroms (The passivity of chromium), *Z. Elektrochem.* **43**, 42–52 (1937).

1938

J. N. Agar and F. P. Bowden, The kinetics of electrode reactions, *Proc. R. Soc.* **169A**, 206–234 (1938).

R. H. Roberts and W. J. Shutt, The anodic behavior of chromium, *Trans. Faraday Soc.* **34**, 1455–1469 (1938).

R. H. Brown and R. B. Mears, The electrochemistry of corrosion, *Trans. Electrochem. Soc.* **74**, 495 (1938).

C. Wagner and W. Traud, Über die Deutung von Korrosionsvorgängen durch Überlagerung von Elektrochemischen Teilvorgängen und über die Potentialbildung an Mischelektroden (On the interpretation of corrosion processes through the superposition of electrochemical partial processes and on the potential of mixed electrodes), *Z. Elektrochem.* **44**, 391 (1938).

1940

M. A. Ryan and H. Heinrich, A theory for the passivity of chromium, *Trans. Electrochem. Soc.* **77**, 427–450 (1940).

U. R. Evans, Researches into the electrochemical character of corrosion, *J. Iron Steel Inst.* **141**, 219P–224P (1940).

1942

A. Hickling, Studies in electrode polarisation. Part IV—The automatic control of the potential of a working electrode, *Trans. Faraday Soc.* **38**, 27–33 (1942).

1944

H. H. Uhlig, Passivity in copper–nickel and molybdenum–nickel–iron alloys, *Trans. Electrochem. Soc.* **85**, 307–333 (1944).

1945

M. Pourbaix, Thesis, Delft and Brussels, 1945; CEBELCOR Reports: E2 (1949) and PD 2 (1951).

1947

D. C. Grahame, The electrical double layer and the theory of electrocapillarity, *Chem. Rev.* **41**, 441, (1947).

1948

E. M. Mahla and N. A. Nielsen, A study of films isolated from passive stainless steels, *Trans. Electrochem. Soc.* **93**, 1–16 (1948).

N. Hackerman and D. I. Marshall, Passivity of chromium, *Trans. Electrochem. Soc.* **93**, 49–54 (1948).

U. R. Evans, *Metallic Corrosion, Passivity and Protection,* Longmans, Green, New York (1948).

1949

M. Pourbaix, Corrosion, passivity and passivation from the thermodynamic point of view, *Corrosion* **5**, 121–133 (1949).

1950

M. V. Lomonosov, *Collection, Vol. 1,* Izv. Akad. Nauk SSSR, Moscow (1950).

1951

D. R. Turner, Anode polarization effects of nickel in sulfuric acid, *J. Electrochem. Soc.* **98**, 434–442 (1951).

W. G. Renshaw and J. A. Ferree, Passivating characteristics of the stainless steels, *Corrosion* **7**, 353–360 (1951).

K. J. Vetter, Über den Zustand des passiven Eisens, insbesondere Salpetersäure (On the condition of passive iron, particularly nitric acid), *Z. Elektrochem.* **55**, 274–280 (1951).

J. J. Lander, Anodic corrosion of lead in H_2SO_4 solutions, *J. Electrochem. Soc.* **98**, 213–219 (1951).

1952

H. C. Gatos and H. H. Uhlig, Passivity of iron in nitric acid, *J. Electrochem. Soc.* **99**, 250–258 (1952).

D. C. Grahame, Mathematical theory of the Faradaic admittance, *J. Electrochem. Soc.* **99**, 370C–385C (1952).

T. P. Hoar and U. R. Evans, Some factors in anodic processes on corroding metals, *J. Electrochem. Soc.* **99**, 212–218 (1952).

J. H. Bartlett and L. Stephenson, Anodic behavior of iron in H_2SO_4, *J. Electrochem. Soc.* **99**, 504–512 (1952).

D. Schlain and J. S. Smatko, Passivity of titanium in hydrochloric acid solutions, *J. Electrochem. Soc.* **99**, 417–422 (1952).

J. V. Petrocelli, The electrochemical behavior of aluminum—III. In buffered and alkaline solutions of potassium ferricyanide and in sodium hydroxide, *J. Electrochem. Soc.* **99**, 513–519 (1952).

1953

O. Kubaschewski and B. E. Hopkins, *Oxidation of Metals and Alloys*, Academic, New York, and Butterworths, London (1953).

J. E. Lilienfeld and C. Miller, Distribution of conductivity within dielectric films on aluminum, *J. Electrochem. Soc.* **100**, 222–226 (1953).

H. H. Uhlig and S. S. Lord, Amount of oxygen on the surface of passive stainless steel, *J. Electrochem. Soc.* **100**, 216–221 (1953).

J. B. Burbank and A. C. Simon, The relation of the anodic corrosion of lead and lead-antimony alloys to microstructure, *J. Electrochem. Soc.* **100**, 11–14 (1953).

M. Pourbaix, The electrochemical behavior of metals and corrosion, *Chem. Appl. Chem.* **30**, 780 (1953).

1954

K. Hauffe and I. Pfeiffer, Über Passivitätserscheinungen an Nickel (Passivation phenomena on nickel), *Z. Metallkd.* **45**, 554–562 (1954).

M. Pourbaix, The utility of thermodynamic interpretation of polarization curves, *J. Electrochem. Soc.* **101**, 9, 217C (1954).

C. Edeleanu, Method for the study of corrosion phenomena, *Nature* **173**, 739 (1954).

R. Landsberg and M. Hollnagel, Das anodische Verhalten des Nickels in Schwefelsäre (The characteristics of nickel as anode in sulfuric acids), *Z. Elecktrochem.* **58**, 680–685 (1954).

M. Pourbaix, The utility of thermodynamic interpretation of polarization curves, *J. Electrochem. Soc.* **101**, 217C–221C (1954).

H. H. Uhlig and A. Geary, Potentials of iron, 18–8, and titanium in passivating solutions, *J. Electrochem. Soc.* **101**, 215–224 (1954).

N. Hackerman and C. D. Hall, Jr. Electrochemical polarization of titanium in aqueous solutions of sodium chloride, *J. Electrochem. Soc.* **101**, 321–327 (1954).

M. Maraghini, G. B. Adams, Jr., and P. V. Rysselberghe, Studies on the anodic polarization of zirconium and zirconium alloys, *J. Electrochem. Soc.* **101**, 400–409 (1954).

N. Hackerman and O. B. Cecil, The electrochemical polarization of zirconium in neutral salt solutions, *J. Electrochem. Soc.* **101**, 419–425 (1954).

M. H. Roberts, A potentiostat for corrosion study, *Br. J. Appl. Phys.* **5**, 351, 352 (1954).

C. Edeleanu, Corrosion control by anodic protection, *Metallurgia* **50**, 113–116 (September 1954).

1955

M. Eisenberg, C. W. Tobias, and C. R. Wilke, Application of backside luggin capillaries in the measurement of nonuniform polarization, *J. Electrochem. Soc.* **102**, 415–419 (1955).

H. B. Bomberger, F. H. Beck, and M. G. Fontana, Polarization studies of copper, nickel, titanium, and some copper and nickel alloys in 3% sodium chloride, *J. Electrochem. Soc.* **102**, 53–58 (1955).

M. Stern, The electrochemical behavior, including hydrogen overvoltage of iron in acid environments, *J. Electrochem. Soc.* **102**, 609–616 (1955).

H. H. Uhlig and T. L. O'Connor, Nature of the passive film on iron in concentrated nitric acid, *J. Electrochem. Soc.* **102**, 562–572 (1955).

A. C. Makrides, N. M. Komodromos, and N. Hackerman, Dissolution of metals in aqueous acid solutions, *J. Electrochem. Soc.* **102**, 363–369 (1955).

G. B. Adams, P. V. Rysselberghe, and M. Maraghini, Anodic polarization of zirconium at low potentials—formation rates, formation field, electrolytic parameters, and film thickness of very thin oxide films, *J. Electrochem. Soc.* **102**, 502–511 (1955).

R. Olivier, thesis, Leyden, 1955; The passivity of Fe–Cr alloys, *International Committee of Electrochemical Thermodynamics and Kinetics, Proceedings of Sixth Meeting,* Butterworth Scientific, London, p. 314 (1955).

C. Carius, Passivierung von Chromstahlen in 6.5 Prozentiger Saltpetersäure, Schwefelsaurer Kupfersulfatlösung und Leitungswasser (Passivation of chrome steel in 6.5% nitrous acid, a sulfuric acid–copper sulfate solution and tap water), *Arch. Eisenhuettenwes.* **26**, 769–776 (1955).

H. Rocha and G. Lennartz, Die Aktivierungspotentiale von Eisen–Chrom-Legierungen und ihre Beziehungen zu der chemischen Beständigkeit in Schwefelsäure (The activation potentials of iron–chromium alloys and their relations to the chemical resistance in sulfuric acid), *Arch. Eisenhuettenwes.* **26**, 117–123 (1955).

S. Morioka and K. Sakiyama, Electrochemical properties of alloys (V). On the anodic behavior of iron–nickel alloys in sulfuric acid solutions, *Nippon Kinzoku Gakkai* **19**, 31–34 (1955).

G. W. Akimov, Electrode potentials, *Corrosion* **11**, 477t–486t (1955).

G. W. Akimov, Electrode Potentials, *Corrosion* **11**, 515t–534t (1955).

R. Olivier, Passiviteit van Ijzer en Ijzer–Chroom Legeringen (Passivity of iron in ferro–chrome alloys), Doctorate dissertation, University of Leiden (1955).

K. F. Bonhoeffer, On the passivity of iron, *Corrosion* **11**, 304t–308t (1955).

M. Stern, The effect of alloying elements in iron on hydrogen overvoltage and corrosion rate in acid environments, *J. Electrochem. Soc.* **102**, 663–668 (1955).

1956

M. Prazak, Anodicka antikorosni ochrana oceli (Anodic anticorrosion of steel), *Hutn. Listy* **11**, 644–648 (1956).

M. Parazak and V. Parazak, O pasivite a korosni odolnosti nerezavejicich occli (On the passivation and corrosion resistance of stainless steel), *Hutn. Listy* **11**, 91–97 (1956).

R. Landsberg and M. Hollnagel, Das anodische Verhalten des Nickels in Schwefelsäure (The anodic behavior of nickel in sulfuric acid), *Z. Elektrochem.* **60**, 1098–1102 (1956).

M. Prazak and V. Prazak, Korrosionsstudium V. Mechanismus der Chemischen Passivierung und Korrosion der Metalle (Corrosion study V. Mechanism of chemical passivation and corrosion of metals), *Collect. Czech. Chem. Commun.* **21**, 564–570 (1956).

W. A. Mueller, Corrosion studies of carbon steel in alkaline pulping liquors by the potential-time and polarization curve methods, I. Theory, methods and selected results, *Can. J. Technol.* **34**, 162–181 (1956).

D. R. Turner, The anode behavior of germanium in aqueous solutions, *J. Electrochem. Soc.* **103**, 252–256 (1956).

1957

M. Stern and R. M. Roth, Anodic behavior of iron in acid solutions, *J. Electrochem. Soc.* **104**, 390–392 (1957).

M. Stern, Electrochemical polarization, *J. Electrochem. Soc.* **104**, 559–563 (1957).

R. A. Covert and H. H. Uhlig, Chemical and electrochemical properties of FeSn₂, *J. Electrochem. Soc.* **104**, 537–542 (1957).

C. Edeleanu, Method for the study of corrosion phenomena, *Nature* **173**, 739 (1957).

M. Stern and A. L. Geary, Electrochemical polarization: I. A theoretical analysis of the shape of polarization curves, *J. Electrochem. Soc.* **104**, 56–63 (1957).

G. Okamoto, H. Kobayashi, N. Sato, and M. Nagayama, Effect of temperature on the passivity of nickel, *J. Electrochem. Soc. Jpn.* **25**, 199–203 (1957).

P. Hancock and J. E. O. Mayne, Anodic polarization as a possible rapid method of deciding whether a given solution is corrosive or inhibitive, *J. Appl. Chem.* **7**, 700–708 (1957).

W. A. Mueller, Corrosion studies of carbon steel in alkaline pulping liquors by the potential-time and polarization curve methods, II. Mixtures of white with oxidized or nonoxidized black liquor, *Tappi* **40**, 129–140 (1957).

G. Okamoto, N. Sato, and M. Nagayama, An application of rapid method to the measurement of polarization characteristics of iron in acid, *J. Electrochem. Soc. Jpn.* **25**, 166–174 (1957).

M. Stern, Electrochemical polarization—II. Ferrous–ferric electrode kinetics on stainless steel, *J. Electrochem. Soc.* **104**, 559–563 (1957).

M. Stern, Electrochemical polarization—III. Further aspects of the shape of polarization curves, *J. Electrochem. Soc.* **104**, 645–650 (1957).

H. A. Johansen, G. B. Adams, Jr., and P. V. Rysselberghe, Anodic oxidation of aluminum, chromium, hafnium, niobium, tantalum, vanadium and zirconium at very low current densities, *J. Electrochem. Soc.* **104**, 339–346 (1957).

E. H. Phelps and D. C. Vreeland, Corrosion of austenitic stainless steels in sulfuric acid, *Corrosion* **13**, 619t–624t (1957).

M. Stern and E. D. Weisert, Experimental observations on the relation between polarization resistance and corrosion rate, *Proc. ASTM* **59**, 1280 (1957).

C. Edeleanu, The potentionstat as a metallographic tool, *J. Iron Steel Inst.,* **185**, 482 (1957).

J. Besson and W. Kunz, The electrochemical behavior of iron–silicon alloys, *Ann. Univ. Sarav. Sci.* **6**, 17–36 (1957).

1958

M. Stern, The mechanism of passivating-type inhibitors, *J. Electrochem. Soc.* **105**, 638–647 (1958).

H. H. Uhlig, The absorption theory of passivity and the flade potential, *Z. Elektrochem.* **62**, 626–632 (1958).

K. G. Weil, Experimentelle Befunde am passiven Eisen (Experimental observations on passive iron), *Z. Elektrochem.* **62**, 638–641 (1958).

H. H. Uhlig, Electron configuration in alloys and passivity, *Z. Elektrochem.* **62**, 700–707 (1958).

U. R. Evans, Protective films in passivity, *Z. Elektrochem.* **62**, 619–626 (1958).

C. Edeleanu, A potentiostat technique for studying the acid resistance of alloy steels, *J. Iron Steel Inst.* **188**, 122–132 (1958).

N. Hackerman, Sorption oxidation and passivity, *Z. Elektrochem.* **62**, 632–637 (1958).

V. Cihai, M. Prazak, and M. Holinka, Potenciostat v metalografii (Potentiostat in metallography), *Hutn. Listy* **13**, 496–501 (1958).

U. F. Franck, Instabilitätserscheinungen an passivierbaren Metallen (Instability phenomena on passivatable metals), *Z. Elecktrochem.* **62**, 649–655 (1958).

C. H. Cartledge, The passivation process in the presence of XO_4^{n-} inhibitors, *Z. Elektrochem.* **62**, 684–690 (1958).

M. J. Pryor, Electrode reactions on oxide covered aluminum, *Z. Elektrochem.* **62**, 782–794 (1958).

H. Gerischer, Passivität der Metalle (Passivity of metals), *Angew. Chem.* **70**, 285-298 (1958).

G. Okamoto, T. Takaishi, and N. Sato, On the flade potential of passive nickel, *Denki Kagaku*, **26**, 615-619 (1958).

T. Heumann and F. W. Dickotter, Über die passivierung des Chroms in verdünnter Schwefelsäure (The passivity of chromium in dilute sulfuric acid), *Z. Elektrochem.* **62**, 745-750 (1958).

J. J. Lingane, *Electroanalytical Chemistry*, Interscience, New York, p. 308 (1958).

Y. M. Kolotyrkin, Electrochemical behavior and anodic passivity mechanism of certain metals in electrolyte solutions, *Z. Elektrochem.* **62**, 664-669 (1958).

G. Okamoto, H. Kobayashi, M. Nagayama, and N. Sato, Effect of temperature on the passivity of nickel, *Z. Elektrochem,* **62**, 775-782 (1958).

K. Schwabe and G. Dietz, Zur Passivität des Nickels (On the passivity of nickel), *Z. Elektrochem.* **62**, 751-759 (1958).

R. Piontelli and G. Serravalle, A contribution to the knowledge of the passivation and passivity of nickel, part I (Aqueous, solutions), *Z. Elektrochem.* **62**, 759-772 (1958).

N. D. Tomashov, Methods for increasing the corrosion resistance of metal alloys, *Corrosion* **14**, 229t-236t (1958).

P. Ruetschi and B. D. Cahan, Electrochemical properties of PbO_2 and the anodic corrosion of lead and lead alloys, *J. Electrochem. Soc.* **105**, 369-377 (1958).

N. Ohtani and H. Sugawara, Effect of halogen ions on the anodic passivation of 13Cr stainless steel, *J. Electrochem. Soc. Jpn.* **26**, E63-E65 (1958).

S. E. S. El Wakkad, A. M. Shams El Din, and H. Kotb, The anodic oxidation of zinc and zinc-tin alloys at very low current density, *J. Electrochem. Soc.* **105**, 47-51 (1958).

M. Eisenberg and R. E. DeLaRue, Anodic polarization of titanium in nonaqueous base etching solutions, *J. Electrochem. Soc.* **105**, 162-169 (1958).

W. A. Mueller, Anodic protection of alkaline pulping digesters, Pulp and Paper Research Institute of Canada Technical Report Series No. 91, Montreal, Quebec (1958).

1959

M. Stern, Evidence for a logarithmic oxidation process for stainless steel in aqueous systems, *J. Electrochem. Soc.* **106**, 376-381 (1959).

H. H. Uhlig and P. F. King, The flade potential of iron passivated by various inorganic corrosion inhibitors, *J. Electrochem. Soc.* **106**, 1-7 (1959).

P. F. King and H. H. Uhlig, Passivity in the iron-chromium binary alloys, *J. Phys. Chem.* **63**, 2026-2032 (1959).

S. Barnartt, The oxygen evolution reaction at gold anodes, II. Overpotential measurements and reaction mechanism in sulfuric acid solutions, *J. Electrochem. Soc.* **106**, 991-994 (1959).

S. Barnartt, The oxygen-evolution reaction at gold anodes, I. Accuracy of overpotential measurements, *J. Electrochem. Soc.* **106**, 772-729 (1959).

V. Cihal and M. Prazak, Corrosion and metallographic study of stainless steels using potentiostatic techniques, *J. Iron Steel Inst.* **193**, 360 (1959).

T. P. Hoar, in *Modern Aspects of Electrochemistry, Vol. 2*, J. O'M Bockris, ed. Butterworths, London (1959).

V. V. Romanov, Effect of polarization on stress corrosion of brass, *J. Appl. Chem. USSR* **32**, 1424-1427 (1959).

E. Lange and H. Brunner, Concerning anodic processes on metal electrodes, especially in chemical and total anodic polishing, *Z. Elektrochem.* **59**, 638-646 (1959).

M. J. Pryor, The significance of the flade potential, *J. Electrochem. Soc.* **106**, 557-562 (1959).

U. F. Franck, Corrosion and Passivity, *Corros. Anticorros.* **7**, 83-97 (1959).

M. Stern and H. Wissenberg, The electrochemical behavior and passivity of titanium, *J. Electrochem. Soc.* **106**, 755-759 (1959).

M. Stern and H. Wissenberg, The influence of noble metal alloy additions on the electrochemical and corrosion behavior of titanium, *J. Electrochem. Soc.* **106**, 759–764 (1959).

G. Okamoto and N. Sato, Measurement of flade potential of the passive nickel in sulfuric acid solutions, *J. Electrochem. Soc. Jpn.* **27**, E125–E128 (1959).

G. Okamoto and N. Sato, Effect of the concentration of hydrogen-ion on the flade potential of nickel, *Nippon Kinzoku Gakkai* **23**, 662–666 (1959).

H. Phisterer, A. Politycki, and E. Fuchs, Zur Natur von Passivschichten auf diinnen Nickelfilmen (Nature of the passive layer on thin nickel films), *Z. Elektrochem.* **63**, 257–261 (1959).

N. D. Greene, The classical potentiostat: Its application to the study of passivity, *Corrosion* **15**, 369t–372t (1959).

G. Okamoto and N. Sato, Self-passivation of nickel in aerated aqueous solutions, *Nippon Kinzoku Gakki,* **23**, 725–728 (1959).

W. A. Mueller, Anodic protection of alkaline pulping digesters, *Tappi* **42**, 179–184 (1959).

W. A. Mueller, Anodic protection of alkaline pulping digesters, *Pulp Pap. Mag. Can.* **60**(1), T3–T8 (1959).

J. K. Higgins, The anodic dissolution and electrolytic polishing of metals, *J. Electrochem. Soc.* **106**, 999–1005 (1959).

J. V. Petrocelli, Anodic behavior of aluminum at low potentials, *J. Electrochem. Soc.* **106**, 566–570 (1959).

A. Bewick, M. Fleischmann, and M. Liler, Some factors in potentiostat design, *Electrochim. Acta,* **1**, 83–105 (1959).

P. Berge and P. S. Jacquet, Quelques exemples d'application de potentiostat aux études sur le comportement des métaux dans les milieux liquides (Some examples of the use of the potentiostat in the study of the behavior of metals in liquid media), 3 Colloq. Annuel Met.—Corros., Saclay, p. 175, June 1959.

H. Fry and M. Whitaker, The anodic oxidation of zinc and a method of altering the characteristic of the anodic films, *J. Electrochem. Soc.* **106**, 606–611 (1959).

P. E. Lake and E. J. Casey, The anodic oxidation of cadmium—II. Electrical properties of the film, *J. Electrochem. Soc.* **106**, 913–916 (1959).

S. Morioka and A. Umezono, The effect of halogen ions on the anodic passivation of titanium, *J. Jpn. Inst. Met.* **23**, 185–189 (1959)

1960

D. A. Shock, O. L. Riggs, and J. D. Sudbury, Application of anodic protection in the chemistry industry, *Corrosion* **16**, 55t–58t (1960).

G. Okamoto and N. Sato, Passivation of nickel in acid solution—Higher oxide film theory, *Nippon Kinzoku Gakkai* **24**, 105–109 (1960).

H. Weidinger and E. Lange, Über Electktrodenreaktionen am aktiven, passiven und transpassiven Chrom (On the subject of electrode reactions of active, passive, and transpassive chromium), *Z. Elektrochem.* **64**, 468–477 (1960).

W. A. Mueller, A model of the mechanism of electrochemical conversion from active to passive states, *J. Electrochem. Soc.* **107**, 157–164 (1960).

A. Rahmel and W. Schwenk, Elektrochemische Messungen über die Passivierung eines Stahles mit 18% Cr und 8% Ni (Electrochemical measurements on the passivation of a steel with 18% Cr and 8% Ni) *Arch. Eisenhuettenwes.* **31**, 189–193 (1960).

N. D. Greene, Effect of oxygen on the active–passive behavior of stainless steel, *J. Electrochem. Soc.* **107**, 457–459 (1960).

K. Arnold and K. J. Vetter, Zum Flade-Potential des passiven Nickels (In regard to the flade-potential of passive nickel) *Z. Elektrochem.* **64**, 407–413 (1960).

H. G. Feller and H. H. Uhlig, Relation of electron configuration to passivity in Cr–Ni–Fe alloys, *J. Electrochem. Soc.* **107**, 864–868 (1960).

B. Corradi and E. Casperini, Impiego del potenziostato per determinare la resistenza degli acciai inossidabili alla corrosione intercristallina (Utilization of potential state to determine inoxidizable steels' resistance to intercrystalline corrosion), *Metall. Ital.* **52**, 249–254 (1960).

A. P. Bond and H. H. Uhlig, Corrosion behavior and passivity of nickel–chromium and cobalt–chromium alloys, *J. Electrochem. Soc.* **107**, 488–493 (1960).

J. D. Sudbury, O. L. Riggs, and D. A. Shock, Anodic passivation studies, *Corrosion* **16**, 47t–54t (1960).

M. Stern and A. C. Makrides, Electrode assembly for electrochemical measurements, *J. Electrochem. Soc.* **107**, 282 (1960).

V. Cihal and M. Prazak, A contribution to the explanation of intergranular corrosion of chromium–nickel steel, *Corrosion* **16**, 530t–532t (1960).

C. E. Locke, M. Hutchison, and N. L. Conger, Now: Anodic corrosion control, *Chem. Eng. Prog.* **56**(11), 50–55 (November 1960).

K. J. Vetter and K. Arnold, Korrosion und Sauerstoffüberspannung des passiven Nickels in Schwefelsäure (Corrosion and oxygen overvoltage of passive nickel in sulfuric acid), *Z. Elektrochem.* **64**, 244–251 (1960).

A. C. Makrides, Dissolution of iron in sulfuric acid and ferric sulfate solutions, *J. Electrochem. Soc.* **107**, 869–877 (1960).

O. L. Riggs, M. Hutchison, N. L. Conger, Anodic control of corrosion in a sulfonation plant, *Corrosion* **16**, 58t–62t (1960).

N. Sato and G. Okamoto, Mechanism of the dissolution of passive nickel in sulfate solutions, *Nippon Kinzoku Gakkai* **24**, 735–739 (1960).

W. A. Mueller, Theory of the polarization curve technique for studying corrosion and electrochemical protection, *Can. J. Chem.* **38**, 576–587 (1960).

I. Sanghi, S. Visvanathan, and S. Ananthanarayanan, Polarization studies on cadmium by galvanostatic and potentiostatic techniques, *Electrochim. Acta* **3**, 65–74 (1960).

R. Bakish, On the anodic oxidation of columbium, *J. Electrochem. Soc.* **107**, 653,654 (1960).

1961

W. A. Mueller, Passivating effect of elemental sulfur on steel in alkaline pulping liquors, *Corrosion* **17**, 557t,558t (1961).

C. Edeleanu and J. G. Gibson, Anodic protection, *Chem. Ind.* 301–308, March 11 (1961).

G. J. Janz and F. Saegusa, Anodic polarization curves in molten carbonate electrolysis, *J. Electrochem. Soc.* **108**, 663–669 (1961).

S. Barnartt and D. van Rooyen, Anodic behavior of austenitic stainless steels and susceptibility to stress corrosion cracking, *J. Electrochem. Soc.* **108**, 222–229 (1961).

L. Young, *Anodic Oxide Films,* Academic, London and New York (1961).

N. D. Tomashov, R. M. Altovsky, and G. P. Chernova, Passivity and corrosion resistance of titanium and its alloys, *J. Electrochem. Soc.* **108**, 113–119 (1961).

W. Schwenk and A. Rahmel, Experimentelle Befunde und Diskussion über den Mechanismus der Passivschichtbildung auf Chrom–Nickel–Stahlen (Experimental findings and discussion of the mechanism of the formation of a passive layer on chrome–nickel steels), *Electrochim. Acta.* **5**, 180–201 (1961).

G. Economy, R. Speiser, F. H. Beck, and M. G. Fontana, Anodic polarization behavior of iron–nickel alloys in sulfuric acid solutions, *J. Electrochem. Soc.* **108**, 337–343 (1961).

J. Osterwald and H. H. Uhlig, Anodic polarization and passivity of Ni and Ni–Cu alloys in sulfuric acid, *J. Electrochem. Soc.* **108**, 515–519 (1961).

H. H. Uhlig, Critical pH and critical current density for passivity in metals, *J. Electrochem. Soc.* **108**, 327–330 (1961).

S. Barnartt, Magnitude of IR-drop corrections in electrode polarization measurements made with a Luggin–Haber capillary, *J. Electrochem. Soc.* **108**, 102–104 (1961).

C. Edeleanu, Anodic protection, *Chem. Ind.* **279**, March 4, 1961.

M. Prazak and J. Spanily, Korrosionsstudium XXIV. Einfluss der Temperatur auf die Passivierungs-Charakteristik Korrosionsbestandiger Stahle (Corrosion study XXIV. The influence of temperature on the passivation characteristics of stainless steels), *Collec. Czech. Chem. Commun.* **26**, 2828–2837 (1961).

N. Sato and G. Okamoto, A mechanism of the anodic passivation of nickel in acid solution, *Trans. Jpn. Inst. Met.* **2**, 113–119 (1961).

N. D. Greene, C. R. Bishop, and M. Stern, Corrosion and electrochemical behavior of chromium–noble metal alloys, *J. Electrochem. Soc.* **108**, 836–841 (1961).

F. H. Giles and J. H. Bartlett, Anodic behavior of copper in phosphoric acid, *J. Electrochem. Soc.* **108**, 266–272 (1961).

G. M. Schmid and N. Hackerman, Anodic polarization of stainless steel in chloride solutions, *J. Electrochem. Soc.* **108**, 741–744 (1961).

M. Pourbaix, Prediction of corrosion and noncorrosion conditions for metals and alloys, *Corros. Anticorros.* **9**, 47–62 (1961).

S. Evans and E. L. Koehler, Use of polarization methods in the determination of the rate of corrosion of aluminum alloys in anaerobic media, *J. Electrochem. Soc.* **108**, 509–514 (1961).

D. J. G. Ives and G. J. Janz, *Reference Electrodes—Theory and Practice,* Academic, New York (1961).

A. Hickling, A simple potentiostat for general laboratory use, *Electrochim. Acta* **5**, 161–168 (1961).

W. A. Mueller, On the passivating effect of elementary sulfur added to alkaline pulping liquors, Pulp and Paper Research Institute of Canada Research Note No. 26, Montreal, Quebec, July 1961.

T. R. B. Watson, Electrolytic corrosion protection of paper mill equipment, *Tappi,* **44**, 208–210 (1961).

G. Wranglen and A. Warg, Determination of the relative effective area of electrodes by polarization measurements, *Alta Chem. Scand.* **15**, 1411, 1412 (1961)

1962

N. D. Greene, Predicting behavior of corrosion resistant alloys by potentiostatic polarization methods, *Corrosion* **18**, 136t–142t (1962).

A. C. Makrides, Hydrogen overpotential on nickel in alkaline solution, *J. Electrochem. Soc.* **109**, 977–984 (1962).

A. C. Makrides, Some electrochemical methods in corrosion research, *Corrosion* **18**, 338t (1962).

W. A. Mueller, The polarization curve and anodic protection, *Corrosion* **18**, 359t–367t (1962).

W. A. Mueller, Derivation of anodic dissolution curves of alloys from those of metallic components, *Corrosion* **18**, 73t–79t (1962).

M. S. Petit and I. Epelboin, A new method for study of passivity in application to 18–8 stainless steels, *Corros. Anticorros.* **10**, 32–38 (1962).

Z. A. Iofa and W. Pao-Ming, The influence of pH on the rate of corrosion and anodic dissolution of cobalt, *Zh. Fix. Khim.* **36**, 1395–1397 (1962).

D. A. Shock, J. D. Sudbury, and O. L. Riggs, Use of anodic passivation for corrosion mitigation of iron and alloy steels, *First International Congress on Metallic Corrosion, 1962,* pp. 363–367, Butterworths, London (1962).

J. W. Ward and J. T. Waber, The electrochemical behavior of uranium, *J. Electrochem. Soc.* **109**, 76–81 (1962).

R. M. Hurd, Polarization curves of redox systems involving consecutive electron transfers: Some theoretical aspects, *J. Electrochem. Soc.* **109**, 327–332 (1962).

M. Nagayama and M. Cohen, The anodic oxidation of iron in a neutral solution. I. The nature and composition of the passive film, *J. Electrochem. Soc.* **109**, 781–790 (1962).

M. Pourbaix and F. Vandervelden, Intentionstatic and potentiostatic studies, CEBELCOR Report E44 (1962).

M. Froment, P. Morel, and I. Epelboin, Etude potentiocinétique des phénomènes de passivité du fer et des aciers, avec observation microscopique simultanée (Potentiokinetic study of passivity phenomena of iron and steel with simultaneous microscopic observation), *Mem. Sci. Rev. Metall.* **49**, 225 (1962).

M. L. Kronenberg, Polarization studies on high-temperature fuel cells, *J. Electrochem. Soc.* **109**, 753–757 (1962).

R. F. Steigerwald and N. D. Greene, The anodic dissolution of binary alloys, *J. Electrochem. Soc.* **109**, 1026–1034 (1962).

L. M. Dvoracek and L. L. Neff, Use of the polarization technique to study corrosion in aqueous ammonia systems, *Corrosion* **18**, 85t–90t (1962).

J. L. Weininger and W. R. Grams, An all-Teflon electrochemical cell assembly, *J. Electrochem. Soc.* **109**, 984, 985 (1962).

A. Ragheb and L. A. Kamel, Anodic behavior and passivation of tin in solutions containing phosphoric acid, *Corrosion* **18**, 153t–157t (1962).

S. J. Acello and N. D. Greene, Anodic protection of austenitic stainless steels in sulfuric acid–chloride media, *Corrosion* **18**, 286t–290t (1962).

V. P. Batrakov, Corrosion diagrams of iron and steel in oxidizing media, *Corrosion* **18**, 437t–439t (1962).

A. Rius and R. Lizarbe, Study of the anodic behaviour of iron at high potentials in solutions containing chloride ions, *Electrochim. Acta.* **7**, 513–522 (1962).

I. Epelboin, M. Froment, and P. Morel, Sur l'etude potentiocinétique de la transpassivité du fer et des aciers (Potentiokinetic study of the transpassivity of iron and steels), *Electrochim. Acta.* **6**, 51–58 (1962).

L. Cavallaro, L. Felloni, F. Pulidori, and G. Trabanelli, Potentiodynamic investigation on the influence of phenylthiourea on the anodic and cathodic polarization curves of iron in acid solution, *Corrosion* **18**, 396t–400t (1962).

N. D. Greene, S. J. Acello, and A. J. Greif, An electrode mount for electrochemical studies of stressed metal specimens, *J. Electrochem. Soc.* **109**, 1001–1002 (1962).

A. Betti, L. Cavallaro, G. Trabanelli, and F. Zucchi, Potentiodynamic analysis of the influence of surface active agents on the corrosion resistance of common and stainless steels in aqueous solutions, *Corrosion* **18**, 351t–358t (1962).

N. D. Greene, The passivity of nickel and nickel-base alloys *First International Congress on Metallic Corrosion,* pp. 113–117, Butterworths, London (1962).

G. M. Schmid and N. Hackerman, Electrical double layer capacities of iron during forced cathodic decay of passivity, *J. Electrochem. Soc.* **109**, 1096–1099 (1962).

A. Desestret, Application du potentiostat a l'etude de la corrosion intergranulaire des aciers inoxydables austenitiques (Application of the potentiostat to the study of intergranular corrosion of austenitic stainless steels), *Mem. Sci. Rev. Metall.* **LIX**, 553 (1962).

S. J. Acello and N. D. Greene, Anodic protection of austenitic stainless steels in sulfuric acid–chloride media, *Corrosion* **18**, 286t (1962).

F. A. Posey and R. F. Sympson, Effect of adsorbed anions on reduction processes on passive stainless steel, *J. Electrochem. Soc.* **109**, 716–723 (1962).

E. H. Phelps, Electrochemical techniques for measurement and interpretation of corrosion, *Corrosion* **18**, 239t–246t (1962).

M. Prazak and V. Cihal, Die Potentiostatische Untersuchung des Einflusses Einiger Legierungselemente auf die Elektrochemischen und Korrosionseigenschaften Nischtrostender Stahle (The potentiostatic investigation of the effect of some alloying elements on the electrochemical and corrosion properties of stainless steels), *Corros. Sci.* **2,** 71–84 (1962).

R. N. Younger, R. G. Baker, and R. Littlewood, The relationship between microstructure and intercrystalline corrosion in an 18Cr–12Ni–1Nb austenitic steel, *Corros. Sci.* **2,** 157–161 (1962).

M. Nagayama and G. Okamoto, The anodic behaviour of passive iron in chromic acid–chromate solutions, *Corros. Sci.* **2,** 203–210 (1962).

R. Juchniewicz, Influence of alternating current on anodic protection, *Corros. Sci.* **2,** 225 (1962).

R. S. Rajogopalan, K. Venu, and K. Balakrishman, Anodic polarization studies in neutral and alkaline solutions containing corrosion inhibitors—I. NaOH–NaCl system, *J. Electrochem. Soc.* **109,** 81–87 (1962).

M. Smialowski and Z. Szklarska–Smialowska, Corrosion of iron in solutions containing ammonium nitrate, *Corrosion* **18,** 1t–4t (1962).

J. M. Matsen and H. B. Linford, A potentiostat for amperometric kinetic studies, *Anal. Chem.* **34,** 142–145 (1962).

I. G. Murgulescu and O. Radovici, The potentiostatic behaviour of some Cr–Ni–Mn stainless steels, *First International Congress on Metallic Corrosion,* pp. 109–112, Butterworths, London (1962).

U. F. Franck, Electrochemical studies of pitting corrosion of passive metals, *First International Congress on Metallic Corrosion,* pp.120–126, Butterworths, London (1962).

E. Uusitalo, Establishing the effect of corrosion inhibitors by potential and polarization measurements with particular reference to operative equipment, *First International Congress on Metallic Corrosion,* pp.138–143, Butterworths, London (1962).

U. R. Evans, Factors deciding between Corrosion and protective film formation, *First International Congress on Metallic Corrosion,* pp. 3–9, Butterworths, London (1962).

J. Voelzel and J. Plateau, Etude electrochimique, au moyen du potentiostat, de la corrosion intergranulaire d'un acier inoxydable du type 18–8 (Electrochemical study, by means of a potentiostat, of intergranular corrosion of a stainless steel of the type 18–8), *C. R. Acad. Sci.* **254,** 1972 (1962).

Y. M. Kolotyrkin, Electrochemical behaviour of metals during anodic and chemical passivation in electrolytic solutions, *First International Congress on Metallic Corrosion,* pp. 10–19, Butterworths, London (1962).

H. H. Uhlig, The advancing frontiers of corrosion science, *First International Congress on Metallic Corrosion,* pp. 36–42, Butterworths, London (1962).

M. G. Hollo, Electronoptical investigations on the pore structure of anodic oxide layers in aluminum, *First International Congress on Metallic Corrosion,* pp. 45–51, Butterworths, London (1962).

J. J. McMullen and M. J. Pryor, The relation between passivation corrosion, and the electrical characteristics of aluminum oxide films, *First International Congress on Metallic Corrosion,* pp. 52–61, Butterworths, London (1962).

M. Pourbaix and G. Govaerts, The work of Cebelcor's *Commission des Fondamentales et Applications* (CEFA), *First International Congress on Metallic Corrosion,* pp. 96–103, Butterworths, London (1962).

T. R. B. Watson, Anodic protection of alkaline pulping digesters, *Pulp Pap. Mag. Can.* **63,** T247, T248 (1962).

1963

M. Prazak, Evaluation of corrosion-resistant steels using potentiostatic polarization curves, *Corrosion* **19,** 75t–80t (1963).

P. F. King, Magnesium as a passive metal, *J. Electrochem. Soc.* **110**, 1113–1116 (1963).

N. E. Wisdom and N. Hackerman, Surface studies on passive iron, *J. Electrochem. Soc.* **110**, 318–325 (1963).

K. J. Vetter, A general thermodynamic theory of the potential of passive electrodes and its influence on passive corrosion, *J. Electrochem. Soc.* **110**, 597–605 (1963).

G. Aronowitz and N. Hackerman, The passivity of iron–chromium alloys, *J. Electrochem. Soc.* **110**, 663–640 (1963).

K. G. Weil, The influence of film thickness on the thermodynamic properties of thin oxide layers on iron, *J. Electrochem. Soc.* **110**, 640–644 (1963).

J. Kruger, Optical studies of the formation and breakdown of passive films formed on iron single crystal surfaces in inorganic inhibitor solutions, *J. Electrochem. Soc.* **110**, 654–663 (1963).

K. Schwabe, Investigation into the nature of anodic passive and barrier coatings, *J. Electrochem. Soc.* **110**, 663–670 (1963).

Y. M. Kolotyrkin, Pitting of metals, *Corrosion* **19**, 261t (1963).

M. Nagayama and M. Cohen, The anodic oxidation of iron in a neutral solution. II. Effect of ferrous ion and pH on the behavior of passive iron, *J. Electrochem. Soc.* **110**, 670–680 (1963).

H. S. Isaacs and J. S. L. Leach, Valency changes in the surface oxide films on metals, *J. Electrochem. Soc.* **110**, 680–687 (1963).

M. Fleischmann and H. R. Thirsk, The growth of thin passivating layers on metallic surfaces, *J. Electrochem. Soc.* **110**, 688–697 (1963).

W. A. Mueller, Throwing power of the current in anodic and cathodic protection, *J. Electrochem. Soc.* **110**, 698–703 (1963).

N. D. Stolica and H. H. Uhlig, Critical and passive current densities for copper–nickel–zinc alloys in sulfuric acid, *J. Electrochem. Soc.* **110**, 1215–1218 (1963).

K. E. Heusler, On the passivity of manganese in acid solutions, *J. Electrochem. Soc.* **110**, 703–708 (1963).

R. Littlewood, The effect of variations in technique on polarization curves obtained with a potentiostat, *Corros. Sci.* **3**, 99–105 (1963).

J. D. Sudbury, C. E. Locke, and D. Coldiron, Corrosion blocked in 98%-H_2SO_4 storage tank, *Chem. Process.* 23–26 (February 11, 1963).

J. D. Sudbury and C. E. Locke, Anodic protection against corrosion, *Chem. Eng.* **70**, 268–272 (1963).

J. L. Weininger and M. W. Breiter, Effect of crystal structure on the anodic oxidation of nickel, *J. Electrochem. Soc.* **110**, 484–490 (1963).

T. Heumann and H. S. Panesar, Contribution to the electrochemical behavior of chromium and iron–chromium alloys in the transpassive region, *J. Electrochem. Soc.* **110**, 628–663 (1963).

M. L. Kronenberg, J. C. Banter, E. Yeager, and F. Hovorka, The electrochemistry of nickel, II. Anodic polarization of nickel, *J. Electrochem. Soc.* **110**, 1007–1013 (1963).

H. H. Uhlig, P. Bond, and H. Feller, Corrosion and passivity of molybdenum–nickel alloys in hydrochloric acid, *J. Electrochem. Soc.* **110**, 650–653 (1963).

W. P. Banks and J. D. Sudbury, Anodic protection of carbon steel in sulfuric acid, *Corrosion* **19**, 300t–307t (1963).

J. D. Sudbury and C. E. Locke, Now anodic control of corrosion is practical for field applications, *Oil Gas J.* **61**(43), 111–113, October (1963).

D. A. Vermilyea, in *Advances in Electrochemistry and Electrochemical Engineering, Vol. 3,* (P. Delahay, ed.), Interscience, New York (1963).

A. O. Fisher and J. F. Brady, Anodic passivation of steel in 100 percent sulfuric acid, *Corrosion* 19, 37t–44t (1963).

N. Sato and G. Okamoto, Anodic passivation of nickel in sulfuric acid solutions, *J. Electrochem. Soc.* 110, 605–614 (1963).

M. Pourbaix, L. Klimizack-Mathieiu, C. Mertens, J. Meunier, C. Vanleugenhaghe, L. deMunck, J. Laureys, L. Neelemans, and M. Warze, Potentiokinetic and corrosimetric investigations of the corrosion behaviour of alloy steels, *Corros. Sci.* 3, 239–259 (1963).

O. L. Riggs, Jr., Anodic protection prevents phosoporic acid tank explosion caused by corrosion-generated hydrogen, *Mater. Prot.* 2(8), 63, 64 (August 1963).

Staff, Anodic systems reduces iron pickup during storage of acid and oleum, *Mat. Prot.* 2(9), 69–71, September (1963).

O. L. Riggs, Jr., Effects of hydrogen halides on anodic polarization of stainless steel, *Corrosion* 19, 180t–185t (1963).

N. D. Tomashov and R. M. Al'Tovskii, Effect of platinum, copper and iron ions on corrosion and passivity of titanium in 15 percent hydrochloric acid, *Corrosion* 19, 217t–221t (1963).

J. F. Brady, Anodic passivation of steel in 100 percent sulfuric acid, *Corrosion* 19, 238t–242t (1963).

A. O. Fisher and J. F. Brady, Anodic passivation of steel in 100 percent sulfuric acid, *Corrosion* 19, 37t–44t (1963).

W. P. Banks and J. D. Sudbury, Anodic polarization of carbon steel in sulfuric acid, *Corrosion* 19, 300t–307t (1963).

M. Prazak, Evaluation of corrosion-resistant steels using potentiostatic polarization curves, *Corrosion* 19, 75t–80t (1963).

F. P. A. Robinson and F. A. Frost, Anodic polarization characteristics of gold and silver in chloride and sulfate media, *Corrosion* 19, 115t–119t (1963).

D. Eurof Davies and S. N. Shah, The anodic behaviour of tin in alkaline solutions—II. 0.1 *M* sodium carbonate and 0.005 *M* potassium chromate solutions, *Electrochim. Acta* 8, 703–808 (1963).

S. N. Shah and D. Eurof Davies, The anodic behaviour of tin in alkaline solutions—I. 0.1 *M* sodium borate solution, *Electrochim. Acta* 8, 663–678 (1963).

T. Hurlen, Anodic behaviour of iron in alkaline solutions, *Electrochim. Acta* 8, 609–619 (1963).

W. Schwenk, Beobachtungen über die Korrosion nichtrostender Stahle in Schwefelsaure unter potentiostatischen Bedingungen (Observations on the corrosion of stainless steels in sulfuric acid under potentiostatic conditions), *Werkst. Korros.* 14, 646 (1963).

A. Bewick and M. Fleischmann, The design and performance of potentiostats, *Electrochim. Acta* 8, 89–106 (1963).

L. Cavallaro, L. Felloni, G. Trabanelli, and F. Pulidori, Potentiodynamic measurements of polarization curves on armco iron in acid medium in the presence of thiourea derivatives, *Electrochim. Acta* 8, 521–527 (1963).

D. N. Staicopolus, The role of cementite in the acidic corrosion of steel, *J. Electrochem. Soc.* 110, 1121–1124 (1963).

J. H. Greenblatt and A. F. McMillan, Polarization studies of aluminum alloys in water at 200C and 300C, *Corrosion* 19, 146t–155t (1963).

Y. M. Kolotyrkin, Pitting corrosion of metals, *Corrosion* 19, 261t–268t (1963).

F. H. Haynie and S. J. Ketcham, Electrochemical behavior of aluminum alloys susceptible to intergranular corrosion. II. Electrode Kinetics of oxide-covered aluminum, *Corrosion* 19, 403t–407t (1963).

S. J. Ketcham and F. H. Haynie, Electrochemical behavior of aluminum alloys susceptible to intergranular corrosion. I. Effect of cooling rate on structure and electrochemical behavior in 2024 aluminum alloy, *Corrosion* **19**, 242t–246t (1963).

T. P. Hoar, Corrosion, *Int. Sci. Technol.* 78–85, December (1963).

1964

N. Sato and G. Okamoto, Kinetics of the anodic dissolution of nickel in sulfuric acid solutions, *J. Electrochem. Soc.* **111**, 897–903 (1964).

A. Desestret and M. Froment, Influence du carbone et du traitement thermique sur la passivité secondaire d'aciers inoxydables austénitiques du type Cr18/Ni12 contenant ou non du titane (Effect of carbon and heat treatment on the secondary passivity of austenitic stainless steels of the 18Cr–12Ni type containing or not-containing titanium, *Corros. Anticorros.* **12**(1), 3–8 (1964).

M. Boyer, M. Keddam, and P. Morel, Sur l'étude de la transpassivité du chrome dans les solutions sulfuriques (Study of the transpassivity of chromium in sulfuric solutions), *C. R. Acad. Sci.* **259**, 1409–1412 (1964).

I. Epelboin and M. Keddam, Sur la dissolution anodique de nickel dans les solutions sulfuriques concentrées (Anodic dissolutions of nickel in concentrated sulfuric solutions), *C. R. Acad. Sci.* **258**, 137–140 (1964).

C. E. Locke, W. P. Banks, and E. C. French, Anodic protection of carbon steel in black sulfuric acids, *Mater. Prot.* **3**(6), 50–53, June (1964).

R. B. Leonard, Developing a new alloy, *Chem. Eng.*, 150 (October 26, 1964).

T. Ishikawa and G. Okamoto, Potentiostatic response of passive metals to the rate of temperature change, *Electrochim. Acta* **9**, 1259–1268 (1964).

N. D. Greene and R. B. Leonard, Comparison of potentiostatic anodic polarization methods, *Electrochim. Acta* **9**, 45–54 (1964).

E. Kunze and K. Schwabe, Beitrag zur Passivität des Nickels (Contribution to the passivity of nickel), *Corros. Sci.* **4**, 109–136 (1964).

S. Tajima and N. Baba, Anodic polishing and passivation of iron in the non-aqueous sulphamic-acid–formamide system, *Electrochim. Acta* **9**, 1509–1519 (1964).

N. D. Greene and G. A. Saltzman, Effect of plastic deformation on the corrosion of iron and steel, *Corrosion* **20**, 293t–298t (1964).

E. Kunze and K. Schwabe, Beitrag zur Passivität des Nickels (Contribution to the passivity of nickel), *Corros. Sci.* **4**, 109–136 (1964).

K. J. Cathro and D. F. A. Koch, The anodic dissolution of gold in cyanide solutions, *J. Electrochem. Soc.* **111**, 1416–1420 (1964).

T. R. B. Watson, Anodic protection of kraft digesters, *Mater. Prot.* **3**(6), 54–56 (June 1964).

L. Cavallaro, G. Trabanelli, L. Felloni, and F. Zucchi, The influence of surfactants on the passivity of 13Cr stainless steel in the presence of Cl⁻ ions and H_2O_2, *Corros. Sci.* **4**, 81–88 (1964).

F. Barbesino, E. Brutto, R. DiPietro, and R. Sesini, Study of corrosion films on zircaloy-2 by impedance measurements, *Energ. Nucl. II,* 435 (1964).

J. G. Hines and R. C. Williamson, Anodic behaviour of mild steel in strong sulphuric acid—I. Steady state conditions, *Corros. Sci.* **4**, 201–210 (1964).

R. C. Williamson and J. G. Hines, Anodic behaviour of mild steel in strong sulphuric acid—II. The sulphation and passivation process, *Corros. Sci.* **4**, 221–235 (1964).

J. Mieluch and M. Smialowski, The behaviour of grain boundaries in iron during anodic polarization in ammonium nitrate solution, *Corros. Sci.* **4**, 237–243 (1964).

W. Schwenk, Theory of stainless steel pitting, *Corros.* **20**, 129t (1964).

J. C. Redden, Metal-loss rate cut 90% by anodic system, *Chem. Process.*, reprint dated October 1964.

P. Neufeld, Application of the polarization resistance technique to corrosion monitoring, *Corros. Sci.* **4**, 245–251 (1964).

R. E. Meyer, The electrochemistry of the dissolution of zirconium in aqueous solutions of hydrochloric acid, *J. Electrochem. Soc.* **111**, 147–155 (1964).

R. Parson, Electrode double layer, *Encyclopedia of Electrochemistry*, C. A. Hampel, ed., Reinhold, New York, p. 404, (1964).

I. Dugdale and J. B. Cotton, The anodic polarization of titanium in halide solutions, *Corros. Sci.* **4**, 397–411 (1964).

Z. Szklarska–Smialowska, Effect of potential of mild steel on stress corrosion cracking in ammonium nitrate solutions, *Corrosion* **20**, 196t–202t (1964).

T. R. B. Watson, Experience with anodic protection of kraft digesters, *Pulp Pap. Mag. Can.* **65**, No. 10, T415–T418 (October 1964).

G. J. Janz and A. Conte, Potentiostatic polarization studies in fused carbonates—II. Stainless steel, *Electrochim. Acta* **9**, 1297–1287 (1964).

G. J. Janz and A. Conte, Potentiostatic polarization studies in fused carbonates—I. The noble metals, silver and nickel, *Electrochim. Acta* **9**, 1269–1278 (1964).

A. M. Shams El Din and F. M. Abd El Wahab, On the anodic passivity of tin in alkaline solutions, *Electrochim. Acta* **9**, 883–896 (1964).

J. Llopis and L. Jorge, Passivation of iridium in hydrochloric acid solutions, *Electrochim. Acta* **9**, 103–111 (1964).

J. Llopis and M. Vasques, Passivation of rhodium in hydrochloric acid solutions, *Electrochim. Acta* **9**, 1655–1663 (1964).

N. D. Tomashov, Passivity and corrosion resistance of metal systems, *Corros. Sci.* **4**, 315–334 (1964).

D. E. Davies and W. Barker, Influence of pH on corrosion and passivation of nickel, *Corrosion* **20**, 47t–53t (1964).

V. D. Kashcheev, B. N. Kabanov, and D. I. Leikis, Anodic activation of iron, *Corrosion* **20**, 54t–56t (1964).

N. D. Tomashov, G. P. Chernova, and O.N. Marcova, Effect of supplementary alloying elements on pitting corrosion susceptibility of 18 Cr–14Ni stainless steel, *Corrosion* **20**, 166t–173t (1964).

C. C. Seastrom, Passivity of chromium and stainless steel in hydrofluoric acid, *Corrosion* **20**, 179t–183t (1964).

F. P. A. Robinson and L. Golante, Minimizing corrosion in uranium extraction solutions by application of anodic protection, *Corrosion* **20**, 239t–244t (1964).

L. M. Dvoracek and L.L. Neff, Effect of pH on anodic behavior of carbon steel in sodium chloride solutions containing ammonia, *Corrosion* **20**, 303t–306t (1964).

K. Nobe and R. F. Tobias, Anodic potentiostatic polarization of iron in sulfuric acid: Effect of chloride ions, *Corrosion* **20**, 263t–266t (1964).

M. Froment, Observation microscopique des electrodes métalliques au cours de leur corrosion (Microscopic observation of metallic electrodes during corrosion), *Mem. Sci. Rev. Metall.* **61**, 283 (1964).

O. L. Riggs, Jr., Sulfates in the passive iron layer, *Corrosion* **20**, 275t–281t (1964).

A. C. Makrides, Kinetics of the Fe^{+++}/Fe^{++} reaction on Fe–Cr alloys, *J. Electrochem. Soc.* **111**, 400–407 (1964).

T. P. Hoar and J. C. Scully, Mechanochemical anodic dissolution of austenitic stainless steel in hot chloride solution at controlled electrode potential, *J. Electrochem. Soc.* **111**, 348–352 (1964).

O. L. Riggs, Jr., Iron analysis with a potentiostat, *Corrosion* **20**, 367t–369t (1964).

B. Baranowski and Z. Szklarska–Smialowska, A galvanostatic and potentiostatic study of the nickel–hydrogen system, *Electrochim. Acta* **9**, 1497–1507 (1964).

L. Cavallaro, L. Felloni, G. Trabanelli, and F. Pulidori, The anodic dissolution of iron and the behaviour of some corrosion inhibitors investigated by the potentiodynamic method, *Electrochim. Acta* **9**, 485–494 (1964).

T. Ishikawa and G. Okamoto, Potentiostatic response of passive metals to the rate of temperature change, *Electrochim. Acta* **9**, 1259–1268 (1964).

K. Kaesche, The passivity of zinc in aqueous solutions of sodium carbonate and sodium bicarbonate, *Electrochim. Acta* **9**, 383–394 (1964).

A. C. Makrides, Kinetics of redox reactions on passive electrodes, *J. Electrochem. Soc.* **111**, 392–400 (1964).

J. Giner, A practical reference electrode, *J. Electrochem. Soc.* **111**, 376,377 (1964).

J. L. Weininger and M. W. Breiter, Hydrogen evolution and surface oxidation of nickel electrodes in alkaline solution, *J. Electrochem. Soc.* **111**, 707–712 (1964).

E. L. Littauer, Impressed currents to curbe corrosion, *Chem. Eng.* 156–164 (September 28, 1964).

Z. A. Foroulis, Fundamental studies of anodic protection, *Ind. Eng. Chem. Process Des. Dev.* **3**, 84–88 (1964).

N. D. Tomashov, Development of the electrochemical theory of metallic corrosion, *Corrosion* **20**, 7t–14t (1964).

1965

J. L. Ord and J. H. Bartlett, Electrical behavior of passive iron, *J. Electrochem. Soc.* **112**, 160–166 (1965).

M. N. Fokin and V. A. Timonin, Extended electrode, partially passivated in a tube, and its ability to stabilize current, *Dok. Akad. Nauk SSSR* **164**, 150 (1965).

J. D. Sudbury, W. P. Banks, and C. E. Locke, Anodic protection of carbon steel in fertilizer solutions, *Mater. Prot.* **4**(6), 81 (1965).

G. Bianchi, A. Barosi, G. Faita, and T. Mussini, A new assemblage for hydrogen electrodes, *J. Electrochem. Soc.* **112**, 921–932 (1965).

M. Daguenet and M. Foment, Influence de la diffusion convective sur le polissage électrolytique du nickel dans les solutions sulfuriques concentrées (Effect on the convective diffusion on electrolytic polishing of nickel in concentrated sulfuric acid solutions), *C. R. Acad. Sci.* **260**, 5534–5537 (1965).

P. Neufeld and R. C. Williamson, The anodic protection of vessels, *Corros. Sci.* **5**, 605–612 (1965).

P. E. Morgan and L. S. Evans, Anodic protection for a titanium heat exchanger, *Mater. Prot.* **4**(1) 60,62 (January 1965).

G. Okamoto, N. Sato, and H. Ohashi, An application of exoelectron emission measurements to the study of stressed surface layer of pure iron, *J. Electrochem. Soc. Jpn.* **33**, 11–18 (1965).

N. Sato and G. Okamoto, Reaction mechanism of anodic oxygen evolution on nickel in sulphate solutions, *Electrochim. Acta* **10**, 495–502 (1965).

G. Okamoto and T. Shibata, Desorption of titrated bound water from the passive film formed on stainless steels, *Nature* **206**, 1350 (1965).

A. R. Tourky, A. A. Abdul Azim, and M.M. Anwar, Effect of carbon content on the corrosion and passivity of iron, *Corros. Sci.*, **5**, 301–307 (1965).

N. D. Greene, W. D. France, Jr., and B. E. Wilde, Electrode mounting for potentiostatic anodic polarization studies, *Corrosion* **21**, 275,276 (1965).

E. P. Koutsoukos and K. Nobe, Corrosion studies, part VI, passivity of inconel in acidic chloride solutions, Department of Engineering, UCLA, Report No. 65-41 (September 1965).

P. W. Bolmer, Polarization of iron in H_2S–NaHs buffers, *Corrosion* **21**, 69–75 (1965).

J. E. Reinoehl, F. H. Beck, and M. G. Fontana, Corrosion, immunity and passivation from an engineering viewpoint, *Corrosion* **21**, 379–381 (1965).

C. E. Locke, Tank trailer gets anodic protection, *Mater. Prot.* **4**(3) 59,60 (March 1965).

Z. A. Foroulis, Fundamental studies on anodic protection: carbon steel in sulfuric acid, *Ind. Eng. Chem. Process Des. Dev.* **4**, 20–23 (1965).

G. Bianchi, A. Barosi, and S. Trosatti, Anodic protection of stainless steel by galvanic coupling with platinum, *Electrochim. Acta* **10**, 83–95 (1965).

J. M. West, *Electrodeposition and Corrosion Processes*, Van Nostrand, New York (1965).

J. J. Podesta and A. J. Arvia, Kinetics of the anodic dissolution of iron in concentrated ionic media: Galvanostatic and potentiostatic measurements, *Electrochim. Acta* **10**, 171–182 (1965).

M. Prazak and K. Barton, The influence of alloying elements on the corrosion behaviour of single-phase alloys, *Corros. Sci.* **5**, 377–382 (1965).

A. M. Sukhotin and K. M. Kartashova, The passivity of iron in acid and alkaline solutions, *Corros. Sci.* **5**, 393–407 (1965).

D. Shaw and A. M. Edwards, A transistorized potentiostat system for corrosion studies, *Corros. Sci.* **5**, 413–425 (1965).

D. deG. Jones, E. G. Kingham, and H. G. Masterson, An automatic voltage programme unit for potentiodynamic polarization studies, *Corros. Sci.* **5**, 503–511 (1965).

D. deG. Jones, A logarithmic current unit for recording electrochemical polarization measurements, *Corros. Sci.* **5**, 559–564 (1965).

R. R. Sayano and K. Nobe, Corrosion studies, part III, anodic polarization of impure and higher purity nickel in H_2SO_4, Department of engineering, UCLA, Report No. 65-38 (September 1965).

I. G. Murgulescu, O. Radovici, and M. Borda, Studies of the mechanism of anodic dissolution of Al–Zn binary alloys in alkaline solutions by potentiodynamic and potentiostatic pulse methods, *Corros. Sci.* **5**, 613–622 (1965).

M. Pourbaix and F. Vandervelden, Intensiostatic and potentiostatic methods. Their use to predetermine the circumstances for corrosion or non-corrosion of metals and alloys, *Corros. Sci.* **5**, 81–111 (1965).

K. Venu, K. Balakrishnan, and K.S. Rajagopalan, A potentiokinetic polarization study of the behaviour of steel in NaOH–NaCl system, *Corros. Sci.* **5**, 59–69 (1965).

A. R. Piggott, H. Leckie, and L.L. Shrier, Anodic polarization of Ti in formic acid—I. Anodic behavior of Ti in relation to anodizing conditions, *Corros. Sci.* **5**, 165–184 (1965).

G. Trabanelli, F. Zucchi, and L. Felloni, Behaviour of nickel and its alloys in acidic media, *Corros. Sci.* **5**, 211–224 (1965).

A. J. Johnson and L. L. Shreir, The anodic behaviour of V, Ti, Zr, Nb and Ta in 3 M $AlCl_3$–diethyl ether solution, *Corros. Sci.* **5**, 269–278 (1965).

T. P. Hoar, D. C. Mears, and G. P. Rothwell, The relationships between anodic passivity, brightening and pitting, *Corros. Sci.* **5**, 279–289 (1965).

H. W. Pickering and R. P. Frankenthal, A transmission electron microscope study of the breakdown of passivity on Fe–24% Cr, *J. Electrochem. Soc.* **112**, 761–767 (1965).

K. Shiobara, Y. Sawada, and S. Morioka, Potentiostatic study on the anodic behavior of iron–chromium alloys, *Trans. Jpn. Inst. Met.* **6**, 58 (1965).

Z. A. Foroulis, Anodic protection—theoretical considerations, *Corros. Sci.* **5**, 383–391 (1965).

I. I. Tingley and R. R. Rogers, Corrosion of niobium and tantalum in alkaline media, *Corrosion* **21**, 132–136 (1965).

L. I. Freiman and Ya. M. Kolotyrkin, Pitting corrosion of iron by perchlorate ions, *Corros. Sci.* **5**, 199–202 (1965).

G. A. DiBari and J. V. Petrocelli, The effect of composition and structure on the electrochemical reactivity of nickel, *J. Electrochem. Soc.* **112**, 99–104 (1965).

J. R. Myers, F. H. Beck, and M. G. Fontana, Anodic polarization behavior of nickel–chromium alloys in sulfuric acid solutions, *Corrosion* **21**, 277–287 (1965).

L. R. Scharfstein, Corrosion resistance of alloy 20Cb in sulfuric acid, *Corrosion* **21**, 254–259 (1965).

M. Pourbaix, A comparative review of electrochemical methods of assessing corrosion and the behaviour in practice of corrodible material, *Corros. Sci.* **5**, 677–700 (1965).

A. Desestret and M. Froment, Sur la dissolution anodique des aciers inoxydables dans le domaine de la transpassivité (On the anodic dissolution of stainless steels in the area of transpassivity), *Mem. Sci. Rev. Metall.* **62**, 135–141 (1965).

N. D. Greene and G. Judd, Relation between anodic dissolution and resistance to pitting corrosion, *Corrosion* **21**, 15–18 (1965).

B. Lovrecek and K. Moslavac, The anodic dissolution of germanium, *Electrochim. Acta* **10**, 627–635 (1965).

J. Llopis, I. M. Tordesillas, and M. Muniz, Anodic corrosion of rhodium in hydrochloric acid solutions, *Electrochim. Acta* **10**, 1045–1055 (1965).

V. Cihal, Application of potential polarization in the study and in the metallography of corrosion resistant steels, *Hutn. Listy* **11**, 817–840 (1965).

1966

H. P. Leckie and H. H. Uhlig, Environmental factors affecting the critical potential for pitting in 18–8 stainless steel, *J. Electrochem. Soc.* **113**, 1262–1267 (1966).

T. K. Ross, G. C. Wood, and I. Mahmud, The anodic behavior of iron–carbon alloys in moving acid media, *J. Electrochem. Soc.*, **113**, 334–345 (1966).

T. C. Downie and C. W. Goulding, A potentiostatic examination of the properties of aluminum in aqueous solutions containing sulfate and other ions, *Metallurgia* **73**, 93 (1966).

M. Pourbaix, *Atlas of Electrochemical Equilibria in Aqueous Solutions,* Pergamon, London (1966).

G. Trabanelli, L. Felloni, and G. Mantovani, Application of potentiodynamic method to the chromium–nickel electroplating corrosion resistance evaluation, *Proceedings of the Second International Congress on Metallic Corrosion,* NACE, Houston, TX (1966).

T. P. Hoar and D. C. Mears, Corrosion-resistant alloys in chloride solutions: materials for surgical implants, *Proc. R. Soc. London* **294**, 486 (1966).

J. L. Ord, A comparison of the passivity of iron in acid and neutral electrolytes, *J. Electrochem. Soc.* **113**, 213–217 (1966).

L. D. Perrigo, Anodic protection of carbon steel in oxalic acid, *Mater. Prot.* **5**(3) 73–76 (March 1966).

K. Osozawa, K. Bohnenkamp, and H.-J. Engell, Potentiostatic study on the intergranular corrosion of an austenitic chromium–nickel stainless steel, *Corros. Sci.* **6**, 421–433 (1966).

D. deG. Jones, E. G. Kingham, and H. G. Masterson, A step-potential programme unit for electrochemical polarization studies, *Corros. Sci.* **6**, 435–444 (1966).

R. Juchniewicz, T. Pompowski, and J. Walaszkowski, Anodic protection of austenitic stainless steel, *Corros. Sci.* **6**, 25–31 (1966).

F. Duffaut, J.-P. Pouzet, and P. Lacombe, Potentiostatic study of structural modifications caused in a Ni–Cr–Fe alloy by heat treatment at 650°C, *Corros. Sci.* **6**, 183–185 (1966).

A. A. Pozdeeva, E. I. Antonovskaya, and A. M. Sukhotin, Passivity of molybdenum, *Corros. Sci.* **6**, 149–158 (1966).

R. Lebet and A. Piotrowski, Resistance to pitting of types 202 and 321 steels to sulfuric acid and sodium chloride solutions, *Corrosion* **22**, 117–131 (1966).

F. P. A. Robinson and D. J. DuPlessis, Polarization and corrosion of ferrosilicon alloys for iron ore beneficiation media, *Corrosion* **22**, 117–131 (1966).

R. F. Steigerwald, Effect of Cr content on pitting behavior of Fe–Cr alloys, *Corrosion* **22**, 107–112 (1966).

K. Osozawa and H.-J. Engell, The anodic polarization curves of iron–nickel–chromium alloys, *Corros. Sci.* **6**, 389–393 (1966).

J. C. Redden, Anodic protection reduces corrosion in acid tanks, *Mater. Prot.* **5**(2), 51 (February 1966).

J. R. Myers and R. K. Saxer, Anodic polarization behavior of low alloy steels in sulfuric acid environments, *Corrosion* **22**, 346–348 (1966).

E. A. Lizlovs, Effect of Mo, Cu, Si and P on anodic behavior of 17Cr steels, *Corrosion* **22**, 297–308 (1966).

R. R. Sayano and K. Nobe, Continuous and pulse polarization of Ni in H_2SO_4, *Corrosion* **22**, 81–87 (1966).

L. R. Hays, How anodic protection is applied to sulfuric acid tanks, *Mater. Prot.* **5**(9), 46–48 (September 1966).

J. R. Myers, W. B. Crow, F. H. Beck, and R. K. Saxer, Observations on the anodic behavior of nickel and chromium: surface topography and temperature effect, *Corrosion* **22**, 32–38 (1966).

I. A. Ammar and S. Darwish, Potentiostatic behaviour of passive nickel in sulphuric acid, *Electrochim. Acta* **11**, 1541–1552 (1966).

R. P. A. Robinson and L. Golante, Anodic polarization characteristics of special alloys in nitric acid–formic acid–ammonium flouride–sulfur dioxide solutions, *Proceedings of the Second International Congress on Metallic Corrosion*, pp. 290–299, NACE, Houston, TX (1966).

A. Bewick and M. Fleischmann, The design of potentiostats for use at very short times, *Electrochim. Acta* **11**, 1397–1416 (1966).

M. H. Tikkanen, T. Tuominen, and A. Laurila, Electrochemical behavior of some cobalt–chromium alloys, *Proceedings of the Second International Congress on Metallic Corrosion*, pp. 563–568, NACE, Houston, TX (1966).

G. Okamoto, M. Nagayama, T. Ishikawa, and T. Shibata, The existence of bound water in the passive films formed on stainless steels, *Proceedings of the Second International Congress on Metallic Corrosion*, pp. 558–562, NACE, Houston, TX (1966).

V. V. Andreeva, Behavior and nature of thin oxide films on some metals in gaseous media and in electrolyte solutions, *Proceedings of the Second International Congress on Metallic Corrosion*, pp. 535–546, NACE, Houston, TX (1966).

I. G. Murgulescu, O. Radovici, and S. Ciolac, Potentio-kinetic studies on some aluminum binary alloys, *Proceedings of the Second International Congress on Metallic Corrosion*, pp. 942–952, NACE, Houston, TX (1966).

W. Schwenk, Theory of stainless steel pitting, *Proceedings of the Second International Congress on Metallic Corrosion*, pp. 256–265, NACE, Houston, TX (1966).

R. L. Every and W. P. Banks, Reference electrodes in sulfuric acid, *Electrochem. Technol.* **4**, 275, 276 (1966).

M. W. Breiter, Automated recording and processing of galvanostatic and potentiostatic data, *J. Electrochem. Soc.* **113**, 1071–1073 (1966).

B. E. Conway, N. Marincic, D. Gilroy, and E. Rudd, Oxide involvement in some anodic oxidation reactions, *J. Electrochem. Soc.* **113**, 1144–1158 (1966).

J. O'M Bockris, A. K. N. Reddy, and B. Bao, An ellipsometric determination of the mechanism of passivity of nickel, *J. Electrochem. Soc.* **113**, 1133–1144 (1966).

H. P. Leckie and H. H. Uhlig, Environmental factors affecting the critical potential for pitting in 18–8 stainless steel, *J. Electrochem. Soc.* **113**, 1962 (1966).

Ya. M. Kolotyrkin, Pitting corrosion of metals, *Proceedings of the Second International Congress on Metallic Corrosion,* pp. 23, NACE, Houston, TX (1966).

A. C. Makrides, Electrochemistry of surface oxides, *J. Electrochem. Soc.* **113**, 1158–1165 (1966).

J. D. Sudbury, W. P. Banks, and C. E. Locke, Anodic protection of carbon steel in fertilizer solutions, *Proceedings of the Second International Congress on Metallic Corrosion,* pp. 267–274, NACE, Houston, TX (1966).

1967

N. D. Tomashov and G. P. Chernova, *Passivity and Protection of Metals Against Corrosion,* Plenum, New York (1967).

W. D. France, Jr., A specimen holder for precise electrochemical polarization measurements on metal sheets and foils, *J. Electrochem. Soc.* **114**, 818, 819 (1967).

H. W. Pickering and C. Wagner, Electrolytic dissolution of binary alloys containing a noble metal, *J. Electrochem. Soc.* **114**, 698–706 (1967).

R. P. Frankenthal, On the passivity of iron–chromium alloys–I. Reversible primary passivation and secondary film formation, *J. Electrochem. Soc.* **114**, 542–547 (1967).

S. J. Ketcham, Polarization and stress–corrosion studies of an Al–Cu–Mg alloy, *Corros. Sci.* **7**, 305–314 (1967).

J. M. Stammen and C. R. Townsend, Cathode effects in anodic protection, *Corrosion* **23**, 343–348 (1967).

W. P. Banks and E. C. French, Steel in phosphoric acid—anodic protection for corrosion control, *Mater. Prot.* **6**(6), 48–49 (June 1967).

M. Levy, Anodic behavior of titanium and commercial alloys in sulfuric acid, *Corrosion* **23**, 236–244 (1967).

W. P. Banks, M. Hutchison, and R. M. Hurd, Anodic protection of carbon steel alkaline sulfide pulp digesters, *Tappi* **50**, 49–55 (1967).

F. Mazza, Influence of long time polarization on anodic breakdown of titanium in concentrated NaCl solutions, *Corrosion* **23**, 223–230 (1967).

B. E. Wilde, An assembly for electrochemical corrosion studies in aqueous environments at high temperature and pressure, *Corrosion* **23**, 331–334 (1967).

D. Shaw, Two further transistorized potentiostat systems for metallurgical studies, *Corros. Sci.* **7**, 367–371 (1967).

T. K. Ross, D. A. Carter, and D. C. Smith, A potentiostatic study of the corrosion of dental silver–tin amalgam, *Corros. Sci.* **7**, 373–376 (1967).

G. J. Spacpen and M. J. Fevery–DeMeyer, Electrochemical corrosion experiments at temperatures above 100°C, *Corros. Sci.* **7**, 405–412 (1967).

N. D. Tomashov and G. P. Chernova, *Passivity and Protection of Metals Against Corrosion,* Plenum, New York (1967).

K. J. Vetter, *Electrochemical Kinetics Theoretical and Experimental Aspects,* Academic, New York (1967).

Z. Szklarska–Smialowska and M. Janik–Czachor, Pitting corrosion of 13Cr–Fe alloy in Na_2SO_4 solutions containing chloride ions, *Corros. Sci.* **7**, 65–72 (1967).

N. Hackerman, E. S. Snavely, Jr., and L. D. Fiel, The anodic polarization behaviour of metals in hydrogen fluoride, *Corros. Sci.* **7**, 39–50 (1967).

B. E. Wilde and G. A. Teterin, Anodic dissolution of copper–zinc alloys in alkaline solutions, *Br. Corros. J.* **2**, 125–128 (1967).

A. Piotrowski and R. Lebet, The electrochemical parameters of stainless steel passivation by some reversible redox systems, *Corros. Sci.* **7**, 231–237 (1967).

T. P. Hoar, The production and breakdown of the passivity of metals, *Corros. Sci.* **7**, 341–355 (1967).

M. J. Humphries and R. N. Parkins, Stress–corrosion cracking of mild steels in sodium hydroxide solutions containing various additional substances, *Corros. Sci.* **7**, 747–761 (1967).

M. Pugh, L. M. Warner, and D. R. Gabe, Some passivation studies on tin electrodes in alkaline solutions, *Corros. Sci.* **7**, 807–820 (1967).

L. E. Kindlimann and N. D. Greene, Dissolution kinetics of nuclear fuels: 1. Uranium, *Corrosion* **23**, 29–34 (1967).

J. Postelthwaite and L. B. Freese, Effect of halide additions on anodic behavior of nickel in sulfuric acid solutions, *Corrosion* **23**, 109–114 (1967).

N. L. Conger and O. L. Riggs, Jr., A precision high current potentiostat, *Corrosion* **23**, 181–184 (1967).

G. Trabanelli, F. Zucchi, and A. Betti, The effect of organic acids on the anodic dissolution of nickel alloys and stainless steels, *Corros. Sci.* **7**, 432–434 (1967).

J. Postlethwaite and D. R. Hurp, Some effects of flow and aeration on the corrosion behaviour of nickel in sulphuric acid solutions, *Corros. Sci.* **7**, 435–445 (1967).

T. P. Hoar, On the relation between corrosion rate and polarization resistance, *Corros. Sci.* **7**, 455–458 (1967).

R. L. Every and W. P. Banks, Reference electrodes in acid and base systems, *Corrosion* **23**, 151–153 (1967).

J. M. Peters and J. R. Myers, Anodic polarization behavior of titanium and titanium alloys in sulfuric acids, *Corrosion* **23**, 326–330 (1967).

Ya. M. Kolotyrkin, Effect of the nature of the anions on the kinetics and mechanism of the dissolution (corrosion) of metals in electrolyte solutions, *Prot. Met.* **3**, 101 (1967).

G. T. Seaman, J. R. Myers, and R. K. Saxer, Anodic polarization behaviour of cobalt–chromium alloys in sulphuric acid solutions, *Electrochim. Acta* **12**, 855–871 (1967).

K. Koslavac and B. Lovrecek, Some aspects of anodic passivity of silicon single crystals, *Corros. Sci.* **7**, 545–551 (1967).

E. A. Lizlovs, Mechanism of the corrosion inhibition of stainless steel in sulfuric acid by sodium molybdophosphate, *J. Electrochem. Soc.* **114**, 1015–1018 (1967).

N. Sato and T. Notoya, Measurement of the anodic oxide film growth on iron for hours, *J. Electrochem. Soc.* **114**, 585–587 (1967).

C. A. Youngdahl and R. E. Loess, Instrumentation for potentiostatic corrosion studies in distilled water, *J. Electrochem. Soc.* **114**, 489–492 (1967).

J. Kruger and J. P. Calvert, Ellipsometric-potentiostatic studies of passivity: I. Anodic film growth in slightly basic solutions, *J. Electrochem. Soc.* **114**, 43–49 (1967).

S. Schuldiner, T. B. Warner, and B. J. Piorsma, Potentiostatic current—potential measurements on a platinum electrode in a high-purity closed system, *J. Electrochem. Soc.* **114**, 343–349 (1967).

1968

N. A. Hampson and N. E. Spencer, Anodic behaviour of tin in potassium hydroxide solution, *Br. Corros. J.* **3**, 1–6 (1968).

J. C. Griess, Jr., Crevice corrosion of titanium in aqueous salt solutions, *Corrosion* **24**, 96–109 (1968).

H. P. Leckie, Effect of pH on the stable passivity of stainless steels, *Corrosion* **24**, 70–74 (1968).

M. E. Komp and H. E. Trout, Jr., Effect of microstructure on anodic polarization of carbon steel in sulfuric acid, *Corrosion* **24**, 11–16 (1968).

O. L. Riggs, Jr. and W. P. Banks, Corrosion problems in phosphoric acid production, *Mater. Prot.* **7**(5) 39–42 (May 1968).

R. F. Steigerwald, Electrochemistry of corrosion, *Corrosion* **24**, 1–10 (1968).

T. P. Dirkse, D. DeWitt, and R. Shoemaker, The anodic behavior of zinc in KCH solutions, *J. Electrochem. Soc.* **115**, 442–444 (1968).

D. C. Mears and G. P. Rothwell, Effects of probe position on potentiostatic control during the breakdown of passivity, *J. Electrochem. Soc.* **115**, 36–38 (1968).

H. J. Cleary, Microelectrodes for corrosion studies, *Corrosion* **24**, 169 (1968).

J. Horvath and H. H. Uhlig, Critical potentials for pitting corrosion of Ni, Cr–Ni, Cr–Fe, and related stainless steels, *J. Electrochem. Soc.* **115**, 791–795 (1968).

W. P. Banks and M. Hutchison, Preventing pitting and general corrosion of stainless steel in phosphoric acid, *Mater. Prot.* **7**, 37–39 (September 1968).

G. H. Cartledge, Passivation and activation of iron in the presence of molybdate ions, *Corrosion* **24**, 223–236 (1968).

W. D. France, Jr., and N. D. Greene, Passivation of crevices during anodic protection, *Corrosion* **24**, 247–251 (1968).

W. D. France, Jr., and N. D. Greene, Predicting the intergranular corrosion of austenitic stainless steels, *Corros. Sci.* **8**, 9–18 (1968).

D. A. Jones, Polarization in high resistivity media, *Corros. Sci.* **8**, 19–27 (1968).

M. Clarke and R. G. P. Elbourne, The passivation potential of 65/35 electrodeposited Sn–Ni alloy, *Corros. Sci.* **8**, 29–39 (1968).

L. Felloni, The effect of pH on the electrochemical behavior of iron in hydrochloric acid, *Corros. Sci.* **8**, 133–148 (1968).

G. C. Wood and G. F. Cammack, The influence of Si and Al on the anodic passivation of Fe–Cr alloys, *Corros. Sci.* **8**, 159–171 (1968).

R. P. Frankenthal, The effect of surface preparation and of deformation on the pitting and anodic dissolution of Fe–Cr alloys, *Corros. Sci.* **8**, 491–498 (1968).

R. F. Ashton and M. T. Hepworth, Effect of crystal orientation on the anodic polarization and passivity of zinc, *Corrosion* **24**, 50–53 (1968).

J. A. S. Green and R. N. Parkins, Electrochemical properties of ferrite and cementite in relation to stress corrosion of mild steels in nitrate solutions, *Corrosion* **24**, 66–69 (1968).

M. M. Mottern and J. R. Myers, Polarization behavior of iron–cobalt alloys in sulfuric acid solutions, *Corrosion* **24**, 197–205 (1968).

J. R. Myers, E. G. Gruenler, and L. A. Smulczenski, Improved working electrode assembly for electrochemical measurements, *Corrosion* **24**, 352,353 (1968).

M. H. Tikkanen and O. Hyvarinen, On the passivity of Ni–Cr alloys, reprint of paper presented at the Fifth Scandinavian Corrosion Congress (1968).

B. E. Wilde and J. S. Armijo, Influence of silicon and manganese on corrosion behavior of austenitic stainless steels, *Corrosion* **24**, 393–402 (1968).

W. D. France, Jr. and N. D. Greene, Interpretation of passive current maxima during polarization of stainless steels, *Corrosion* **24**, 403–406 (1968).

B. E. Wilde, The influence of hydrogen, oxygen and ammonia on the corrosion behavior of plain carbon steel in high temperature water, *Corrosion* **24**, 338–343 (1968).

S. R. Maloof, Dislocation etch pitting studies on tungsten single crystals under potentiostatic conditions, *Corrosion* **24**, 283–290 (1968).

W. D. France, Jr. and R. W. Lietz, Improved data recording for automatic potentiodynamic polarization measurements, *Corrosion* **24**, 298–300 (1968).

R. B. Leonard, The electrochemical behavior of some nickel base and cobalt base alloys, *Corrosion* **24**, 301–307 (1968).

A. B. Ijzermans and A. J. van der Krogt, Pitting corrosion of an austenitic Cr–Ni stainless steel in H_2SO_4 containing H_2S, *Corros. Sci.* **8**, 679–687 (1968).

J. S. Armijo and B. E. Wilde, Influence of Si content on the corrosion resistance of austenitic Fe–Cr–Ni alloys in oxidizing acids, *Corros. Sci.* **8**, 649–664 (1968).

G. Bianchi, G. A. Camona, G. Fiori, and F. Mazza, Galvanic platinizing and passivity of stainless steel in H_2SO_4, *Corros. Sci.* **8**, 751–757 (1968).

W. E. Cowley, F. P. A. Robinson, and J. E. Kerrich, Anodic protection for the control of corrosion fatigue, *Br. Corros. J.* **3**, 223–237 (1968).

J. M. Stammen, Is anodic protection effective in preventing steel corrosion in H_2SO_4? *Mater. Prot.* **7**(12) 33–35 (December 1968).

C. M. Shepherd and S. Schuldiner, Potentiostatic current–potential measurements of iron and platinum electrodes in high-purity closed alkaline systems, Naval Research Laboratory Report No. 6748, Washington, D.C. (August 26, 1968).

1969

I. D. Dirmeik, Anodic passivation of metals in fused salt electrolytes, *Corrosion* **25**, 180 (1969).

J. M. Stammen and H. A. Webster, Anodic protection, Preliminary Technical Practices Committee Report, NACE, Houston, TX (1969).

D. A. Jones, Anodic polarization of anodized aluminum, *Corrosion* **25**, 187 (1969).

W. E. Henry and B. E. Wilde, An automatic polarization apparatus for electrochemical corrosion studies, *Corrosion* **25**, 515–518 (1969).

J. E. Reinoehl and F. H. Beck, Passivity and anodic protection, *Corrosion* **25**, 233–242 (1969).

W. A. Mueller, Theoretical considerations in the measurements of polarization and corrosion potential curves, *Corrosion* **25**, 473 (1969).

M. Pourbaix, Recent applications of electrode potential measurements in the thermodynamics and kinetics of corrosion of metals, *Corrosion* **25**, 267–281 (1969).

D. Shaw, E. Fletcher, and J. S. Wilde, An automatic potentiostat system for corrosion studies, *Br. Corros. J.* **4**, 249 (1969).

F. L. LaQue, Electrochemistry and corrosion (research and tests), Acheson Metal Address, *J. Electrochem. Soc.* **116**, 73C (1969).

R. L. Horst, Jr., E. H. Hollingsworth, and W. King, A new solution potential measurement for predicting stress corrosion performance of 2219 aluminum alloy, *Corrosion* **25**, 199 (1969).

R. L. Chance and W. D. France, Jr., Anodic polarization characteristics of phosphated steels, *Corrosion* **25**, 329 (1969).

B. F. Wilde and N. D. Greene, The variable corrosion resistance of 18Cr–8Ni stainless steels: Behavior of commercial alloys, *Corrosion* **25**, 300 (1969).

E. Fot and J. Weber, Progress in corrosion research—our new facilty for the acquisition and processing of electrochemical measurement data, *Sulzer Tech. Rev.*, No. 3 (1969).

M. A. Streicher, The time factor in potentiostatic studies of intergranular corrosion of austenitic stainless steels, *Corros. Sci.* **9**, 53 (1969).

W. D. France, Jr., Controlled potential corrosion tests, their applications and limitations, *Mater. Res. Stand.* **9**(8) 21 (1969).

M. Levy and G. N. Sklover, Anodic polarization of titanium and titanium alloys in hydrochloric acid, *J. Electrochem. Soc.* **116**, 323 (1969).

D. A. Jones and N. D. Greene, Electrochemical detection of localized corrosion, *Corrosion* **25**, 367–370 (1969).

F. M. Donahue, Comments on anodic tafel slopes on iron, *Corrosion* **25**, 379 (1969).

A. Piqeaud and H. B. Kirkpatrick, A correlated potentiostatic microscopic study of iron passivation in sulfuric acid, *Corrosion* **25**, 209 (1969).

1970

J. M. West, Applications of potentiostats in corrosion studies, *Br. Corros. J.* **5**, 65–71 (1970).

P. Neufeld and E. D. Queenan, Frequency dependence of polarisation resistances measured with square-wave alternating potential, *Br. Corros. J.* **5**, 72–75 (1970).

J. G. N. Thomas, T. J. Nurse, and R. Walker, Anodic passivation of iron in carbonate solutions, *Br. Corros. J.* **5**, 87–92 (1970).

K. Schwabe and M. Schmidt, Der Einfluss des Wassers auf die Passivierbarkeit von Nickel in schwefelsaurer Lösung, *Corros. Sci.* **10**, 143–155 (1970).

C. M. Chen, F. H. Beck, and M. G. Fontana, Stress corrosion cracking of Ti–8Al–1Mo–1V alloy: Electrochemical behavior in aqueous solutions, *Corrosion* **26**, 135–140 (1970).

J. E. Reinoehl, F. H. Beck, and M. G. Fontana, Effect of polarization rate in potentiokinetic anodic polarization of iron in $1N$ H_2SO_4, *Corrosion* **26**, 141–150 (1970).

T. C. Finley and J. R. Myers, Effect of cold work on anodic polarization of iron in sulfuric acid, *Corrosion* **26**, 150–152 (1970).

T. P. Hoar and J. R. Galvele, Anodic behaviour of mild steel during yielding in nitrate solutions, *Corros. Sci.* **10**, 211–224 (1970).

Ya. Omurtag and M. Doruk, Some investigations on the corrosion characteristics of Fe–Si alloys *Corros. Sci* **10**, 225–231 (1970).

J. A. von Fraunhofer, The polarization behaviour of mild steel in aerated and deaerated $1M$ $NaHCO_3$, *Corros. Sci.* **10**, 245–251 (1970).

G. Trabanelli and F. Zucchi, Measuring the corrosion resistance of four alloys in Cl solutions, *Mater. Prot. Perform.* **9**(7) 16–19 (July 1970).

M. E. Indig and C. Groot, Electrochemical measurement of corrosion, *Corrosion* **26**, 171–176 (1970).

W. D. France, Jr., Effects of stress and plastic deformation on the corrosion of steel, *Corrosion* **26**, 189–199 (1970).

A. B. Ijzermans, Effect of the potential traverse rate on the potentiokinetic polarization curves of Ni in $1N$ H_2SO_4, *Corros. Sci.* **10**, 113–117 (1970).

W. D. France, Jr. and N. D. Greene, Comparison on chemically and electrochemically induced pitting corrosion, *Corrosion* **26**, 1 (1970).

H. W. Leavenworth, Jr., J. P. Carter, and D. Schlain, The mechanism of corrosion of columbium in sodium hydroxide, *Corrosion* **26**, 43–50 (1970).

A. Akiyama, R. E. Patterson, and K. Nobe, Electrochemical characteristics of iron during corrosion: Effect of heat treatment and purity *Corrosion* **26**, 51–57 (1970).

R. T. Foley, Role of the chloride ion in iron corrosion, *Corrosion* **26**, 58–70 (1970).

R. A. Legault, S. Mori, and H. P. Leckie, An electrochemical-statistical study of the effect of the chemical environment on the corrosion behavior of mild steel, *Corrosion* **26**, 121–128 (1970).

G. P. Cammarota, L. Felloni, G. Palombarini, and S. Sostero Traverso, Optical microscopy studies of anodic dissolution of iron in sulfuric and hydrochloric acid solutions: Influence of metal purity, structure, heat treatment, *Corrosion* **26**, 129–140 (1970).

F. G. Hodge and B. E. Wilde, Effect of chloride ion on the anodic dissolution kinetics of chromium–nickel binary alloys in dilute sulfuric acid, *Corrosion* **26**, 146–150 (1970).

O. L. Riggs, Jr., A potentiostatic and galvanostatic study of carbon steel surfaces in NH_3–NH_4NO_3–H_2O solutions, *Corrosion* **26**, 155–157 (1970).

H. Brandt, M. Fischer, and K. Schwabe, Untersuchungen über die Sparbeizwirkung organischer Sulfide im System Eisen–Schwefelsäure, *Corros. Sci.* **10**, 631–639 (1970).

D. Rettig, C. Voigt, and K. Schwabe, Untersuchunger über periodische Potentialschwankungen des Zirkons in NaF–Lösungen—I. Experimentelle Ergebnisse, *Corros. Sci.* **10**, 657–669 (1970).

H. G. Feller, C. Kamp, and M. Kesten, Untersuchungen über den Einfluss der magnetischen Umwandlung auf das anodische Verhalten von Ni–W und Ni–V-Legierungen, *Corros. Sci.* **10**, 687–692 (1970).

R. F. North and M. J. Pryor, The influence of corrosion product structure on the corrosion rate of Cu–Ni alloys, *Corros. Sci.* **10**, 297–311 (1970).

G. Okamoto and T. Shibata, Stability of passive stainless steel in relation to the potential of passivation treatment, *Corros. Sci.* **10**, 371–378 (1970).

N. Azzerri, L. Giuliani, and G. Bombara, Influence of surface finish on passivity retention on stainless steels, *Corrosion* **26**, 381–386 (1970).

M. N. Bentley and A. G. Denham, A simple laboratory potentiostat, *Br. Corros. J.* **5**, 227–229 (1970).

V. K. Gouda, Corrosion and corrosion inhibition of reinforcing steel—I. Immersed in alkaline solutions, *Br. Corros. J.* **5**, 198–203 (1970).

V. K. Gouda and W. Y. Halaka, Corrosion and corrosion inhibition of reinforcing steel—II. Embedded in concrete, *Br. Corros. J.* **5**, 204–208 (1970).

R. Grauer and U. Gut, Thermodynamische und morphologische aspekte der Korrosion in Trinkwasserähnlichen Lösungen—II. Cadmium in Hydrogencarbonatlösungen bei Temperaturen von 25 bis 80°C, *Corros. Sci.* **10**, 503–511

R. Grauer, U. Gut, and K. Blaser, Thermodynamische und Morphologische Aspekte der Korrosion in Trinkwasserahnlichen Losungen—I. Zink in Hydrogencarbonatlosungen bei Temperaturen von 25 bis 80°C, *Corros. Sci.* **10**, 489–502 (1970).

V. Ashworth, P. J. Boden, J. S. L. Leach, and A. Y. Nehru, On the Cl^- breakdown of passive films on mild steel, *Corros. Sci.* **10**, 481–488 (1970).

J. O'M. Bockris and P. K. Subramanyan, Contributions to the electrochemical basis of the stability of metals, *Corros. Sci.* **10**, 435–466 (1970).

A. A. Abdul Azim, Pitting corrosion of Pb in $H_2SO_4 + HCLO_4$, *Corros. Sci.* **10**, 421–433 (1970).

H. Helber and E. L. Littauer, Anodic behaviour of Pb–Pt bielectrodes in low-salinity electrolytes, *Corros. Sci.* **10**, 411–419 (1970).

M. Kesten and H. G. Feller, Über den Einfluss von adsorbierten Anionen auf die magnetische Umwandlung von Ni–Mo-Legierungen, *Corros. Sci.* **10**, 401–409 (1970).

P. Rothenbacher, Zur Entzinkung von rekristallisierten und statisch belasteten Kupfer–Zink-Legierungen mit 30 at.% Zink, *Corros. Sci.* **10**, 391–400 (1970).

H. Fietler, Instantaneous corrosion rate measurement, *Mater. Prot. Perform.* **9**(10), 37–41 (October 1970).

L. Giuliani and G. Bombara, An electrochemical study of copper alloys in chloride solutions, *Br. Corros. J.* **5**, 179–183 (1970).

A. B. Ijzermans, Pitting Corrosion and Intergranular attack of austenitic Cr–Ni stainless steels in NaSCN, *Corros. Sci.* **10**, 607–615 (1970).

P. E. Morris and R. C. Scarberry, Anodic polarization measurements of active–passive nickel alloys by rapid-scan potentiostatic techniques, *Corrosion* **26**, 169–179 (1970).

L. E. Kindlimann and N. D. Greene, Mechanism for the acid corrosion behavior of neutron-irradiated uranium, *Corrosion* **26**, 189–192 (1970).

K. G. Compton, The use of reference electrodes in corrosion studies, *Mater. Res. Stand.* **10**(1), 13 (1970).

G. Baudo, A. Tamba, and G. Bombara, An electrochemical investigation of corrosion of ferritic steels in molten sulfates, *Corrosion* **26**, 193–199 (1970).

J. W. Johnson, C. H. Chi, C. K. Chen, and W. J. James, The anodic dissolution of Molybdenum, *Corrosion* **26**, 238–242 (1970).

O. L. Riggs, Jr., A modified electrochemical expression for corrosion rate measurements, *Corrosion* **26**, 243–245 (1970).

J. Flis, Effect of C on the corrosion of Fe in NH_4NO_3 solution within a wide potential range, *Corros. Sci.* **10**, 745–759 (1970).

C. D. Kim and B. E. Wilde, Influence of cathodic pretreatment on the anodic dissolution kinetics of stainless steels in dilute acid media, *Corros. Sci.* **10**, 735–744 (1970).

V. Ashworth and P. J. Boden, Potential–pH diagrams at elevated temperatures, *Corros. Sci.* **10**, 709–718 (1970).

J. R. Galvele and S. M. de Micheli, Mechanism of intergranular corrosion of Al–Cu alloys, *Corros. Sci.* **10**, 795–807 (1970).

N. Sato, K. Kudo, and T. Noda, Single layer of the passive film on Fe, *Corros. Sci.* **10**, 785–794 (1970).

A. Sh. Valeev, Polarization curves of the anodic dissolution of metals with decreasing characteristics, *Mater. Nauchn. Konf., Inst. Org. Fiz. Khim.*, Akad. Nauk. USSR 1969, 219–221 (1970).

V. A. Suprunov, M. K. Freid, and N. S. Baburina, Anodic protection of stainless steel in scouring baths containing sulfuric acid, *Tr. Ivanov. Khim. Teknol. Inst.* **12**, 180–184 (1970).

E. A. Lizlovs, Polarization cell for potentiostatic crevice corrosion testing, *J. Electrochem. Soc.* **117**, 1235–1237 (1970).

P. E. Morris and R. C. Scarberry, Anodic polarization measurements of active–passive nickel alloys by rapid-scan potentiostatic techniques, *Corrosion* **26**, 169–179 (1970).

O. L. Riggs, Jr., A potentiostatic and galvanostatic study of carbon steel surfaces in NH_3-NH_4NO_3-H_2O solutions, *Corrosion* **26**, 155–157 (1970).

R. L. Jones, L. W. Strattan, and E. D. Osgood, The potentiostatic passivation of mild steel in 300 °C NaOH solution, *Corrosion* **26**, 399–405 (1970).

P. E. Morris and R. C. Scarberry, Anodic polarization measurements of active–passive nickel alloys by rapid-scan potentiostatic techniques, *Corrosion* **26**, 169–179 (1970).

T. C. Finley and J. R. Myers, Effect of temperature on anodic protection of iron in sulfuric acid, *Corrosion* **26**, 544–546 (1970).

M. Pourbaix, Significance of protection potential in pitting and intergranular corrosion, *Corrosion* **26**, 431, 432 (1970).

N. D. Greene and B. E. Wilde, Variable corrosion resistance of 18Cr–8Ni stainless steels: Influence of environmental and metallurgical factors, *Corrosion* **26**, 533–538 (1970).

E. Mansfeld and E. P. Parry, A quantitative electrochemical test for porosity in permalloy plated memory wire, *Corrosion* **26**, 542–544 (1970).

J. W. Johnson, M. S. Lee, and W. J. James, Electrochemical behavior of molybdenum in acid chloride solutions, *Corrosion* **26**, 507–510 (1970).

B. Lovrecek and S. Lipanovic, Investigation of anodic passivity of Ni, *Corros. Sci.* **10**, 865–874 (1970).

1971

P. Forchhammer and H.-J. Engell, Investigations of the corrosion of stainless steel by potentiokinetic measurements, *Corros. Sci.* **11**, 49–53 (1971).

N. Subramanyan, V. Kapali, and S. Venkitakrishna Iyer, Influence of hydroxy compounds on the corrosion and anodic behaviour of Al in 1 N NaOH solutions, *Corros. Sci.* **11**, 115–123 (1971).

I. F. Danzig, R. M. Dempsey, and A. B. LaConti, Characteristics of tantalided and hafnided samples in highly corrosive electrolyte solutions, *Corrosion* **27**, 55–62 (1971).

F. Zucchi and G. Trabanelli, Anodic behaviour of Fe in phosphate solutions, *Corros. Sci.* **11**, 141–151 (1971).

M. B. Rockel, Interpretation of the second anodic current maximum on polarization curves of sensitized chromium steels in 1 N H_2SO_4, *Corrosion* **27**, 95–103 (1971).

Y. Tsuji and M. Ichikawa, Passivity of electrodeposited Sn–Co alloy, *Corrosion* **27**, 168, 169 (1971).

R. K. Flatt and P. A. Brook, The effects of anion-concentration on the anodic polarization of copper, zinc, and brass, *Corros. Sci.* **11**, 185–196 (1971).

R. P. Jackson and D. Van Rooyen, Electrochemical evaluation of resistance of stainless alloys to chloride media, *Corrosion* **27**, 203–210 (1971).

J. P. Hoare, K.-W. Mao, and A. J. Wallace, Jr., ECM behavior in $NaClO_4$ electrolyte, *Corrosion* **27**, 211–215 (1971).

I. Gary and V. Hafke, Der Einfluss der Metallstruktur auf das Passivierungsverhalten von Nickel, *Corros. Sci.* **11**, 329–336 (1971).

L. Giuliani, G. Agabio, and G. Bombara, Breakdown of passivity in low-alloy steels, *Corros. Sci.* **11**, 403–410 (1971).

E. D. Verink, Jr. and M. Pourbaix, Use of electrochemical hysteresis techniques in developing alloys for saline exposures, *Corrosion* **27**, 495–505 (1971).

K. S. Rajagopalan and K. Venu, Activation of passivated steel in sodium chromate solution containing sodium chloride, *Corrosion* **27**, 506–508 (1971).

R. M. Latanision and H. Opperhauser, Passivation of nickel monocrystal surfaces, *Corrosion* **27**, 509–515 (1971).

M. Indig and D. A. Vermilyea, A reference electrode for corrosion studies in aqueous solutions at temperatures to 289°C, *Corrosion* **27**, 312, 313 (1971).

B. J. Bornong and P. Martin, Jr., Ellipsometric-potentiostatic studies of steel corrosion: oxide film development and effects of inhibitors, *Corrosion* **27**, 315–319 (1971).

R. L. Jones and E. W. Steinkuller, Lithium hydroxide effects in the potentiostatic passivation of mild steel at 300°C, *Corrosion* **27**, 353–359 (1971).

L. Troselius, Polarization performance of stainless steels in H_2SO_4 and HCl, *Corros. Sci.* **11**, 473–484 (1971).

C. M. Chen, F. H. Beck, and M. G. Fontana, Crevice effect during polarization of Ti–8 Al–1 Mo–1 V alloy in aqueous and methanol environments, *Corrosion* **27**, 234–238 (1971).

C. P. Kin and K. Nobe, Polarization of copper in acidic and alkaline solutions, *Corrosion* **27**, 382–385 (1971).

T. G. Gooch, J. Honeycombe, and P. Walker, Potentiostatic study of the corrosion behaviour of austenitic stainless steel weld metal, *Br. Corros. J.* **6**, 148–154 (1971).

Z. Szklarska–Smialowska and M. Janik–Czachor, The analysis of electrochemical methods for the determination of characteristic potentials of pitting corrosion, *Corros. Sci.* **11**, 901–914 (1971).

W. B. Crow, J. R. Myers, and B. D. Marvin, Anodic polarization of Ni–Al alloys in sulfuric acid solutions, *Corrosion* **27**, 459–465 (1971).

M. Kesten and H. G. Feller, Über die Rolle der Sulphationen bei der anodischen Auflösung und bei der Passivierung von Nickel, *Electrochim. Acta* **16**, 763–778 (1971).

B. E. Wilde, The role of passivity in the mechanism of stress–corrosion cracking and metal dissolution of 18Cr–8Ni stainless steels in boiling magnesium and lithium chlorides, *J. Electrochem. Soc.* **118**, 1717–1725 (1971).

G. Bombaro, A. Tamba, and N. Azzerri, Potentiostatic anodic pickling of stainless steel, *J. Electrochem. Soc.* **118**(4) 675–681 (April 1971).

S. Asakura and K. Nobe, Kinetics of anodic processes on iron in alkaline solutions, *J. Electrochem. Soc.* **118**, 536–541 (1971).

G. Bombara and G. Agabio, An investigation of anodic behaviour of low alloy steels, *Corrosion* **27**, 26–31 (1971).

W. B. Crow, J. R. Myers, and B. D. Marvin, Anodic polarization behaviour of Ni–Al alloys in sulfuric acid solutions, *Corrosion* **27**, 459–465 (1971).

R. M. Latanision and H. Opperhauser, Passivation of nickel monocrystal surfaces, *Corrosion* **27**, 509–515 (1971).

W. K. Behl and J. E. Toni, Anodic oxidation of cobalt in potassium hydroxide electrolyte, *J. Electroanal. Chem.* **31**, 63–75 (1971).

C. Lamy and P. Malaterre, Analyse théorique du fonctionnement et de la stabilité d'un système potentiostatique à large bande passante avec correction automatique de chute ohmigue, *J. Electroanal. Chem.* **32**, 137–151 (1971).

R. G. Casino, J. J. Podestá, and A. J. Arvia, On the anodic passivity of iron in molten potassium bisulfate, *Electrochim. Acta* **16**, 121–130 (1971).

N. Sato and K. Kudo, Ellipsometry of the passivation film on iron in neutral solution, *Electrochim. Acta* **16**, 447–462 (1971).

R. V. Moshtev, Temperature dependence of passive film growth on iron at constant current, *Electrochim. Acta* **17**, 2401 (December 1972); N. Sato, *Electrochem. Acta* **16**, 659 (1971).

1972

S. Ya Grilikhes, B. S. Krasikov, T. G. Apartsina, and E. D. Levin, Anodic polarization of an iron–nickel alloy in a sulfuric acid solution, *Zh. Prikl. Khim.* **45**, 2252–2256 (October 1972).

F. Mansfeld, Passivity and pitting of aluminum, nickel, titanium, and stainless steel in methanol plus sulfuric acid, *J. Electrochem. Soc.* **120**, 188–192 (February 1972).

R. Bruno and F. Calderon, Anodic behaviour of 17–7 precipitation hardening stainless steel in acid media, *Br. Corros. J.* **7**, 174–179 (April 1972).

N. Barbouth, J. Pagetti, and J. Qudar, Electrochemical study of different iron–chromium alloys in a sodium sulfate medium, *C. R. Acad. Sci. Ser. C.* **275**(18), 997–1000 (1972).

V. M. Novakouskii, and L. L. Faingold, Protection properties of an air passivated film on the iron anode in an acid electrolyte, *Zashch. Met.* **8**, 683–685 (June 1972).

R. V. Moshtev, Temperature dependence of passive film growth on iron at constant current, *Electrochim. Acta* **17**, 2401 (December 1972); N. Sato, *Electrochim. Acta* **16**, 659 (1971).

V. K. Altukhev, I. K. Marshakev, E. S. Vorontsov, and D. E. Emel'yanov, The effect of concentration of chloride ions on the kinetics of anodic dissolution of copper, *Izv. Vyssh. Ucheb. Zaved. Khim. Khim. Teknol.* **15**, 1752–1754 (November 1972).

R. G. Bringham, Pitting of molybdenum bearing austenitic stainless steel, *Corrosion* **28**, 177–179 (1972).

A. J. Stavros and H. W. Paxton, Effects of heat treatment on the potentiostatic polarization behaviour of an 18% Ni maraging steel, *Corros. Sci.* **12**, 739–748 (1972).

C. J. Chatfield and L. L. Shreir, Effect of sweep rate on the active–passive transitions of the Ni/H_2SO_4 systems, *Corros. Sci.* **12**, 563–570 (1972).

T. Suzuki and Y. Kitamura, Critical potential for growth localized corrosion of stainless steel in chloride media, *Corrosion* **28**, 1–6 (1972).

R. Bruno and F. Calderone, Anodic behavior of 17–7 precipitation hardening stainless steel in acid media, *Br. Corros. J.* **7**, 174–179 (April 1972).

N. Subramanyan and M. Krishman, Corrosion and anodic polarisation behaviour of aluminum in sodium hydroxide solution containing sucrose or allied products, with and without calcium, *Br. Corros. J.* **7**, 184–188 (1972).

B. L. Trout and R. D. Daniels, Potentiostatic polarization of inconel X-750, *Corrosion* **28**, 331–336 (1972).

F. Zucchi, G. Gilli, P. A. Borea, and G. Trabanelli, Anodic behaviour of Fe/Ni alloys in concentrated H_2SO_4, *Corros. Sci.* **12**, 699–711 (1972).

W. B. Crow, J. R. Myers, and J. V. Jeffreys, Anodic polarization behavior of Fe–Si alloys in sulfuric acid solutions, *Corrosion* **28**, 77–82 (1972).

D. O. Condit, Effect of chloride ion concentration on the polarization behavior of an Fe–25Ni alloy, *Corrosion* **28**, 95–100 (1972).

A. A. Abdul Azim and K. M. El-Sobki, Corrosion and passivity of Pb in strongly alkaline solutions, *Corros. Sci.* **12**, 207–215 (1972).

A. J. Sedriks, J. A. S. Green, and K. D. L. Novak, Electrochemical behavior of Ti–Ni alloys in acidic chloride solutions, *Corrosion* **28**, 137–142 (1972).

E. Elayaperumal, P. K. De, and J. Balachandra, Passivity of type 304 stainless steel—effect of plastic deformation, *Corrosion* **28**, 269–273 (1972).

G. Jouve, M. Patriciu, and M. Aucouturier, C and Cr segregation and precipitation in inconel alloys (77Ni–16Cr–7Fe) and electrochemical behavior of these passivated alloys in H_2SO_4 solution, *Corros. Sci.* **12**, 537–541 (1972).

R. Grauer ana H. Kaesche, Elektronenmikroskopische und elektrochemische Untersuchungen über die Passivierung von Zink in 0.1 *M* Natronlauge, *Corros. Sci.* **12,** 617–624 (1972).

S. Lipanovic, Investigation of anodic passivity of nickel in approximately neutral solution of NiSO₄, *Corros. Sci.* **12,** 637–649 (1972).

P. Neufeld and A. F. Bromley, Anodic passivation of mild steel in hot concentrated sodium hydroxide solution, *Br. Corros. J.* **7,** 285, 286 (June 1972).

T. M. Alkhazishvili, V. A. Dolidze, R. K. Ralitchev, L. A. Danielyan, V. A. Makarov, and Ya. M. Kolotyrkin, Monitoring the potential of chemical equipment, *Zashch. Met.* **8,** 742–744 (June 1972).

G. Bouyssoux and J. De Becdelievre, Anodic behaviour of single and polycrystalline chromium electrodes in sulfuric acid, *J. Solid State Chem.* **5,** 346–353 (May 1972).

H. Kaesche, Electrochemical methods of corrosion protection, *Proceedings of the International Congress on Metallic Corrosion, 1969,* 4th edition, pp. 15–25, NACE, Houston, TX (1972).

A. K. Hanna, P. Novak, F. Franz, and J. Korritta, Corrosion of anodically passivated iron–chromium nickel alloys, *Sb. Vys. Sk. Chem. Technol. Praze. Anorg. Chem. Technol.* **B15,** 207–217 (1972).

P. I. Zabotin and A. A. Tibekina, Corrosion and potentiodynamic characteristics of chromium behaviour in some solutions of different nature, *Tr. Inst. Org. Katal. Elektrokhim., Akad. Nauk. Kaz. SSR* **3,** 97–106 (1972).

K. I. Noninski, R. G. Raichev, Z. L. Georgiev, Corrosion behaviour of electrolytically deposited tin–nickel alloys, *Khim. Ind.* **44,** 456–459 (October 1972).

T. Markovic and E. G. Leonlopez, Flade potential consideration reversible potential from a thermodynamic viewpoint, *Met. Corros. Ind.* **47**(567), 379–382 (1972).

I. Eppelboin, C. Gabrielli, M. Keddam, and J. C. Lestrade, Passivation of iron in sulfuric acid medium, *J. Electrochem. Soc.* **119,** 1632–1637 (December 1972).

H. Saito, K. Tachibana, and G. Okamoto, Effect of potential of passivation treatment on the pit formation in stainless steel, *Boshoku Gijutsu* **22,** 18–22 (January 1972).

V. Brusil, Passivation and passivity, *Oxides and Oxide films* **1,** 1–89 (1972).

J. Siejka, C. Cherki, and J. Yahalom, Study of nickel passivity by nuclear microanalysis of oxygen-16 and oxygen-18 isotopes, *Electrochim. Acta* **17,** 2371–2380 (December 1972).

C. L. McBee and J. Kruger, Nature of passive films on iron–chromium alloys, *Electrochim. Acta* **17,** 1337–1341 (1972).

A. J. Arvia, J. J. Podestá, and R. C. V. Piatti, Kinetics of iron passivation and corrosion in molten alkali nitrates, *Electrochim. Acta* **17,** 33–44 (1972).

J. Siejka, C. Cherki, and J. Yahalom, A study of nickel passivity by nuclear microanalysis of O¹⁶ and O¹⁸ isotopes, *Electrochim. Acta* **17,** 161–170 (1972).

R. D. Armstrong and H. R. Thirsk, The effect of thin anodic films on metal dissolution, *Electrochim. Acta* **17,** 171–176 (1972).

I. Eppelboin and M. Keddan, Kinetics of formation of primary and secondary passivity in sulphuric aqueous media, *Electrochim. Acta* **17,** 177–186 (1972).

H. G. Feller, H. J. Ratzer–Schiebe, and W. Wendt, Anodic dissolution of Ni in 1*N* H₂SO₄ studied by dynamic impedance measurements, *Electrochim. Acta* **17,** 187–196 (1972).

R. Knoedler and K. E. Heusler, The mechanism of oxidation of chromium to chromate, *Electrochim. Acta* **17,** 197–212 (1972).

P. E. Morris and R. C. Scarberry, Predicting corrosion rates with the potentiostat, *Corrosion* **28,** 444–452 (December 1972).

K. Schwabe, U. Ebersbach, and W. Leimbrick, Zur Kinetic der Passivierung von Eisen, Kobalt und Nickel—V, *Electrochim. Acta* **17,** 957–963 (1972).

U. Ebersbach, G. Kreysa, and K. Schwabe, Zur Kinetic der anodischen Passivierung von Eisen, Kobalt, und Nickel—IV, *Electrochim. Acta* **17,** 445–450 (1972).

G. M. Bulman and A. C. C. Tseung, The kinetics of the anodic formation of the passive film on stainless steel, *Corros. Sci.* **12**, 415–432 (1972).

D. O. Condit, Potentiodynamic polarization studies of Fe–Ni binary alloys in sulfuric acid solution at 25°C, *Corros. Sci.* **12**, 451–462 (1972).

W. B. Crow, J. R. Myers, and J. V. Jeffreys, Anodic polarization behaviour of Fe–Si alloys in sulfuric acid solutions, *Corrosion* **28**, 77–82 (1972).

B. L. Trout and R. D. Daniels, Potentiostatic polarization of inconel X-750, *Corrosion* **28**, 331–336 (1972).

A. A. Aboul Axim and S. H. Sanad, The anodic dissolution of Fe and C steel in acidic sulfate solution, *Electrochim. Acta* **17**, 1699–1704 (September 1972).

T. M. Alkhazishvili, V. A. Dolidze, R. K. Kalitchev, L. A. Danielyan, V. A. Makarov, and Ya. M. Kolotyrkin, *Zashch. Met.* **8**(6), 742–744 (1972).

M. K. Freid and V. A. Suprunov, Anodic dissolution of chromium carbide in a sulfuric acid solution, *Tr. Ivanov. Khim. Teknol. Inst.* **12**, 168–172 (1972).

E. T. Shapovalov, Potentiostatic and potentiodynamic studies of hard tungsten–cobalt and titanium–cobalt alloys in Na_2SO_4 solutions, *Sb. Tr. Vses. Nauchno.-Issled. Instrum. Inst.*, No. 1, 55–60 (1972).

V. M. Filimonenko and B. M. Kreichman, Passivation during anodic dissolution of tungsten carbide–cobalt alloys, *Tr. Novosib. Elektrotekh. Inst.*, No. 2, 71–74 (1972).

R. P. Jackson and D. Van Rooyen, Crevice corrosion of nickel–chromium–molybdenum–iron alloys in laboratory tests, *Amer. Soc. Test. Mater.*, Spec. Tech. Publ., **516**, 210–221 (1972).

V. N. Filimonenko and B. M. Kreichman, Passivation during anodic dissolution of tungsten carbide–cobalt alloys, *Tr. Novosib. Elektrotekh. Inst.*, No. 2, 71–74 (1972).

1973

C. Wagner, Models for lattice defects in oxide layers on passivated iron and nickel, *Ber. Bunsenges. Phys. Chem.* **77**(12) 1090–1097 (1973).

A. Caprani, I. Epelboin, and P. Morel, Dissolution value of titanium in hydrofluoric and sulfuric acid media, *J. Electroanal. Chem. Interfacial Electrochem.* **43** (January 1973).

D. Sinigalia, G. Taccani, and B. Vilentini, Behaviour of titanium in acidic solutions. I. Anodic behaviour and behaviour in abandoned conditions, *Werkst. Korros.* **24**, 1027–1036 (December 1973).

T. Notaya, Anodic behavior of copper in molten alkali metal nitrates, *Denki Kagaku* **41**, 779–783 (October 1973).

G. M. Bulman and A. C. C. Tseung, Ellipsometric study of passive film growth on stainless steel, *Corros. Sci.* **13**, 531–544 (July 1973).

P. I. Buler, G. A. Toporishchev, D. A. Esin, V. B. Lepinsiich, and V. A. Kopysov, Anodic behaviour of cobalt in molten sodium tetraborate, *Zashch. Met.* **9**, 724–727 (June 1973).

K. Ogura, *Passivity of Iron,* University of Texas, Austin (1973).

C. Gabrielli, Regulation and analysis of systems having multiple steady states. Application to the identification of the process of electrochemical passivation of iron, *Met. Corros. Ind.* **48**(573), 171–184 (1973); *Met. Corros. Ind.* **48**(577), 309–327 (1973); *Met. Corros. Ind.* **48**(578), 356–369 (1973).

S. Colac and P. Mateo, Anticorrosive protection methods based on metal passivity property, *Rev. Chim.* **24**, 285–289 (April 1973).

Q. K. Ju, F. Bourelier, E. Chassaing, and J. Montueller, Role of molybdenum and chromium in the corrosion resistance of nickel based alloys in hydrochloric acid, *Mem. Sci. Rev. Metall.* **70**, 417–421 (May 1973).

S. E. Rauch, Jr. and R. N. Steinbicker, Surface chromium on tin plate, *J. Electrochem. Soc.* **120**, 735–738 (June 1973).

W. R. Roser and F. E. Rizzo, Corrosion engineer's look at passive alloys, *Mater. Prot. Perform.* **12**, 51–54 (April 1973).

K. Elayaperumal, S. S. Choutai, and J. Balachanora, Passivity of zirconium in sulfuric acid solutions: Effect of cathodic pretreatment, *Corrosion* **29**, 59–63 (February 1973).

A. A. Voronin, A. G. Efimov, G. I. Korneev, E. K. Kudryavstev, M. S. Grilikhes, M. A. Sokolov, B. P. Artomonov, and L. A. Backholdina, Modular electronic potentiostat, *Zaschch. Met.* **9**, 122–124 (January 1973).

A. A. Yun, I. B. Murashova, and A. V. Pomosov, Nature of the anodic passivity of copper, *Electrokhimiya* **9**, 465–469 (April 1973).

C. Gabrielli and M. Koddam, Heterogeneous reactions coupled by diffusion, multiple stationary states, impedance and stability, *J. Electroanal. Chem. Interfacial Electrochem.* **45**, 267–277 (February 1973).

I. Epelboin, C. Gabrielli and P. Morel, Etude de la passivation du cobalte en milieu sulfurique, *Electrochim. Acta* **18**, 509–513 (July 1973).

V. A. Makarov, Anodic behavior of thallium in solutions containing chloride ions, *Elektrokhimiya* **9**, 1291–1293 (September 1973).

Ubirasjara, Q. Cabral, and L. Sathler, Electrochemical and corrosion behaviour of some low-alloyed steels in acidic medium in commercial manufacture, *Met. ABM* **29**(183), 91–96 (1973).

V. A. Timonin, *Passivity of metals and Alloys,* Moskovskii Institut Stali Splavov, Moscow, USSR (1973).

N. S. D. Elayathu and J. Balachandra, Passivation behaviour of zirconium base alloys in nitric acid solutions containing fluoride, *J. Electrochem. Soc. India* **22**, 243–250 (March 1973).

J. B. Gnanamoorthy and J. Balachandra, Anodic behaviour of commercial stainless steels in sulfuric acid containing chloride ions, *J. Electrochem. Soc. India* **22**, 257–261 (March 1973).

F. F. Faizullin, G. I. Yakhvarov, and V. V. Mosolov, Electrochemical kinetics of the anodic oxidation of zirconium in some salt solutions, *Elektrokhimiya* **9**, 1508–1510 (October 1973).

I. M. Novosel'skii and M. G. Kahkimov, Kinetic theory of the passivation of anodically soluble metal. XIII. Impedance measurements capacitance of the electric double layer, *Elektrokhimya* **9**, 999–1002 (July 1973).

M. Turner, G. E. Thompson, and P. A. Brook, Anodic behaviour of nickel in sulfuric acid solutions, *Corros. Sci.* **13**, 985–991 (December 1973).

K. I. Noninski, R. G. Rakchev, and Z. L. Georgiev, Investigation of the passivity of chromium steels in an acid medium with the acid of a self-cleaning rotating electrode, *Khim. Ind. Sofia* **45**, 311–315 (July 1973).

G. C. Paut, P. K. De, and K. Elayaperumal, Chromium induced passivity of a zirconium–chromium alloy in a chloride solution, *Electrochim. Acta* **18**, 927–931 (December 1973).

T. R. Beck, Electrochemistry of freshly generated titanium surfaces. II. Rapid fracture experiments, *Electrochim. Acta* **18**, 851–937 (November 1973).

D. R. Gabe and P. Sripatr, Anode behaviour of tin during alkaline stannate plating, *Trans. Inst. Metal Finish* **51**, Part 4, 141–144 (1973).

M. Kanda and A. Akiyama, Anodic polarization of mild steel in condensed phosphoric acids and their mixtures with sulfuric acid, *Denki Kagaku* **41**, 818–822 (November 1973).

K. I. Noninski, R. G. Raichev, and Z. L. Georgiev, Anodic behaviour of self-passivating steel studied with a self-cleaning rotating electrode, *Khim. Ind. Sofia* **45**, 407–411 (September 1973).

1974

M. Mito and H. Sugawara, Anodic polarization of iron–chromium binary alloys embrittled at 475° and iron–chromium coupled electrodes in N H_2SO_4, *Nippon Kinzoku Gakkaishi* **38**(1) 22–28 (1974).

R. M. Horton and K. Kim, Anodic polarization behaviour of nickel–manganese alloys in N H_2SO_4 solution, *Corrosion* **30**(1) 13–17 (1974).

R. Lebet, K. Schwabe, and H. Worch, Electrochemical behaviour of the iron–chromium alloys GX40Cr30 and GX120Cr30 in sulfate—and chloride—containing electrolytes, *Werkst. Korros.* **25**, 32–39 (January 1974).

R. D. Armstrong and A. C. Coates, Passivation of iron in carbonate/bicarbonate solutions, *J. Electroanal. Chem. Interfacial Electrochem.* **50**, 303–313 (March 1974).

K. Sugimote, K. Kishi, S. Ikeda, and Y. Sawada, X-ray photoelectron spectra of passivated chromium–nickel (18Cr–8Ni) stainless steel, *Nippon Kinzoku Gakkaishi* **38**, 54–62 (January 1974).

M. Mito and H. Sugawara, Anodic polarization of iron–chromium binary alloys embrittled at 475° and of iron–chromium coupled electrodes in N sulfuric acid, *Nippon Kinzoku Gakkaishi* **38**, 22–28 (August 1974).

M. Okuyama and S. Harvyama, Passive film formed on nickel in a neutral solution, *Corros. Sci.* **14**, 1–14 (January 1974).

S. H. Cadle, Ring-disk electrode study of palladium dissolution, *J. Electrochem. Soc.* **121**, 645–648 (May 1974).

M. Zamin and M. B. Ives, The anodic polarization behaviour of nickel in acid chloride solution, *J. Electrochem. Soc.* **121**(9), 1141–1145 (September 1974).

T. Koizumi and H. H. Uhlig, Passivity and pitting of Ni–Cu alloys in NaCl solutions, *J. Electrochem. Soc.* **121**(9), 1141–1145 (September 1974).

M. Okuyama and S. Haruyama, Passive film formed on nickel in a neutral solution, *Corros. Sci.* **14**, 1–14 (January 1974).

J. Tousek, Ursache der Geschwindigkeitsänderungen des Eisenlochfrasses bei zweitem Depassivierungspotential, *Corros. Sci.* **14**, 543–552 (October 1974).

N. D. Tomashov, G. P. Chernova, Yu. S. Ruscol, and G. A. Ayuyan, The passivation of alloys on titanium bases, *Electrochim. Acta* **19**, 159–172 (April 1974).

M. Okuyama and S. Haruyama, Effects of nickel contents on electrochemical properties of passive films formed on iron–nickel alloys, *Nippon Kinzoku Gakkaishu* **38**(4), 358–365 (1974).

R. Bartonicek, Anodic polarization of curves of brass alloys in ammonium chloride solutions, *Werkst. Korros.* **25**(4), 253–261 (1974).

A. Piotrowski and A. Slimak, Anodic potentiostatic phase dissolution of heterogeneous alloys, *Werkst. Korros.* **25**(4), 262–265 (1974).

T. Yoshimura, Y. Imanaka, and M. Yamashita, *Doshisha Daigaku Rikogaku Kenkyu Hokoku* **14**(4), 247–255 (1974).

A. C. Hart, Anodic passivation and dissolution of nonactivated nickel in electrolytic solutions of nickel chloride–nickel sulfate, *Galvanotecnica* **25**(11), 189–197 (1974).

V. B. Lepinskikh, P. I. Buler, G. A. Toporishev, and O. Yu. Esin, Anodic passivation of iron in a borate melt, *Tr. Ural. Nauchno-Issled. Inst. Chern. Met.* **21**, 57–60 (1974).

S. Nishiyama, K. Tachibana, G. Okamoto, and T. Sugita, The analysis of passive current noise of stainless steels under potentiostatic conditions. II. Changes in anodic noise patterns of passivated 18-8 stainless steels caused by the addition of chloride ions during the induction period of pitting, *Boshoku Gijutsu* **23**(9), 445–451 (1974).

K. Sugimoto and Y. Sawada, Interfacial impedance of austenitic stainless steel under anodic polarization, *Boshoku Gijutsu* **23**(2), 63–71 (1974).

G. Okamoto, T. Sugita, S. Nishiyama, and K. Tachibana, The analysis of passive current noise of stainless steel under potentiostatic conditions. I. Anodic noise patterns of passivated 18-8 stainless steel in sulfuric acid obtained by low noise electronic system with amplifiers, *Boshoku Gijutsu* **23**(9), 439–444 (1974).

N. Sato, Anodic passivation films on iron, *Boshoku Gijutsu* **23**(11), 535–544 (1974).

H. Abo, M. Ueda, and S. Noguchi, Corrosion resistance of various stainless steels to sulfuric acid solution, *Boshoku Gijutsu* **23**(7), 341–346 (1974).

S. Ishankhodzhaev and G. A. Tsyganov, Anodic dissolution of a molybdenum–tungsten alloy in an ammonium nitrate solution in the presence of hydrogen peroxide, *Dok. Akad. Nauk. Uzb. SSR* **31**(10), 30–31 (1974).

P. I. Buler, G. A. Toporishchev, O. A. Esin, V. B. Lepinskikh, and V. A. Kopysov, Anodic passivation of metals in oxide melts, *Ionnye Rasplavy* **2**, 159–176 (1974).

F. F. Faizullin and V. K. Saifullin, Anodic passivation of iron in a eutectic mixture of a potassium nitrate–sodium nitrate melt, *Prikl. Elektrokhim.* **3–4**, 32–33 (1974).

A. M. Shams El Din and F. M. Abe El Wahab, Behavior of copper–zinc alloys in alkaline solutions upon alternate anodic and cathode polarization, *Proceedings of the Fifth International Congress on Metallic Corrosion,* pp. 189–191, Butterworths, London (1974).

A. A. Tibekina and V. M. Tsokalo, Role of weak coordination reactions in the anode dissolution of chromium and in other electrochemical reactions, *Tezisy Dokladov Vsesoyuznoi Soveshchanii Elektrokhimiya, 5th* **2**, 259–261 (1974).

F. F. Faizullin and G. A. Kamalova, Effect of pH and anions of an electrolyte on anodic behavior of cadmium, *Issled. Elektrokhim. Magnetokhim. Elektrokhim. Metodam Anal.* **4**, Ch. 1, 35–40 (1974).

F. F. Faizullin, D. A. Baitalov, and K. A. Zakirova, Temperature—kinetic study of the kinetics of anodic passivation of titanium in 1.0 M potassium hydroxide, *Issled. Elektrokhim. Magnetokhim. Elektrokhim. Metodam Anal.* **4**, Ch. 1, 178–185 (1974).

P. I. Buler, V. B. Lepinskikh, and G. A. Toporishchev, Effect of water on anodic behavior of passive nickel in a borosilicate melt, *Tr. Inst. Metall. Akad. Nauk SSSR Ural. Nauchn. Tsentr* **28**, 97–100 (1974).

E. G. Ivanov, E. A. Berkman, and G. M. Petrova, Negative differential effect and passivation during anodic dissolution of aluminum and its alloys, *Sb. Rab. Khim. Istochnikam Toka* **9**, 186–193 (1974).

A. P. Brynza, V. S. Buglakova, A. N. Stepanchuk, and E. V. Koblyanskaya, Anodic behavior of molten titanium carbide in sulfuric acid, *Vopr. Khim. Khim. Tekhnol.* **36**, 64–9 (1974).

T. R. Jayaraman, V. K. Venkatesan, and H. V. K. Udupa, Anodic behavior of electroless cobalt in sodium hydroxide, *Proc. Semin. Electrochem., 14th*, pp. 274–82 (1974).

H. Yamamoto and S. Ito, Anodic dissolution behaviors of stainless steel in concentrated sodium chloride and sodium nitrate aqueous solutions, *Denki Kagaku Oyobi Kogyo Butsuri Kagaku* **42**(10), 535–539 (1974).

P. I. Buler, V. B. Lepinskikh, and G. A. Toporishchev, Rate of anodic dissolution of passive nickel in a borate melt, *Elektrokhim. Rasplavy,* pp. 82–5 (1974).

I. A. Atanasiu and M. Constantinescu, Anodic dissolution of nickel using varying current. I. Effect of electrolyte nature and concentration, *Rev. Chim. Bucharest* **25**(9), 730–734 (1974).

A. M. Azzam, S. S. Abdel Rehim, and M. H. Fawzy, Alkaline stannate baths. I. Anodic and cathodic polarization, *Indian J. Technol.* **12**(11), 492–496 (1974).

T. Noda and N. Sato, Ionic current in the passive oxide film on iron, *Nippon Kinzoku Gakkaishi* **38**(12), 1143–1149 (1974).

V. M. Novakovskii and A. A. Sokolov, Process for increasing and reducing the passivity of a titanium anode, *Zashch. Met.* **10**(5), 520–526 (1974).

K. Toda and K. Tomikashi, High current density electrolytic refining of copper with anodic passivation avoided by using intermittent D. C., Japan Patent No. 74-18896, May 19 (1974).

T. Yamashita and S. Nonaka, Anodic behavior of zinc. I. Effect of temperature of the passivation of zinc in alkaline solution, *Nippon Kagaku Kaishi* **12**, 2287–2290 (1974).

V. V. Gerasimova and V. V. Gerasimov, Effect of mechanical stresses on the kinetics of the anodic process of chromium–nickel steel KH18N9 and its stress corrosion cracking resistance, *Fiz.-Khim. Mekh. Mater.* **10**(5), 29–32 (1974).

G. G. Lopovok, S. D. Kokin and M. P. Kalyanova, Effect of molybdenum on the anodic behavior of stainless steels in thiocyanate solutions, *Zashch. Met.* **10**(5), 557–559 (1974).

N. D. Tomashov, Yu. S. Ruskol, and G. A. Ayuyan, Titanium passivity in sulfuric acid, *Zashch. Met.* **10**(5), 515–519 (1974).

V. S. Kuzub, V. S. Novitskii, L. B. Golovneva, and V. P. Rebrunov, Anodic protection of reaction vessels for the production of a nitrogen–phosphorus–potassium fertilizer, *Khim. Promst. Moscow* **8**, 609, 610 (1974).

A. D. Davydov, A. D. Romashkan, M. A. Monina, and V. D. Kashcheev, Anodic dissolution of cobalt at high current densities, *Elektrokhimiya* **10**(11), 1681–1685 (1974).

N. Sato, T. Noda, and K. Kudo, Thickness and structure of passive films on iron in acidic and basic solution, *Electrochim. Acta* **19**(8), 471–475 (1974).

N. Sato and K. Kuko, Ellipsometric study of anodic passivation of nickel in borate buffer solution, *Electrochim. Acta* **19**(8), 461–470 (1974).

A. I. Tsinman, L. M. Pishchik, and G. L. Makovei, Corrosion—electrochemical behavior of metals in organic media. I. Passivation and anodic dissolution of molybdenum in anhydrous acetic acid solutions, *Elektrokhimiya* **10**(9), 1321–1327 (1974).

J. Yahalom, Anodic oxidation of gallium phosphide semiconductor disks, *Ger. Offen.* **14**, September 26 (1974).

U. Ebersbach, Kinetics of the anodic passivation of metals. II, *Zashch. Met.* **10**(4), 374–381 (1974).

B. MacDougall and M. Cohen, Anodic oxidation of nickel in neutral sulfate solution, *J. Electrochem. Soc.* **121**(9), 1152–1159 (1974).

V. I. Vigdorovich, L. E. Tsygankova, and N. V. Filippova, Anodic behavior of chromium–nickel–titanium steel KH18N10T in $1N$ aqueous alcohol solutions of hydrogen chloride, *Zashch. Met.* **10**(4), 427–431 (1974).

P. I. Buler, V. B. Lepinskikh, G. A. Toporishchev, and O. Yu. Esin, Reasons for the inhibition of the anodic oxidation of iron in a borate melt, *Elektrokhimiya* **10**(7), 1153–1155 (1974).

B. N. Afanas'ev, V. I. Bukarinov, and N. N. Milyutin, Kinetics of the anodic oxidation of a cadmium electrode in 1 N potassium hydroxide, *Zh. Prikl. Khim. Leningrad* **47**(6), 1312–1316 (1974).

J. Krupski and H. G. Feller, Study of the anodic dissolution of nickel with the radioisotope sulfur-35, *Z. Metallkd.* **65**(5), 401–407 (1974).

A. Piotrowski and A. Slimak, Anodic potentiostatic phase dissolution of heterogeneous alloys, *Werkst. Korros* **25**(4), 262–265 (1974).

R. Bartonicek, Anodic polarization curves of brass alloys in ammonium chloride solutions, *Werkst. Korros* **25**(4), 253–261 (1974).

J. Schich, Passivation and repassivation of aluminum, *Prax. Naturwiss., Teil 3* **23**(3), 57–63 (1974).

A. C. Hart, Anodic passivation and dissolution of unactivated nickel in nickel sulfate and chloride electrolytes, *Fachber. Oberflaechentech.* **12**(2), 35–40 (1974).

V. N. Filimonenko, B. M. Kreichman, and G. A. Iskhakova, Anodic polarization and passivity of a cobalt-activated tungsten carbide heterogeneous system, *Fiz. Khim. Obrab. Mater.* **1**, 143–146 (1974).

I. N. Frantsevich, L. N. Yagulpol'skaya, V. A. Lavrenko, O. K. Teodorovich, and G. N. Braterskaya, Corrosion properties of nickel–silver pseudoalloys and principle of electrochemical reaction independences, *Dok. Akad. Nauk. SSSR* **214**(5), 1128–1130 (1974).

R. C. Bracken, J. G. Harper, and A. D. Paddock, Anodic passivating of integrated circuits, U. S. Patent No. 3,798,135 January 19 (1974).

S. O. Izidinov, A. P. Blokhina, and M. N. Sinyushin, Anodic passivation of silicon in alkali solutions during galvanostatic polarization, *Elektrokhimiya* **10**(2), 244–250 (1974).

R. M. Horton and K. Kim, Anodic polarization behavior of nickel–manganese alloys in $1N$ sulfuric acid solution, *Corrosion* **30**(1), 13–17 (1974).

H. Vaidyanathan, M. E. Straumanis, and W. J. James, Polarization characteristics and anodic disintegration of beryllium in nonaqueous solutions, *J. Electrochem. Soc.* **121**(1), 7–12 (1974).

1975

M. MacDougall and M. Cohen, The effect of cathodic treatment on nickel dissolution, *J. Electrochem. Soc.* **122**, 383–385 (March 1975).

O. L. Riggs, Jr., The second anodic current maximum, *Corrosion* **31** (1975).

R. D. Armstrong, T. Dickinson, B. MacFarlane, and H. R. Thirsk, The anodic behavior of indium and indium/bismuth alloys in alkaline solutions, *Power Sources Symp. Proc., 9th,* 479–502 (1975).

O. Atdaev and K. N. Nikitin, Passivation of zirconium during anodic dissolution in acid alcoholic solutions of chlorides, *Deposited Doc. Viniti* 70–79, 1626–1675 (1975).

A. M. Sukhotin, E. A. Gankin, and A. I. Khentov, Anodic behavior of passive iron and ferriferrous oxide in solutions of perchloric and sulfuric acid, *Passivnost Korroz. Met.,* 4–15 (1975).

P. L. Bonora, G. P. Ponzano, and M. Bassoli, Anodic behavior of copper alloys in alkaline solutions, *Ann. Chim. Rome* **65**, 11–12, 677–686 (1975).

V. M. Blinov, Effect of forming protective films on the kinetics of an anodic process, *Vopr. Khim. Biokhim. Sistem Soderzh. Marganets Polifenoly* **3**, 67–72 (1975).

M. Maja, P. Spinelli, and B. De Benedetti, Anodic behavior of iron–chromium alloys in borate solutions, *Atti Accad. Sci. Torina, Cl. Sci. Fis., Mat. Nat.* **109**, 1–2, 217–227 (1975).

H. Saito, K. Tachibana, and G. Okamoto, Anodic current characteristics of 18–8 stainless steels in chloride media and the effect of passivation treatment temperature on anodic pitting, *Boshoku Gijutsu* **24**(9), 465–470 (1975).

A. F. Bogoyavlenskii and I. P. Oranskaya, Some properties of anodic germanium oxide formed in an acetate electrolyte, *Tr. Kazan. Aviats. In-ta.* **185**, 3–5 (1975).

P. Juzikis, T. Jankauskas, and V. Kaikaris, Anodic dissolution of silver. III. Formation of passivating films, *Liet. Tsr. Aukst. Mokyklu Mokslo Darb., Chem. Chem. Technol.* **17**, 217–220 (1975).

Y. Kikuchi and E. Sato, Anodic passivation behavior of lead alloy and its platinum coating in electrodes in artificial sea water, *Boshoku Gijutsu* **24**(9), 459–463 (1975).

N. Sato, K. Kudo, and T. Noda, Anodic passivating films on iron in phosphate and borate solutions, *Z. Phys. Chem. Frankfurt Am Main* **98**, 1–6, 271–284 (1975).

K. Schwabe and G. Fuhrmann, On the anodic behavior of armco iron in trifluoroacetic acid, *Z. Phys. Chem. Frankfurt Am Main* **98**, 1–6, 261–270 (1975).

L. V. Companys, Effect of chloride ions in the process of anodic passivation of palladium, *An. Quim.* **71**, 11–12, 1056–1060 (1975).

I. N. Ozeryanaya and I. V. Zauzolkov, Anodic behavior of zirconium in alkali chloride melts, *Tr. Inst. Elektrokhim. Ural. Nauchn. Tsentr. Akad. Nauk SSSR* **22**, 80–83 (1975).

I. M. Tordesillas, L. Victori, and J. M. Chao, Passivation of palladium in neutral solutions of sodium chloride, *An. Quim.* **71**, 7–8, 642–646.

M. Morita, C. Iwakura, H. Yoneyama, and H. Tamura, Electrode characteristics of tantalum carbide, *Denki Kagaku Oyobi Kogyo Butsuri Kagaku* **43**(12), 740–745 (1975).

F. F. Faizullin and G. M. Abdulkadyrov, Anodic oxidation of the tantalum–titanium alloy in aqueous solutions of some salts, *Elektrokhimiya* **11**(12), 1905 (1975).

T. Yamashita, Effect of sodium(+) and potassium(+) on the anodic passivation of zinc in alkaline solutions, *Kinzoku Hyomen Gijutsu* **26**(12), 602–603 (1975).

T. Yoshimura, Y. Imanaka, and M. Yamashita, Anodic behavior of copper in alkaline solution. 1. On the mechanism of passive film formation as revealed by ellipsometry and the rotating disk electrode method, *Boshoku Gijutsu* **24**(3), 117–122 (1975).

A. G. Akimov, M. G. Astaf'ev, and I. L. Rozenfel'd, Electrochemical properties of a titanium electrode studied by electroabsorption, *Elektrokhimiya* **11**(19), 1578–1581 (1975).

A. M. Sukhomin and O. S. Andreeva, Nonsteady-state kinetics of dissolution of passive metals, *Elektrokhimiya* **11**(11), 1698–1701 (1975).

L. M. Danilov and A. P. Brynza, Anodic behavior of alloys of the titanium–oxygen system in sulfuric acid. II. Range of potentials for the complete passivation of titanium, *Vopr. Khim. Khim. Tekhnol. Resp. Mezhved. Temat. Nauchn. Tekhnd. Sb.* **37**, 28–30 (1975).

R. W. Lewis and J. Turner, Effect of silicate ion on the anodic behavior of zinc in concentrated alkali, *J. Appl. Electrochem.* **5**(4), 343–349 (1975).

A. A. Ravdel, N. A. Novikova, and A. A. Musakina, Composition of a passivating film formed on a zinc electrode during its anodic polarization in an alkaline solution, *Zh. Prikl. Khim. Leningrad* **48**(10), 2303–2304 (1975).

N. N. Milyutin and I. F. Danilenko, Equation of an anodic potentiostatic curve of a metal undergoing passivation, *Zh. Prikl. Khim. Leningrad* **48**(10), 2211–2216 (1975).

G. Reinhard and J. Pirrwitz, Determination of water constituents in anodic passive layers of iron by a tritium labeling technique, *Isotopenpraxis* **11**(4), 134–140 (1975).

Z. A. Foroulis and M. J. Thubrikar, Kinetics of the breakdown of passivity of preanodized aluminum by chloride ions, *J. Electrochem. Soc.* **122**(10), 1296–1301 (1975).

O. Tajima and K. Nakao, Anodic polarization behavior of nickel–copper alloys in sulfuric acid solutions, *Technol. Rep. Kansai Univ.* **16**, 123–132 (1975).

A. I. Rusin, B. P. Yur'ev, and O. V. Potokova, Anodic behavior of a copper electrode in a lead–copper battery, *Sb. Rab. Khim. Istochnikam Toka* **10**, 57–62 (1975).

J. B. Cotton, Practical use of anodic passivation for the protection of chemical plants, *Br. Corros. J.* **10**(2), 66–68 (1975).

S. O. Izidinov, Principles of silicon passivation and photoactivation during chemical dissolution of a passivating oxide film studied by chronopotentiometry, *Elektrokhimiya* **11**(3), 420–427 (1975).

V. N. Kovtun, V. F. Mogilenko, and A. M. Greshchik, Kinetic scheme of reactions during anodic dissolution of nickel in sulfuric acid, *Elektrokhimiya* **11**(2), 2, 277–280 (1975).

E. J. Casey and C. L. Gardner, Anodic passivation by cadmium oxide studied by ESR, *J. Electrochem. Soc.* **122**(7), 851–854 (1975).

L. N. Yagupol'skaya, Nature of current peaks in the anodic polarization curve of nickel in $1N$ sulfuric acid, *Zashch. Met.* **11**(3), 338–341 (1975).

S. O. Izidinov, Two stages in the formation of a passivating anodic oxide film at constant potential and mechanisms of partial photoactivation of silicon in concentrated alkali, *Elektrokhimiya* **11**(5), 730–738 (1975).

S. O. Izidinov and Yu. Ya. Vinnikov, Ellipsometric study of passivation and kinetics of complete photoactivation of passive N-type silicon in a concentrated alkali, *Elektrokhimiya* **11**(3), 414–419 (1975).

E. A. Lizlovs and A. P. Bond, Anodic polarization behavior of high-purity 13 and 18% chromium stainless steels, *J. Electrochem. Soc.* **122**(6), 719–722 (1975).

A. Caprani, Parameters determining the anodic behavior of titanium in fluorinated sulfuric acid solutions, *J. Chim. Phys. Phys. Chim. Biol.* **72**(2), 171–182 (1975).

G. Pourcelly and M. Rolin, Electrolysis of liquid mixtures of ammonia and hydrofluoric acid in the hydrofluoric acid-rich range of composition. Anodic behavior of metals and carbon, *J. Chim. Phys. Phys. Chim. Biol.* **72**(3), 283–289 (1975).

G. P. Chernova, G. S. Kalnina, and N. D. Tomashov, Effect of structure and composition of alloys in the titanium–chromium–tantalum system on their anodic behavior, *Zashch. Met.* **11**(2), 159–164 (1975).

I. L. Serushkin, G. A. Tedoradze, G. I. Kaurova, G. P. Il'inskaya, and T. L. Razmerova, Anodic behavior of nickel in liquid hydrogen fluoride studied by a chronopotentiometric method, *Elektrokhimiya* **11**(5), 705–710 (1975).

A. Mihajlovic, A. Mance, and B. Djuric, Effect of microstructure on the anodic behavior of a copper–beryllium–nickel–zirconium alloy system, *J. Appl. Electrochem.* **5**(2), 121–124 (1975).

A. I. Tsinman and L. M. Pischik, Corrosion–electrochemical behavior of metals in organic media. III. Anodic behavior and corrosion resistance of molybdenum-containing alloys and their components in an acetic acid solution of a chloride, *Elektrokhimiya* **11**(1), 127–132 (1975).

J. Mieluch, Ellipsometric studies of the anodic passivity of nickel in 0.5 M sulfuric acid, *Rocz. Chem.* **49**(2), 365–378 (1975).

A. I. Tsinman, L. M. Pischik, and G. L. Makovei, Corrosion—electrochemical behavior of metals in organic media. II. Passivation and anodic dissolution of molybdenum-containing alloys in acetic acid solutions of sodium acetate, *Elektrokhimiya* **11**(1), 101–103 (1975).

N. N. Bibikov, L. V. Povarova, and E. A. Kashcheeva, Region of anodic passivity of titanium in sea water, *Zashch. Met.* **11**(1), 77–79 (1975).

T. Ohtsuka and N. Sato, Anodic passivation of cobalt in neutral solutions, *Nippon Kinzoku Gakkaishi* **39**(1), 60–67 (1975).

S. O. Izidinov and V. I. Fedotova, Elimination of anodic passivation of N–silicon during steady-state illumination under conditions of elevated temperature and galvanostatic polarization in 10N potassium hydroxide, *Elektrokhimiya* **11**(1), 23–28 (1975).

F. F. Faizullin and K. V. Egorova, Kinetics of anodic dissolution and passivation of the intermetallic compounds indium antimonide, *Elektrokhimiya* **11**(1), 93–95 (1975).

B. Wallen and J. E. Andersson, Electrode for potential measurements, Swed. 393404, 6 pp. 5/9/77 Pat App/Prty = 7512691, 11/12/75. PCL C23F-013/00 P (patent).

1976

N. Azzerri and A. Tamba, Potentiostatic pickling: A new technique for improving stainless steel processing, *J. Appl. Electrochem.* **6**, 347–352 (1976).

O. L. Riggs, Jr., and L. S. Surtees, Potentiostatic cleaning and conditioning of the cathodes used in electrodeposition cells, *Metall. Trans. B* **73**, 245–251 June (1976).

R. D. Armstrong and M. Reid, The effect of electrode orientation upon the anodic behavior of the vitreous carbon–sodium polysulfide interphase at 350 degree C, *Electrochim. Acta* **21**(11), 1105–1106 (1976).

V. A. Khadeev and A. M. Gevorgyan, Characteristics of nonplatinum metallic electrodes used in anodic amperometry, *Dokl. Akad. Nauk Uzb. SSR* **7**, 47–48 (1976).

V. V. Gneushev and V. A. Panin, Anodic behavior of metals in fused thiocyanates, *Issled. Obl. Khim. Istochnikov Toka* **4**, 15–22 (1976).

L. A. Kamel, Studies on the anodic and cathodic polarization of amalgams. The behavior of thallium amalgams in alkaline solutions, *Egypt. J. Chem.* **17**(5), 535–545 (1976).

J. P. Hoare, R. F. Paluch, and S. G. Meibuhr, On the differential thermal analysis of the platinum–oxygen system, *J. Electrochem. Soc.* **123**(12), 1821–1824 (1976).

F. Zucchi, G. Brunoro, G. Trabanelli, and M. Zucchini, Semiconducting characteristics of anodic passive films on iron, *Surf. Technol.* **4**(6), 497–504 (1976).

S. Feliu and M. Morcillo, The behavior of lamellar second phases in lead matrix under anodic action, *Electrochim. Acta* **21**(11), 1035–1039 (1976).

H. Hasegawa and T. Suzuki, On the behaviors of the cell voltage during anodic oxidation of gallium arsenide under dark and illuminated conditions, *Jpn. J. Appl. Phys.* **15**(12), 2489–2490 (1976).

M. M. Pecherskii, V. V. Gorodetskii, V. M. Pulina, and V. V. Losev, Effect of acidity on the anodic behavior of ruthenium, *Electrokhimiya* **12**(9), 1445–1448 (1976).

E. Paczoska, Effect of temperature and current density on cell voltage and anodic passivation in the copper refining process, *Pr. Inst. Met. Niezelaz.* **5**(1), 13–16 (1976).

A. G. Atanasyants and T. P. Mavrenkova, Anodic behavior of vanadium in alkali. II. Transition to a passive state, *Zh. Fiz. Khim.* **50**(10), 2644–2646 (1976).

A. F. Mazanko, I. M. Osadchenko, B. L. Klyuev, and A. P. Tomilov, Copper oxide by anodic dissolution of copper in a solution of alkali with subsequent removal of the copper oxide precipitated from the solution, *U.S.S.R* August 5 (1976).

B. N. Stirrup and N. A. Hampson, Anodic passivation of tin in buffered phosphate electrolyte, *J. Electroanal. Chem. Interfacial Electrochem.* **73**(2), 189–206 (1976).

R. A. H. Penne and H. Jullien, Talc and anodic passivation, *FATIPEC Congr.* **13**, 496–503 (1976).

S. O. Izidinov and V. P. Osipov, Anodic passivation and potential distribution on passive silicon in alkali solutions, *Elektrochim.* **12**(7), 1081–1088 (1976).

Yu. A. Popov, A. A. Vasil'ev, and Ya. M. Kolotyrkin, Theory of crystallization of a protective layer on a dissolving metal. Comparison with experimental data, *Elecktrokhimiya* **12**(8), 1298–1301 (1976).

G. W. Simmons, E. Kellerman, and H. Leidheiser, Jr., *In situ* studies of the passivation and anodic oxidation of cobalt by emmission moessbauer spectroscopy. I. Theoretical background, experimental methods, and experimental results for borate solution (pH 8.5), *J. Electrochem. Soc.* **123**(9), 1276–1284 (1976).

P. C. Catagnus and C. T. Baker, Passivation of mercury cadmium telluride semiconductor surfaces by anodic oxidation, *U.S.* **9** (August 24, 1976).

S. Casadio, Cyclic voltammetry for mixed diffusion–charge transfer control of simple anodic dissolution and passivation of the electrode material, *J. Electroanal. Chem. Interfacial Electrochem.* **72**(2), 243–250 (1976).

K. Szabo and K. Solymos, Passivation kinetics during anodic plarization on a liquid gallium electrode in hydrochloric acid potassium chloride solutions, *Magy. Kem. Foly.* **82**(7), 341–346 (1976).

Z. Szklarska–Smialowska, Conclusions on the critical potential of passivation to be drawn from ellipsometric studies of anodic films on iron, *Passivity Its Breakdown Iron Iron Base Alloys. U.S.A.–Japan Seminar,* pp. 60–64 NACE, Houston, TX (1976).

A. I. Tsinman and L. A. Danielyan, Possible anodic protection of carbon steel in alkali solutions, *Zashch. Met.* **12**(4), 450–452 (1976).

V. I. Spitsyn, M. M. Kurtepov, and R. A. Baru, Electrochemical and corrosion characteristics of technetium and stainless steel alloyed with technetium, *Dokl. Adad. Nauk SSSR* **229**(3), 673–675 (1976).

T. Shibata, T. Takeyama, and G. Okamoto, Plastic deformation and reactivity of passive metals, *Passivity Its Breakdown Iron Iron Base Alloys, U.S.A.–Japan Seminar,* pp. 165–168 NACE, Houston, TX (1976).

A. B. Delgado, D. Posadas, and A. J. Arvia, Kinetics and mechanism of the anodic dissolution of nickel in hydrochloric acid–dimethylsulfoxide solutions, *Electrochim. Acta* **21**(6), 385–393 (1976).

M. Romand, G. Bouyssoux, J. S. Solomon and W. L. Baun, Auger spectrometric study of passive films formed on tantalum in phosphoric acid solution, *J. Electron Spectrosc. Relat. Phenom.* **9**(1), 41–50 (1976).

A. E. Yaniv, J. B. Lumsden, and R. W. Staehle, The composition of passive films on ferritic stainless steels, *Passivity Its Breakdown Iron Iron Base Alloys, U.S.A.–Japan Seminar*, pp. 72–74 NACE, Houston, TX (1976).

A. Jouanneau, M. Keddam, and M. C. Petit, A general model of the anodic behavior of nickel in acidic media, *Electrochim. Acta.* **21**(4), 287–292 (1976).

T. M. Devine and L. Wells, Anodic polarization behavior of an amorphous metal, *Scr. Metall.* **10**(4), 309–310 (1976).

Yu. A. Popov, Crystallization of a protective salt layer on anodically dissolving metals. I. Structure of a porous layer, *Elektrokhimiya* **12**(5), 817–820 (1976).

E. L. Littauer and K. C. Tsai, Anodic behavior of lithium in aqueous electrolytes. II. Mechanical passivation, *J. Electrochem. Soc.* **123**(7), 964–969 (1976).

R. P. Frankenthal and D. E. Thompson, The anodic behavior of gold in sulfuric acid solutions. Effect of chloride and electrode potential, *J. Electrochem. Soc.* **123**(6), 799–804 (1976).

F. C. Sessa and N. D. Greene, Influence of deionized water on the anodic behavior of active-passive metals, *Corrosion* **32**(5), 203–204 (1976).

P. Novak, J. Novohradsky, and F. Franz, Crevice corrosion during the anodic passivation of metals, *Sb. Vys. Sk. Chem. Technol. Praze Anorg. Chem. Technol.* **B21**, 77–91 (1976).

D. W. Shoesmith, T. E. Rummery, D. Owen, and W. Lee, Anodic oxidation of copper in alkaline solutions. I. Nucleation and growth of cupric hydroxide films, *J. Electrochem. Soc.* **123**(6) 790–799 (1976).

E. L. Littauer and K. C. Tsai, Anodic behavior of lithium in aqueous electrolytes. I. Transient passivation, *J. Electrochem. Soc.* **123**(6), 771–776 (1976).

K. Szabo and K. G. Solymos, The anodic behavior of a liquid gallium electrode in hydrochloric acid solutions, *Magy. Kem. Foly.* **82**(3), 139–145 (1976).

A. Jouanneau and M. C. Petit, Simulation of the anodic behavior of nickel in acid media, *J. Chim. Phys. Phys. Chim. Biol.* **73**(1), 82–88 (1976).

W. D. Johnston, Jr., Anodic passivation and coating of aluminum arsenide in aqueous solutions, *J. Electrochem. Soc.* **123**(3), 442,443 (1976).

T. Ohtsuka, K. Kudo, and N. Sati, Thickness and layer structure of anodic passivating films on cobalt in neutral solutions, *Nippon Kinzoku Gakkaishi* **40**(2), 124–132 (1976).

R. Nishimura, K. Kudo, and N. Sato, The potential dependence of the the compostion of anodic passive films on iron in neutral solutions, *Nippon Kinzoku Gakkaishi* **40**(2), 118–124 (1976).

M. Jaenchen and K. Schwabe, The anodic behavior of iron in nonaqueous solvents having different water contents, *Z. Phys. Chem. Leipzig* **257**(1), 129–144 (1976).

E. Kelly, Anodic dissolution of titanium in acidic sulfate solutions. II. Effects of titanium (III) and titanium (IV) ions, *J. Electrochem. Soc.* **123**(2), 162–170 (1976).

V. P. Kochergin, I. N. Vinyarskaya, and Yu. M. Kovalev, Anodic behavior of nickel and nickel-containing materials in melted sodium and potassium phosphates and borates, *Zashch. Met.* **12**(1), 79–82 (1976).

N. A. Balashova, G. D. Zakumbaeva, and L. A. Beketaeva, Effect of anions on the anodic behavior of nickel in alkaline solutions, *Zashch. Met.* **12**(1), 39–41 (1976).

B. N. Stirrup and N. A. Hampson, Anodic passivation of tin in sodium hydroxide solutions, *J. Electroanal. Chem. Interfacial Electrochem.* **67**(1), 45–56 (1976).

V. G. Moisa and V. S. Kuzub, Corrosion protection of articles which come into contact with an electrically conducting medium, Ger. Offen. 2637633, 13 pp., 2/23/78 Pat App/Prty = 2637633, 8/20/76. PCL C23F-013/00 P (patent).

Yu. Ya. Nichaenko, B. A. Gru, and V. S. Kuzub, Reference electrode for anodic protection of a container for chemical nickel plating bath, Ger. Offen. 2655679, 11 pp., 6/15/78 Pat App/Prty = 3655679, 12/8/76. PCL C23C-003/02 P (patent).

R. Walker and L. L. Ed Shreir, *Anodic protection corrosion, Vol.2,* 2nd ed., University of Surrey, Department of Technology, Guilford, pp. 11:112–11:125, Butterworth, London,

David J. Palmer, Use anodic protection with care, *Can. Chem. Process. Ind.,* **60**(8) 35, 37–9 (1976).

1977

W. A. Szymanski, Anodic protection of carbon steel in aqueous ammonia solution, *Mater. Perform.,* 16–18, November (1977).

T. Yamashita, Anodic dissolution behavior of zinc studied with a rotating disk electrode, *Kinzoku Hyomen Gijutsu* **28**(9), 464–467 (1977).

A. D. Daydov, R. A. Mirzoev, and A. N. Kamkin, Role of oxide film defects in the anodic-anionic activation of niobium, *Elektrokhimiya* **13**(8), 1271 (1977).

Y. Kawashima, T. Kishi, and T. Nagai, Anodic behavior of iron– nickel ferrite electrodes, *Denki Kagaku Oyobi Kogyo Butsuri Kagaku* **45**(5), 299–303 (1977).

L. A. Medvedeva, V. M. Knyazheva, Ya. M. Kolotyrkin, and S. G. Babich, Problem of choosing a standard corrosion medium for obtaining anodic potentiodynamic curves for steel 18–10, *Zashch. Met.* **13**(4), 400–405 (1977).

G. N. Trusov, E. P. Gonchalieva, B. A. Goncharenko, and V. S. Mikheev, Anodic behavior of niobium–nickel and tantalum–nickel alloys in acid solutions, *Fiz. Khim. Inst. Im. Karpova Moscow* (1977).

T. Mitamura, H. Kobayashi, F. Noguchi, and M. Kikuchi, Studies on lead alloy anodes in artificial sea water. 1. Anodic behavior of lead–silver alloy in sodium sulfate solutions, *Denki Kagaku Oyobi Kogyo Butsuri Kagaku* **45**(3), 176–180 (1977).

M. Koudelkova, J. Augustynski, and H. Berthou, On the composition of the passivating films formed on aluminum in chromate solutions, *J. Electrochem. Soc.* **124**(8), 1165–1168 (1977).

B. Prakash, Effect of tin on the metallurgical and corrosion properties of iron–carbon–silicon (alloys), *J. Electrochem. Soc. India* **26**(1), 37–47 (1977).

R. G. Barradas, S. Fletcher, and J. D. Porter, The anodic behavior of lead amalgam electrodes in hydrochloric acid solution, *J. Electroanal. Chem. Interfacial Electrochem.* **80**(2), 295–304 (1977).

E. E. Milam, W. B. Crow, and J. R. Myers, Anodic polarization behavior of iron–aluminum alloys in sulfuric acid solutions, *Corrosion* **33**(7), 240–243 (1977).

J. J. Podesta, R. C. V. Piatti, and A. J. Arvia, Potentiostatic curves of iron, nickel, and cobalt in the passivity and transpassivity regions in fused lithium nitrate–sodium nitrate–potassium nitrate eutectic, *Corros. Sci.* **17**(3), 225–235 (1977).

G. Bouyssoux, M. Romand, H. D. Polaschegg, and J. T. Calow, XPS and AEC studies of anodic passive films grown on chromium electrodes in sulfuric acid baths, *J. Electron Spectrosc. Relat. Phenom.* **11**(2), 185–196 (1977).

J. Perkins, W. H. Luebke, K. J. Graham, and J. M. Todd, Anodic corrosion of zinc alloys in seawater, *J. Electrochem. Soc.* **124**(6), 819–826 (1977).

A. V. Nikolaev, D. P. Semchenko, and I. V. Borisova, Effect of an alternating current on the anodic polarization of cobalt in sulfuric acid solutions, *Zh. Prikl. Khim. Leningrad* **50**(2), 331–334 (1977).

A. M. Shams El Din and F. M. Abd El Wahab, The behavior of copper–zinc alloys in alkaline solutions upon alternate anodic and cathodic polarization, *Corros. Sci.* **17**(1), 49–58 (1977).

A. V. Merkulov, Yu. I. Naumov, and V. N. Flerov, Anodic processes on copper in a 5 *N* solution of sodium hydroxide with additives of polyhydric alchohols, *Izv. Vyssh. Uchebn. Zaved., Khim. Khim. Tekhnol.* **20**(1), 104–106 (1977).

T. M. Devine, Anodic polarization and localized corrosion behavior of amorphous nickel–iron–chromium–phosphorus–boron alloy (N135FE30CR15P14B6) in near-neutral and acidic chloride solutions, *J. Electrochem. Soc.* **124**(1), 38–44 (1977).

W. A. Mueller, Corrosion rates of carbon steel tubes in kraft liquors with and without anodic or cathodic protection, *Pulp Pap. Res. Inst.* **2**, 140–146 (1977).

C. K. Chang, Electrochemical corrosion prevention—cathodic and anodic protection, *K'o Hsueh Shih Yen* **5**, 163–165 (1977).

V. A. Makarov, P. I. Zarubin, Yu. M. Mironov, L. A. Poluboyartseva, L. N. Yurlova, and E. A. Sidel'nikova, Study and modeling of anodic protection of heat exchangers from steel, *Zashch. Met.* **13**(2), 181–183 (1977).

V. S. Kuzub, A. L. Anokhin, R. N. Akbasheva, L. D. Timofeeva, and S. G. Sabirova, Anodic protection of apparatus for the evaporation of sodium thiocyanate, *Khim. Promst.* **2**, 129–130 (1977).

1978

C. K. Dyer and J. S. L. Leach, Reversible reactions within anodic oxide films on titanium electrodes, *Electrochim. Acta* **23**(12), 1387–1394 (December 1978).

D. D. MacDonald and B. Roberts, A potentiostatic transient study of the passivation of carbon steel in 1M NaOH, *Electrochim. Acta* **23**(6), 557–564 (June 1978).

G. Beck-Nielsen, The anodic dissolution of iron—VIII: The influence of ionic strength on reaction orders with respect to anions, *Electrochim. Acta* **23**(5), 425–431 (May 1978).

T. E. Cummings, M. A. Jensen, and P. J. Elving, Construction, operation and evaluation of a rapid-response potentiostat, *Electrochim. Acta* **23**(11), 1173–1184 (November 1978).

I. R. McGill and B. McEnaney, A novel reference electrode arrangement for high temperature polarization studies, *Corros. Sci.* **18**(3), 257–259 (March 1978).

K. Asami, K. Hashimoto, and S. Shimodaira, An XPS study of the passivity of a series of iron–chromium alloys in sulfuric acid, *Corros. Sci.* **18**(3), 151–160 (March 1978).

A. Cigada, G. Re, D. Sinigaglia, and F. Borile, Contribution to the interpretation of current maxima in the passivity range of austenitic stainless steels, *Corrosion* **34**(11), 407–410 (1978).

Z. Szklarska-Smialowska and N. Lukomski, Ellipsometric study of surface films grown on austenitic stainless steel in chloride solutions, *Corrosion* **34**(5), 177–182 (1978).

H. Ogawa, H. Omata, I. Itoh, and H. Okada, Auger electron spectroscopic and electrochemical analysis of the effect of alloying elements on the passivation behavior of stainless steels, *Corrosion* **34**(2), (1978).

T. Sydberger and S. Nordin, Corrosivity of wet process phosphoric acid, *Corrosion* **34**(1), 16–22 (1978).

V. S. Muralidharan and K. S. Rajagopalan, Kinetics and mechanism of passivation of steel in sodium phosphate solutions, *Electrochim. Acta* **23**(12), 1297–1302 (December 1978).

K. Ogura and T. Majina, Formation and reduction of the passive film on iron in phosphate–borate buffer solution, *Electrochim. Acta* **23**(12), 1361–1365 (December 1978).

A. Vértes, H. Leidheiser, Jr., M. L. Varsányi, A. W. Simmons, and L. Kiss, Mössbauer studies of the passive film formed on tin in borate buffer, *J. Electrochem. Soc.* **125**(12), 1946–1950 (December 1978).

I. L. Rosenfeld, I. S. Danilou, and R. N. Oranskaya, Breakdown of the passive state and repassivation of stainless steels, *J. Electrochem. Soc.* **125**(11), 1729–1735 (November 1978).

E. N. Paleolog, A. Z. Fedetova, O. G. Derjagina, and N. D. Tomashov, Anodic process kinetics on the passive surfaces of titanium, nickel and titanium–nickel alloys, *J. Electrochem. Soc.* **125**(9), 1410–1415 (September 1978).

J. R. Galvele, J. B. Lumsden, and R. W. Staehle, Effect of molybdenum on the pitting potential of high purity 18% Cr–ferritic stainless steels, *J. Electrochem. Soc.* **125**(9), 1204–1208 (August 1978).

S. P. Napreenko, V. A. Timonin, E. V. Uvarov, A. G. Tokarenko, and V. S. Kuzub, Anodic electrochemical protection of railroad tank cars for transporting extrapure battery sulfuric acid, *Issled. Zashch. Met. Kopr. Khim. Promst.* 55–59 (1978).

S. Mladenovic, Protection of metals from electrochemical corrosion. I. Electrochemical protection of metals and alloys, *Hem. Ind.* **32**(9), 631–634 (1978).

P. Novak, J. Bystriansky, F. Franz, and V. Bartel, Anodic protection of a reactor for the continuous production of formic acid, *Chem. Prum.* **28**(9), 461–464 (1978).

A. Antoniuk, System for the anodic protection of steel tanks, Pol. 97405, 3 pp, 3/31/78 Pat App/Prty = 183574, 9/25/75, PCL C23F-013/00.

J. Montuelle, Possibility of anodic protection in the chemical industry and particularly in heat exchangers, *Rev. Metall. Paris* **76**(11), 641–643 (1978).

W. A. Ashby, L. S. Evans, and W. Shepherd, Anodic protection of cast iron in boiling sulfuric acid under industrial conditions, *Br. Corros. J.* **13**(2), 85–87 (1978).

Yu. Ya. Nikhaenko, Reference electrode for anodic protection of nickel electroless plating vessel, Japan. Kokai 7864623, 4 pp, 6/9/78 Pat App/Prty = 76139625, 11/22/76, PCL C23C-003/02.

1979

F. M. Delnick and N. Hackerman, Passive iron—a semiconductor model for the oxide film, *J. Electrochem. Soc.* **126**(5), 732–741 (May 1979).

N. Pessall, G. P. Airey, and X. Lingenfelter, The influence of thermal treatment on the VCC behaviour of inconel alloy 600 at controlled potentials in 10% caustic soda solutions at 315°C, *Corrosion* **35**(3), 100–107 (March 1979).

D. D. MacDonald, B. C. Syrett, and S. S. Wing, The use of potential-pH diagrams for the interpretation of corrosion phenomena in high salinity geothermal brines, *Corrosion* **35**(1), 1–11 (1979).

E. T. Eisenmann, Rapid data acquisition in the study of slow electrode processes by means of simultaneous potentiostats—the sulfidation of copper, *Corrosion* **35**(1), 12–16 (1979).

F. Mansfeld and R. V. Inman, A versatile inexpensive potentiostat for corrosion studies, *Corrosion* **35**(1), 21–23, (1979).

M. Cohen, D. Mitchell, and K. Hashimoto, The composition of anodically formed iron oxide films, *J. Electrochem. Soc.* **126**(3) (March 1979).

H. Saito, T. Shibata, and G. Okamoto, The inhibitive action of bound water in the passive film of stainless steel against chloride corrosion, *Corros. Sci.* **19**(10), 693–708 (1979).

B. Linder, Electrochemistry of effective corrosion protection, *Kem, Tidskr.* **91**(4), 32–37 (1979).

L. G. Naidorf, V. S. Novitskii and V. S. Kuzub, Determination of conditions and effectiveness of the anodic protection of steel, *Zashch. Met.* **15**(4), 456–458 (1979).

V. S. Kuzub, A. G. Tokarenko, L. G. Kuzub, V. A. Makarov, Yu. M. Ironov, Ya. V. Shvartsshten, A. I. Furman, B. M. Prozorov, and E. S. Martynenko, Anodic protection of shell and tube coolers, *Khim. Promst. Moscow* 4, 224–226 (1979).

P. Novak, L. Vesela, and F. Franz, Anodic protection of heat-exchange apparatus made from corrosion-resistant steel in concentrated sulfuric acid, *Chem. Prum.* **29**(7), 342–344 (1979).

S. Mladenovic, Anodic protection of carbon steel in calcium carbonate–ammonium nitrate solutions, *Hem. Ind.* **33**(6), 223–225 (1979).

Index